# INSECTS

*OF THE*

*LOS ANGELES*

*BASIN*

# INSECTS

# *OF THE LOS ANGELES BASIN*

BY CHARLES L. HOGUE

NATURAL HISTORY MUSEUM
OF LOS ANGELES COUNTY

*This revision is rededicated,*

*in loving memory,*

*to my departed sister*

LEONORA McTERNAN

Natural History Museum of Los Angeles County
Los Angeles, California 90007


Published 1974 as Science Series 27.
Second Edition 1993.
Printed in Hong Kong

LC 93-084264
ISBN 0-938644-29-7 (softcover)
ISBN 0-938644-32-7 (hardcover)

A gift in support of this publication was made by
Mr. and Mrs. James A. Robertson, in memory of
Dr. Charles L. Hogue.

Frontispiece, page 2: A male tree cricket, *Oecanthus* species. See page 56. Photograph by C. Hogue.

# CONTENTS

MOHAVE DESERT
SAN GABRIEL
VERDUGO MOUNTAINS
SAN FERNANDO VALLEY
SANTA MONICA MOUNTAINS
BALDWIN HILLS
LAX
PACIFIC OCEAN

MOUNTAINS

SAN GABRIEL VALLEY

SAN JOS
HILLS

PUENTE HILLS

Civic
Center

Los Angeles River

San Gabriel River

DOMINGUEZ
HILLS

PALOS VERDES HILLS

PREVIOUS PAGES:
Major topographic features of the Los Angeles Basin as seen from a point over the Pacific Ocean. To the north, beyond the Palos Verdes Peninsula, are the San Gabriel Mountains; the Mohave Desert is visible on the far horizon. Computer-generated image made from a Landsat photograph. Courtesy of Image Processing Applications and Development Section, Jet Propulsion Laboratory, Pasadena, Calif., Kevin Hussey.

# PREFACE

THIS BOOK IS FOR THE ANGELENO. Its purpose is to give the resident of the Greater Los Angeles Basin a means of identifying and understanding the insects and related terrestrial arthropods that are most likely to be encountered here. Although the book specifically addresses the insects of the basin and its immediate fringes, it should be useful to residents of the entire coastal area of southern California, from Santa Barbara to San Diego.

One might suppose that the need for such a guide would be negligible today, because few insects would remain in such a heavily urbanized region as Los Angeles and its neighboring cities. But our fauna of these invertebrate animals is extensive: I estimate that we have somewhere between 3,000 and 4,000 species. Not only are there still many natural areas in the basin supporting communities of innumerable species, but remnants of the assemblages of our native insects have found haven in semiimproved areas—in the gardens and vacant lots of the suburbs, in parks, and in many other refuges, even in the midst of civic complexes. And there are some forms that, like the Starling and Opossum, have adapted completely to civilization and are sharing our homes and buildings. Where two very large groups of organisms are in contact—in this case, approximately 8 million human beings and many more millions of insects—there is an obvious need for investigation and record of their interactions and mutual tolerance.

In the time since the publication of the first edition of this book in 1974, the field of urban entomology, which was heralded by Walter Ebeling's pioneering book on the subject, has come of age. New perspectives on "bugs in the big city" have been given in a number of publications, and the subject has now become a university specialty.

And southern California's concrete jungle has continued to grow. More and more natural areas and open spaces are succumbing to development. The human population of the Los Angeles Basin has increased dramatically in size, and there has been a considerable influx of cultures from all parts of the world. These factors have forced changes in the spectrum of insect species: new exotics have appeared, and old indigenous elements have become extinct or rare.

These developments made a revision of *Insects of the Los Angeles Basin* a necessity. Readers will find a completely reworked and expanded text, with all previous errors corrected (I hope!). In all, this new edition covers some 485 types of insects and related terrestrial arthropods (130 more than in the first edition) and includes as many color illustrations as possible to make identifications easier for the novice. As in the first edition, I have tried to answer the questions that I am most often asked in my position as Curator of Entomology at the Natural History Museum; more detailed information is available from the publications cited throughout the book.

I have not provided recommendations for eradication of pest species because there are many handbooks of pest control that have this purpose. However, I have discussed economically important forms, so that the reader will be able to identify and know his insect adversaries and thus plan an intelligent course of action against them.

There is still a great deal to be learned about many of our local insects and arachnids. Practically nothing is known of the habits and venomology of our local spiders, and there is a need for research on such mundane insect types as the Brown Leatherwing Beetle (whose immature stages have not yet been found), the Canyon Fly (whose larval habitat is unknown), and the Jerusalem Cricket (whose native host plants have not been fully documented). I hope that this introduction to our insect fauna will stimulate entomologists—amateur and professional alike—to further study.

# ACKNOWLEDGMENTS

IN THE YEARS since its publication, I have kept my personal copy of the first edition of *Insects of the Los Angeles Basin* close by my desk at the Natural History Museum. It has become filled with notes, annotations, additions, and corrections as my experience with the local insect fauna increased. The literature expanded my knowledge on points of fact and newly appreciated species, and friends, colleagues, and the public provided me with many items of information.

To all who helped with the first edition I repeat my deepest appreciation. Many of the same individuals as well as other entomologists and biologists have contributed to this revision, and to these friends I am greatly indebted: Robert P. Allen (lacewings), Peter F. Bellinger (springtails), John R. Bryce (food inhabiting insects), Joe M. Cicero (beetles), Don G. Denning, deceased (caddisflies), John T. Doyen (beetles), Franklin Ennik (health pests), Arthur V. Evans (general), Joel Floyd (exotic insects), Don C. Frack (moths), Gordon Gordh (parasitoid wasps), Franklin Hall (health pests), David C. Hawks (moths), Blain R. Hebert (arachnids), James N. Hogue (general), Frank T. Hovore (beetles), Wendle R. Icenogle (spiders), David H. Kavanaugh (beetles), Graeme Lowe (scorpions), Robert J. Lyon (gall wasps), Scott E. Miller (general), Alan R. Olsen (food inhabiting insects), John H. Poorbaugh (flies), James A. Robertson (beetles), Rick Rogers (flies), Findley E. Russell (spiders), Rowland M. Shelley (myriapods), Elbert L. Sleeper (beetles), Fred S. Truxal (water bugs), David B. Weissman (Orthoptera and relatives), and Stanley C. Williams (scorpions).

For providing essential information on so many species, I give special thanks to the entomological staffs of the California Department of Agriculture in Sacramento (Fred G. Andrews, George M. Buxton, Eric M. Fisher, Ray J. Gill, Alan R. Hardy, Tokuwo Kono, George T. Okumura, Terry N. Seeno, and Marius M. Wasbauer) and the Orange County Vector Control District (Gilbert L. Challet, Rudy Geck, Stephen G. Bennett, and James P. Webb).

Special appreciation is also due my colleagues in entomology at the Natural History Museum of Los Angeles County for reviewing portions of the book and for generously offering so much to its improvement: Julian P. Donahue, Roy R. Snelling, Fred S. Truxal, and Chris Nagano. Steven R. Kutcher of Pasadena, California, and Rosser W. Garrison of the Los Angeles County Agricultural Commissioner's Office, also reviewed the entire manuscript and provided innumerable suggestions, additions, and corrections. Many of these reviewers made direct contributions of original information. Robin A. Simpson, Head of Museum Publications, is responsible for an outstanding accomplishment in pulling the diverse elements of the work together and creating the final product.

I would like to thank Peggy Zeadow for typing the manuscript, Ruth Ann DeNicola for drawings of ants, and Tina Ross for preparation of art work and her many excellent renderings.

I am also grateful to all those who graciously allowed me use of their photographs for this volume. These are Robert P. Allen, Max E. Badgley, Leland R. Brown, Peter J. Bryant, Elmer P. Catts, Jr., Donald C. Frack, Rosser W. Garrison and the Los Angeles County Agricultural Commissioner's Office (some photographers unknown), David Hawks, James N. Hogue, Frank T. Hovore, Alpha James, Alden Johnson, Jack N. Levy, Susan Marquez, Rudolf Mattoni, Chris Nagano, Phillip D. Nordin, Roy J. Pence, Lawrence S. Reynolds, James A. Robertson, Walter Sakai, Darwin L. Tiemann, and Keith V. Wolfe. Individual acknowledgments are found in the figure legends. "LACM photograph" indicates an illustration (photographer unknown) from the files of the Entomology Section, Natural History Museum of Los Angeles County. Kevin Hussey of the Imaging Processing Applications and Development Section, Jet Propulsion Laboratory of the California Institute of Technology, provided the Landsat and computer images used for the map on pages 10 and 11.

Finally, I acknowledge the support of the County of Los Angeles and the Natural History Museum Foundation for financing this publication.

# Part I
# INTRODUCTION

INSECTS AND THEIR TERRESTRIAL RELATIVES have inhabited the earth for a very long time. Fossils of the earliest, most primitive forms date from the Devonian Period over 350 million years ago. By the Age of Dinosaurs (in the late Paleozoic and Mesozoic Eras 300 million years before the present), insects were already abundant in numbers and kinds. And like the dinosaurs, some prehistoric forms were immense: the dragonfly *Meganeuropsis* had a wing span of 35 inches (75 cm), and the giant Mesozoic relatives of the scorpions, the eurypterids, measured over 6 feet (2 meters) in length.

The origins and evolution of insects and related groups is complex. Insects themselves undoubtedly descended from many-legged animals that were similar to present-day centipedes. The functional effectiveness of six legs was decided early, for this has been a constant feature among the orders of insects. Wings also gave considerable advantage to even the earliest insects, which dominated the air millions of years before the flying reptiles (pterosaurs), birds, or bats. Insects took advantage of multiple developmental stages (metamorphosis) to further diversify and thus increase their chances for survival. All of these characteristics contributed to the overriding success of insects as a group: today insect species outnumber those in all other living groups combined.

The history of the creatures that are the subject of this book is generally well known. But paleoentomologists are still working to unravel the particular lines of descent leading to the myriad species living today, as well as to the species that were present only in the past.

REFERENCES. Hennig, W. 1981. *Insect phylogeny*. New York: Wiley and Sons.

Manton, S. M. 1977. *The Arthropoda*. Oxford: Clarendon.

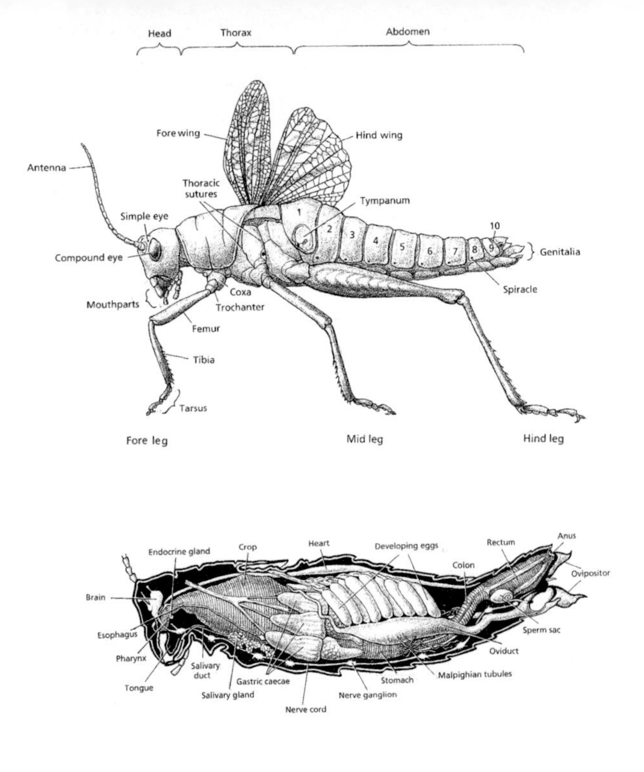


1. Principal external (top) and internal features of the insect body as seen in a common representative type, the grasshopper. Top: Drawing by T. Ross after J. Belkin, 1972, *Fundamentals of entomology,* p. 157 (Baltimore: American Biological Supply). Bottom: Drawing from R. Matheson, 1951, *Entomology for introductory courses,* fig. 52 (Ithaca, N.Y.: Comstock).

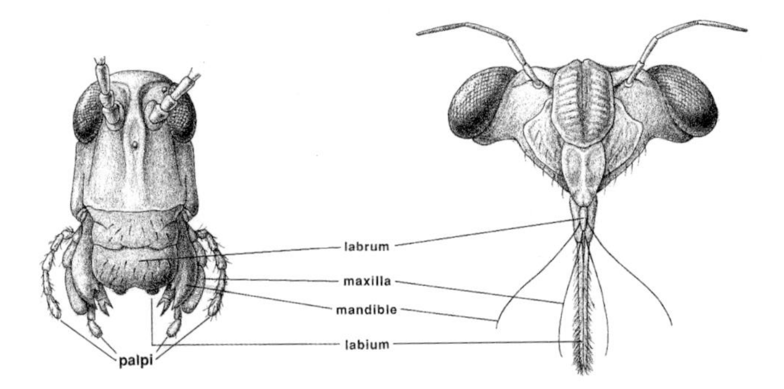


2. Major elements of insect mouthparts in a chewing type (grasshopper; left) and a sucking type (cicada; right). Drawing by T. Ross.

# 1 INSECTS IN GENERAL

## BODY STRUCTURE AND FUNCTION

INSECTS BELONG TO THE PHYLUM ARTHROPODA and therefore exhibit as their most characteristic features a hardened outer shell or *exoskeleton* and a segmented body with jointed appendages. The segments of the body are not all alike but are fused into three regions—the head, thorax, and abdomen, each of which serves its own set of functions. Typical external and internal features of insect anatomy are illustrated in Figure 1.

As with other animals, insects may be grouped according to their food preferences into *carnivores* (flesh feeders), *herbivores* (plant eaters), or *omnivores* (animals with catholic feeding habits). Carnivorous insect *predators* kill and consume their prey. But some carnivorous insects take nutrients from the tissues of other animals without killing them and are true *parasites;* a special category are the *parasitoids,* which live like parasites but eventually kill the host. The anatomy of a given insect reflects its feeding preferences to some degree.

The head is the chief center of orientation, bearing the eyes and supplementary sensory organs, the antennae. The mouth is surrounded by several structures collectively known as the mouthparts: these are the *labrum* (upper lip); *mandibles* (or jaws); *labium* (lower lip); *maxillae* (accessory jaws); and *hypopharynx* (tongue). The arrangement of these elements in chewing insects is illustrated in Figure 2.

Many insects take liquid food and accordingly possess a set of mouthparts modified for sucking or siphoning rather than for chewing. In these, the chewing elements are all present, but they are greatly elongated (for probing and piercing), and they interlock in such a way as to form a *proboscis* or beak with a food channel or canal running through it. An added element is a pumping apparatus located farther back in the head or thorax that, through a system of valves and pressure chambers, promotes the flow of fluid (for example, nectar, sap, or blood) into the digestive tract. The general anatomy of sucking mouthparts is illustrated in Figure 2.

An insect's brain functions much like our own, receiving external stimuli via the eyes and myriads of microscopic sensing structures that are located all over the body surface but are especially numerous on the antennae, *palpi* (antenna-like structures on the mouthparts) and *tarsi* (feet). The brain integrates the stimuli received from these sources and initiates reactions that collectively form the insect's behavior.

Insects have a number of smaller secondary or accessory "brains" (called *ganglia)* located along the ventral nerve cord, which lies in the floor of the body cavity. Each ganglion assumes some of the duties of the true brain and provides a degree of autonomy to the segment in which it lies. This explains the ability of an insect to live and move for some time after decapitation.

The thorax, or middle body region, bears the wings and legs and thus is the center of locomotion. It is composed of three segments *(prothorax, mesothorax,* and *metathorax),* the boundaries of which are not always clearly discernible (one segment may be greatly enlarged and dominate the others). A small triangular area *(scutellum)* on the rear of the thorax is often well developed and is useful in identifying some insects. Internally, the thorax is filled with the muscles and large ganglia needed to operate and control the appendages.

Wings, which many insects lack, are usually broad flattened membranous structures, although they may be greatly modified as in the hard horny outer wings of beetles. They are strengthened by a system of stout veins that diverge from the wing base and run throughout the membrane.

The legs, of which there are never more than three pair,* generally resemble each other; each is divided into four major sections that roughly correspond in function to our own leg regions: a basal coxa with associated trochanter (like the human hip); an elongate femur (thigh); a slender tibia (shin); and a terminal tarsus (foot). The tarsus itself consists of a number of smaller segments, the last of which is tipped with claws or other devices for securing the

---

*Supernumerary legs *(pseudopods)* are found on the abdomen in many immature insects, such as caterpillars, but these are of different origin and structure than the six legs located on the thorax.

insect's hold on the substrate. The abdomen contains vital organs concerned with digestion, excretion, blood circulation, and reproduction. Respiration is performed by a network of tubes (tracheal system) that ramify throughout the body and open to the atmosphere by means of breathing pores called *spiracles.* Oxygen entering these breathing pores passes through the tracheal tubes directly into the tissues; it does not travel in the blood, as it does in our bodies.

The digestive processes of insects are similar to those in the vertebrate body. Food is broken down by digestive fluids in the stomach and front part of the intestine and passes in molecular form through the intestinal wall into the blood.

The blood of an insect circulates freely in the body cavity, bathing the tissues and organs with nutrients, hormones, and other necessary substances. Thus, the circulatory system is of the "open" type—that is, there is no closed system of vessels through which the blood flows in a confined path. The only organ in the insect comparable to our own blood vessels is the so-called heart and its extension, the aorta, which runs the length of the insect in the roof of the body cavity. Blood in the abdomen passes into the heart through perforations and is propelled forward to the head by pulsating contractions of muscles in the walls of the aorta. Flooding from this vessel in the vicinity of the brain, the blood percolates rearward through the entire body, including the legs and wings, to the abdomen, where it is again picked up by the heart. This cycle constitutes blood circulation.

Excretion of metabolic wastes is the job of the insect's kidneys, called Malpighian tubules. These organs are long slender hollow tubes branching off the front part of the intestine. The cells that make up the walls of the tubules have the ability to selectively remove waste substances in solution from the blood and pass them to the intestinal cavity, from which they are later voided with the feces.

During the breeding season, the abdomen of the female of many insects contains eggs, which develop within the paired ovaries. As the eggs pass down the lateral ovarian ducts, they mature and are fertilized by the male sperm previously placed at mating time in a special side pouch off the main duct. Fertilization is usually internal among insects, although many species use alternative systems, some even reproducing entirely in the absence of the male *(parthenogenesis)*.

Before leaving the subject of insect structure and function, we should look at the special nature of the external body wall or exoskeleton, which serves primarily as a protective covering (preventing loss of vital moisture and entry of disease-causing microorganisms) and as a supportive skeleton (giving form to the body and providing a place for muscle attachment). The body wall has a unique chemistry that places it among the toughest and most resistant of all natural products. A principle ingredient is *chitin,* a substance that is made of complex crosslinked chain molecules like those in plastic and that is insoluble in most reagents, including dilute acids, alkalies, and the digestive enzymes of other animals. Chitin does not give the body wall its rigidity, however. Another substance, *sclerotin,* is largely responsible for this characteristic.

The body wall of the insect is firm and rigid only in certain areas. These rigid regions or plates (called *sclerites)* are connected by membranous zones of soft tissue, which permit movement between the plates. The total arrangement is something like the armor suits worn by knights of old.

A more complete discussion of the anatomy and physiology of insects and other terrestrial arthropods can be found in the sources listed below and in works listed in the bibliography at the end of this book.

REFERENCES. Chapman, R. F. 1969. *The insects: Structure and function.* New York: American Elsevier.

Daly, H. V., J. T. Doyen, and P. R. Ehrlich. 1978. *Introduction to insect biology and diversity.* New York: McGraw-Hill.

Kaestner, A. 1968. *Invertebrate zoology,* vol. 2. New York: Interscience, John Wiley.

## GROWTH AND DEVELOPMENT

LIKE MOST ANIMALS, an insect begins its life as an egg. Insect eggs come in many shapes and sizes and are deposited in a wide variety of places—on the surface of leaves, in and on twigs, between flower petals, in the ground, and in water—and almost always in the vicinity of a food source for the newly hatched young. Within the egg shell a period of embryonic development occurs that is analagous to that in the egg of a bird or reptile. The young insect hatches by eating an opening in the shell or by splitting the egg along some line of weakness.

A period of growth follows hatching. Before reaching maturity, the young insect may increase its size and weight many times. Growth does not proceed evenly and continuously as it does in mammals. Rather, because the body wall is nonliving, enlargement can continue only to the point at which the integument will stretch no more. Then, to allow for a further increase in size, the unrelenting skin must be shed by a process called molting. Young insects usually molt a number of times on their way to maturity.

The seemingly impossible trick of shedding its integument is performed routinely by an insect through the following steps: (1) The old outer skin becomes detached from the new skin through the action of enzymes secreted by epidermal cells; these enzymes dissolve the lower layers of the skin to be discarded. (2) Internal pressure, which the insect exerts by making writhing movements and swallowing air to increase its body volume, causes the old skin to split down the back. (3) The insect in its new skin wriggles through the fissure by degrees, stepping from the old skin as a farmer sheds his long underwear in the spring.

The obsolete skin must be firmly anchored to the ground while it is being discarded. As a necessary preliminary to the molting process, some insects embed the tips of their tarsi in the substrate; others spin a special silken pad to which their feet cling as they molt.

As it grows to adulthood, the developing insect may alter not just its size but also its body form and habits, through a phenomenal process of change known as *metamorphosis.* The degree of change varies, and most species exhibit one of three processes of reaching maturity:

■ Complete metamorphosis, or *holometabolism.* Insects such as butterflies and moths change radically as they mature, going through a worm-like caterpillar or larval stage and a mummy-like pupal stage (Figure 3); these species are said to undergo complete metamorphosis. If the species is to have wings as an adult, these develop internally within the larva.

Species that experience complete metamorphosis have four stages of development: egg, followed by larva (caterpillar, grub, maggot, etc.), followed by pupa or chrysalis, followed by adult or *imago.*

Complete metamorphosis is exhibited by ants, bees, and wasps; beetles; fleas; gnats and flies; butterflies and moths; caddisflies; and lacewings and antlions, among others.

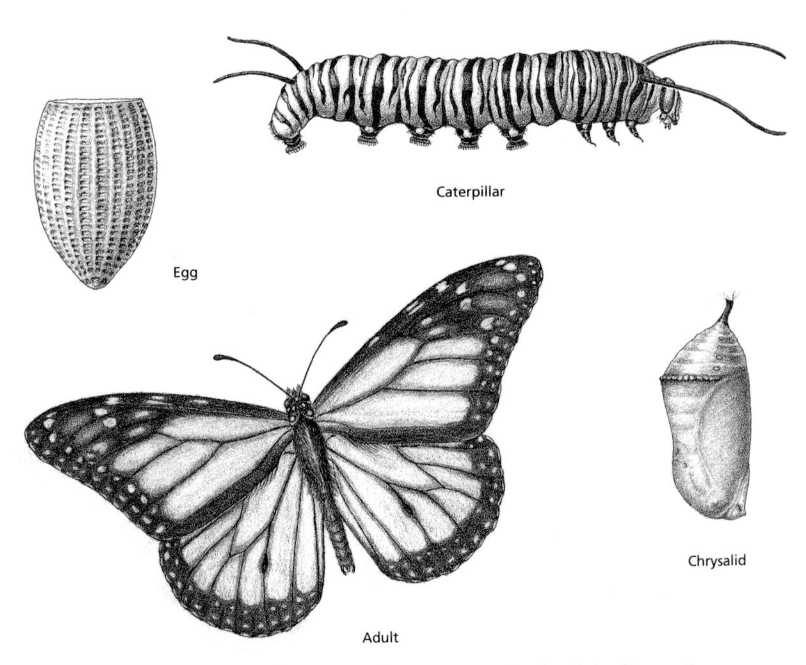


3. Stages in complete metamorphosis of the Monarch butterfly. Drawing by T. Ross after C. Riley, 1873, *Canadian Entomologist,* vol. 5, pp. 5-8.

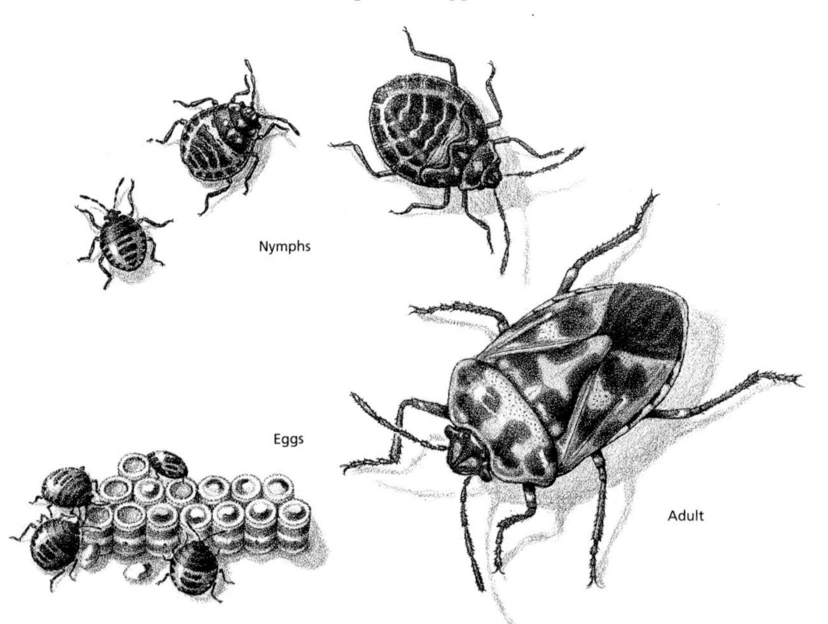


4. Stages in incomplete metamorphosis of the Harlequin Bug. Drawing by T. Ross after F. Chittenden, 1920, U.S. Department of Agriculture Farmer's Bulletins, no. 1061.

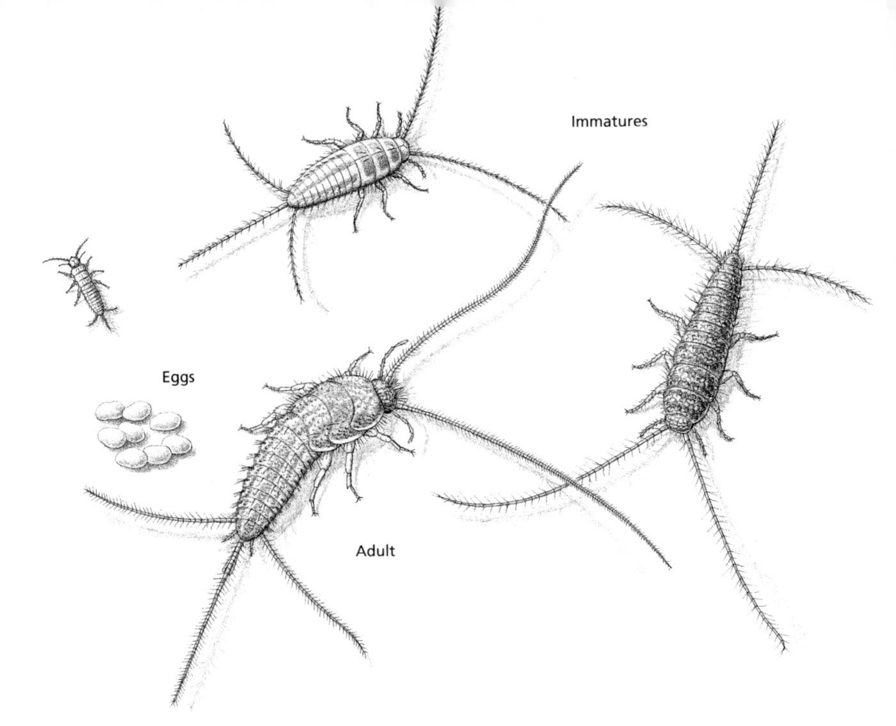


5. Stages in development without metamorphosis in the Silverfish. Drawing by T. Ross after C. Metcalf et al., 1962, *Destructive and Useful Insects,* fig. 5.8 (New York: McGraw-Hill).

■ Incomplete metamorphosis, or *hemimetabolism.* Here there is a moderate or relatively slight change in body form and habits during development, as in true bug nymphs (Figure 4). When wings are present in the adults, they have developed externally in pads or pouches on the nymph.

There are three stages in incomplete metamorphosis: egg, followed by nymph, followed by adult.

Incomplete metamorphosis occurs in thrips; true bugs; sucking and biting lice; bark lice; termites; grasshoppers, cockroaches, mantids, and walkingsticks; earwigs; dragonflies and damselflies; and mayflies.

■ No metamorphosis, or *ametabolism.* There is virtually no change in body form and habits during development of silverfish (Figure 5) and other insects in this category: the young differ very little from the adult except in size. Wings are never present in these species.

The insect develops in three stages: egg, followed by young, followed by adult.

Insects that do not undergo metamorphosis include silverfish and springtails.

As is true in humans, growth processes in insects are under the control and direction of hormones, chemical substances that are manufactured internally and released by special cells and organs into the insect's body and distributed to the target tissues via the blood. The balance of two hormones is responsible for maturation and molting: neotenin, the juvenile hormone, inhibits maturation; and ecdysone, the growth and differentiation hormone, promotes growth and molting of the immature insect and triggers the transformation through the growth stages to adulthood.

## CLASSIFICATION

AS A RESULT OF ORGANIC EVOLUTION, all living animals are believed to be related by heredity, having descended from simpler forms over the course of millions of years. Using the tools of paleontology, comparative anatomy, embryology, physiology, and genetics, the interrelationships of a group of organisms, such as the insects, can be studied, and a hypothetical diagram of descent (a phylogenetic or evolutionary tree) can be constructed. By circumscribing more and more inclusive branches on a phylogenetic tree, a hierarchical classification system can also be developed. The ends of the branches represent the finest divisions, and the more basal branches and the trunk, the higher categories. Such a diagram not only shows closeness or distance in the relatedness of different types of insects but also which insects are more or less complex in body form and function.

At the tips of the branches of the insect classification tree are the species of insects. A species is a group of individual animals that are very similar to each other and that can breed successfully among themselves to produce fertile offspring. To organize the species, those that are closely related are grouped into genera, related genera into families, and families into orders. The various orders of insects make up the class Insecta. At the base of this diagram are the phylum and kingdom to which this group of organisms belong: insects, with spiders and certain other fairly small invertebrates, are arthropods—members of the phylum Arthropoda. Finally, as animals, the Arthropoda are members of the kingdom Animalia.

Insects in a particular order share a few features in common; those in a genus have a greater number of similarities and are more closely related. Two closely related insects—the Convergent Lady Beetle

*(Hippodamia convergens)* and the Five-spotted Ladybird Beetle *(Hippodamia quinquesignata)* would appear in an evolutionary classification as follows:

| | |
|---|---|
| Kingdom | Animalia (animals) |
| Phylum | Arthropoda (arthropods) |
| Class | Insecta (insects) |
| Subclass | Pterygota (advanced insects with wings) |
| Order | Coleoptera (beetles) |
| Family | Coccinellidae (ladybird beetles) |
| Genus | *Hippodamia* |
| Species 1 | *convergens* |
| Species 2 | *quinquesignata* |

Each species is most precisely referred to by its scientific name, which consists of two words—the generic name and the specific name. Recognition of finer divisions, and of relationships that are closer than those among members of a species, is possible with the recognition of subspecies, which are assigned three-part names. For example, a butterfly known as the Square-spotted Blue *(Euphilotes battoides)* seems to have given rise to several subspecies, among them the El Segundo Blue *(Euphilotes battoides allyni)*, which is known only from the coastal sand dunes of southern California, and the Bernardino Blue *(Euphilotes battoides bernardino)*, which is found in the interior mountains of this part of the state (the subspecies of a species are normally geographically separated).

The classification and name of a species may change as more is learned about it and its relatives. The scientists who specialize in making classifications, called taxonomists, usually work in museums (and sometimes in universities) where the large collections of organisms that must be examined in determining relationships are kept.

The linear classification of the orders of insects given in Table 1 is a rudimentary representation of their evolutionary history. The number of groups represented and the name assigned to each group are simplified for our purposes here. Table 2 contains a list of the few other groups of terrestrial arthropods and insect-like creatures discussed in the last chapters of this book.

REFERENCE. Clarke, K. U. 1988. Impact arthropodisation. *Antenna* (Bulletin of the Royal Entomological Society of London), vol. 12, no. 2, pp. 49-54.

**Subclass Apterygota.** Primitive wingless insects with vestiges of legs on some abdominal segments (indicates closer relationship to the centipede-like insect ancestors than to winged insects); no metamorphosis.

| | |
|---|---|
| * Order Protura | Proturans: microscopic soil-inhabiting forms |
| Order Collembola | Springtails |
| * Order Diplura | Diplurans: small moisture-loving insects |
| Order Zygentoma | Silverfish |
| Order Microcoryphia | Bristletails |

**Subclass Pterygota.** Advanced insects with wings (wings secondarily lost in some); without abdominal legs in adult; metamorphosis present.

**Infraclass Paleoptera.** Wings incapable of folding over back of insect; metamorphosis incomplete; immatures aquatic.

| | |
|---|---|
| Order Ephemeroptera | Mayflies |
| Order Odonata | Dragonflies and damselflies |

**Infraclass Neoptera.** Wings capable of folding over back of insect; degree of metamorphosis and habitat of immatures varied.

**Division Hemimetabola.** Metamorphosis incomplete; nymph (with externally developing wings) similar to adult.

| | |
|---|---|
| * Order Plecoptera | Stoneflies: immatures aquatic; adults found near mountain streams |
| * Order Embiidina | Webspinners: small tree-bark and ground dwellers |
| Order Dermaptera | Earwigs |

Table 1. Orders of insects

| | |
|---|---|
| Order Orthoptera | Grasshoppers, crickets, katydids |
| Order Blattodea | Cockroaches |
| Order Mantodea | Mantids |
| Order Phasmatodea | Walkingsticks |
| Order Isoptera | Termites |
| * Order Zoraptera | Zorapterans; rare insects not found in the West |
| Order Psocoptera | Bark lice |
| Order Mallophaga | Biting lice, bird lice |
| Order Anoplura | Sucking lice |
| Order Heteroptera | True bugs |
| Order Homoptera | Homopterans (cicadas, aphids, scale insects) |
| Order Thysanoptera | Thrips |

**Division Holometabola.** Metamorphosis complete; larva (with internally developing wings) very different from adult.

| | |
|---|---|
| Order Neuroptera | Lacewings, etc. |
| * Order Mecoptera | Scorpionflies, not found in the Los Angeles Basin |
| Order Trichoptera | Caddisflies |
| Order Lepidoptera | Butterflies and moths |
| Order Diptera | Gnats, midges, and flies |
| Order Siphonaptera | Fleas |
| Order Coleoptera | Beetles |
| * Order Strepsiptera | Twisted-wing insects: rare, parasitic on other insects |
| Order Hymenoptera | Ants, bees, and wasps |

*Not covered in this book.

Table 2. Terrestrial invertebrates other than insects covered in this book.

| | |
|---|---|
| **Phylum Arthropoda** | Arthropods |
| Class Arachnida | Arachnids |
| Order Scorpionida | Scorpions |
| Order Solpugida | Sun spiders |
| Order Opiliones | Daddy long-legs |
| Order Araneae | Spiders |
| Order Acari | Mites and ticks |
| Class Chilopoda | Centipedes |
| Class Diplopoda | Millipedes |
| Class Crustacea | Crustaceans |
| Order Isopoda | Sowbugs, etc. |
| Order Amphipoda | Sand hoppers, etc. |
| **Phylum Mollusca** | Mollusks |
| Class Gastropoda | Gastropods (snails, slugs) |
| **Phylum Platyhelminthes** | Planarians, flukes, tapeworms |
| Class Turbellaria | Flatworms |

# 2 LOS ANGELES BASIN ENVIRONMENT AND ECOLOGY

THE GREATER LOS ANGELES AREA (pages 10–11) is accurately described as a basin—that is, an area largely enclosed by higher lands—because it is bounded on three sides by mountains. To the north are the San Gabriels; at the northwest are the Santa Susanas; and to the southeast are the Santa Anas. The only gaps in the basin's periphery are the narrow valley at the mouth of the San Gabriel River to the east and the coast of the Pacific Ocean to the southeast.

A complicated system of mountains and hills divides the basin into three main areas—the Los Angeles Plain, the San Fernando Valley, and the San Gabriel Valley. The most prominent barrier is the Santa Monica chain, rising highest in the west and declining eastward to form the Elysian Hills. The Los Angeles River neatly terminates the hills, with a sharp incision at Elysian Park. To the northeast lie the Verdugo, Repetto, and San Raphael hill systems. Completing the fragmentation of the greater basin are the Puente Hills, San Jose Hills, and Santa Ana Mountains to the east. Otherwise flat or gently sloping to the sea, the uniformity of the basin floor is broken only by minor prominences, such as the Baldwin Hills, Dominguez Hills, and Signal Hill.

Numerous small streams cut impressively steep canyons into the San Gabriels and Santa Anas and merge into three major river systems, the Los Angeles (with the Rio Hondo tributary), the San Gabriel (with Big Dalton and Coyote Creek tributaries), and the Santa Ana. None of these channels carries a sizable flow during most of the year, but each may develop raging torrents during winter downpours. Broad alluvial flood plains spreading over the basin floor are evidence of the disruptive force of rainwater falling on the high ridges of the San Gabriels (at times, at the phenomenal rate of up to an inch an hour) and then rushing down to the sea only 40 miles (65 km) distant.

## INSECT HABITATS

### ■ MICROHABITATS AND MACROHABITATS

BECAUSE INSECTS ARE SMALL ANIMALS, their activities are only indirectly influenced by the climate. They are instead most profoundly affected by changes in their

| Microhabitat | Possible Insect Residents |
|---|---|
| On the exposed surfaces of leaves | aphids |
| In contact with the ground under stones and boards | ground beetles |
| Among the hairs on the bodies of mammals | Cat Flea |
| Among feathers on the bodies of birds | Slender Pigeon Louse |
| Within the tubular flowers of many plants | thrips |
| Within tunnels chewed in dead decaying wood | Pacific Dampwood Termite |
| Beneath the surface of loamy soil | Jerusalem crickets |
| In the bottom mud of stagnant ponds | water midge larvae |
| Attached to the surfaces of rocks in swiftly flowing streams | black fly larvae |
| Within the flesh of fresh carrion | flesh fly larvae |
| Within the flesh of old dry carrion | dermestid beetle larvae |
| Within tunnels bored in the stems of living plants | Yucca Weevil larvae |
| Among the hairs on the human head | Human Louse |
| In human hair follicles | Hair Follicle Mite |
| Under the kitchen sink | American Cockroach |
| Under the bark on the trunk of a dead rotting oak tree | Ironclad Beetle |
| In fresh cattle droppings in a pasture | Drone Fly larvae |
| On the surface of still ponds | Common Water Strider |
| In pools of crude oil near oil derricks and natural oil seeps | Petroleum Fly |
| In rotting kelp on the beach | Stable Fly larvae |

Table 3. Twenty selected examples of insect and arachnid microhabitats of the Los Angeles Basin

immediate environment—all the variables of temperature, moisture, space, and food that delimit their "homes" or places of residence in the world. These places are usually referred to as *microhabitats.*

Conditions in a particular microhabitat may differ greatly from those in the general surroundings or *macrohabitat.* For example, loamy soil that is 7 to 10 inches (18 to 25 cm) deep, a microhabitat of many insects, has been shown to have an average minimum winter temperature higher than that of the air at the surface above it. The significance of this is greatest when air temperatures are at or below freezing and soil temperatures are above freezing. It is easy to see how such small differences between macro- and microhabitat conditions may prohibit or permit an insect's winter activities, for example, or its very existence.

Insects frequently live quite well in microhabitats that are completely surrounded by decidedly hostile general environments. Consider, for example, the desert, with summer air temperatures that are commonly above 100° Fahrenheit (38° Celcius) and humidity below 5 percent; this is the home of the soft-bodied, moisture-loving larvae of the Cactus Fly. The larvae could not survive were it not for the cooler, more humid interior of rotting cactus stems, which provides them with a suitable habitat in an otherwise inhospitable environment.

Additional examples of insect microhabitats are given in Table 3. These are the places where one might look to collect specimens and observe the marvels of adaptation so characteristic of this class of animals. If amplified or made more specific, the list would go on almost indefinitely; but twenty examples will suffice to show the variety of microhabitats used by both common and unusual types of insects.

The macrohabitats of insects are not as diverse nor as numerous as the microhabitats, because they are defined by climate, soil, terrain, vegetation, surface water, and human activity, and these factors vary little and over geographic areas that are relatively broad, at least from an insect's perspective. Yet changes in the macrohabitat sometimes have very important specific effects on insects.

The life cycles of almost all species are controlled by climatic factors, principally temperature and moisture. Sharp variations in temperature and humidity resulting from Santa Ana winds over the basin may trigger emergence of some forms, such as owlet moths. The first winter rains bring out others, such as rain beetles. Smog may even disturb insects as it does humans, but there is as yet no scientific knowledge about its effects. Differences between our four climate types—Maritime, Intermediate Valley, Mountain, and Transitional—are not extreme, and there are few insects limited to any one of them: the most distinctively different insect fauna is that of the mountains (the Santa Monicas and the San Gabriel slopes and foothills) because these areas tend to have appreciably higher rainfall and lower temperatures than the others.

Apart from special silts and muds, such as tidal marsh, alkali flat, and river wash, there are four major types of soils in the basin, which are classified according to origin:

■ Residual soils are derived from the disintegration and weathering-in-place of granite ("decomposed granite") and other consolidated rocks.

■ Old valley-filling and coastal plain soils result from the weathering of old sedimentary deposits.

■ Recent alluvial soils are derived from material transported from the various drainage basins.

■ Wind-laid soils are derived from beach sands blown inland into dunes.

Most of our insects are not affected by soil conditions except indirectly, in that the soil dictates the types of vegetation that will be present in an area. However there are species with narrow tolerances of variables in the subterranean realm. Many underground species—Jerusalem crickets and trapdoor spiders, for example—require soil of sufficiently pliable texture or moisture-holding capacity to permit burrowing. Generally, our soils are light-colored, low in organic and mineral content, without profile, and covered with little or no vegetable litter, except in Riparian and Oak-Woodland canyons. Consequently, the rich insect fauna (springtails and mites, for example) typical of humic forest floors is for the most part absent in the basin.

Standing and running surface water is scarce except in impoundments. There are several artificial lakes or reservoirs, but their insect fauna is depauperate because they are paved or covered or lack emergent shore vegetation. Our streams and rivers are, for the most part, channelized and intermittent, flowing only during the rainy months and then with scouring velocity. The best natural ponds and streams for supporting insect life are in the Santa Monica Mountains and San Gabriel foothills. Our once numerous artesian wells, sloughs, and marshes (Madrona and Bixby Sloughs, Bouton Lake), and our estuary salt marshes (Alamitos and Bolsa Chica Bays), have been virtually destroyed by land fills and other construction. They were the habitat of some insects that are now locally extinct or rare.

Although terrain is an important modifier of climate, it has little immediate effect on insects. The hilly and mountainous regions of the basin are presently richest in insect diversity and abundance, mainly because they have remained relatively free of human disturbance. Vestiges of natural surroundings still persist here and there but are even now coming under the modifying influence of property development.

No single component of the environment in itself directly limits the occurrence of any species. Rather, it is the combination of all of them that defines the macrohabitats of the basin and influences the distribution of plants, people, and insects.

The multiplicity of basin macrohabitats may be organized into two general categories: natural and artificial. The examples given in Table 4 were chosen because they are the most familiar; the insects characteristic of each may be considered indicators of the community in the same way as the plants.

Although some tolerant species freely cross habitat boundaries and occur over the entire basin, most are found in only one of the environments described in Table 4. Most herbivores eat only one or a few species of plants and thus are restricted to the community where those plants occur. This is the reason that the California Sister butterfly, for example, whose food is the Canyon Oak, lives only in the Riparian Woodland, which is the habitat of the tree. Carnivores, parasitoids, and parasites that prey on specific herbivores are in turn limited to the ranges of their hosts.

Our region's insect fauna is, of course, not strictly limited by the perimeter of the Los Angeles Basin as defined for this book. Species encroach westward from the neighboring San Gabriel mountains and desert. Many species included in the book are marginal in this respect, being more properly adapted to high elevations or dry conditions but showing up on occasion within the area. Species from the San Gabriel Mountains—the Pine Sawyer, for example—sometimes drift down to the foothills through major river canyons. Pockets of moisture-deficient environments in the basin, such as the El Segundo Sand Dunes, contain isolated populations of

NATURAL

Essentially undisturbed wild or native areas; the original soil and vegetation* are still present.

| Macrohabitat | Example | Characteristic or Indicator Species: Plants | Characteristic or Indicator Species: Insects |
|---|---|---|---|
| **Rocky Seashore:** Rocks between and above tides along the coast | Malibu Beach | Kelp and seaweed | Tidepool Springtail *(Entomobrya laguna)* |
| **Coastal Strand:** Sandy beaches and dunes scattered along the coast; low or prostrate vegetation, commonly succulent perennials | Dockweiler Beach<br>El Segundo Sand Dunes | Shore sandbur *(Ambrosia)*<br>Saltbush *(Atriplex)*<br>Iceplant *(Carpobrotus edulis)*<br>Shore Morning Glory *(Convolvulus soldanella)* | Sand wasps *(Bembix)*<br>Kelp flies *(Fucellia* and *Coelopa)*<br>El Segundo Blue *(Euphilotes battoides allyni)* |
| **Coastal Salt Marsh:** Salt marshes along the coast, from sea level to 10 feet (3 meters)—now almost completely destroyed; low herbs or shrubs, commonly succulents. | Alamitos Bay<br>Seal Beach inlets<br>Sunset Beach<br>Ballona Wetlands<br>Malibu Lagoon | Inkweed *(Suaeda californica)*<br>Salt Grass *(Distichlis spicata)*<br>Sea Heath *(Frankenia grandifolia)* | Pygmy Blue *(Brephidium exilis)*<br>Wandering Skipper *(Panoquina errans)*<br>Oregon Tiger Beetle *(Cicindela oregona)* |
| **Freshwater Marsh:** Inland bodies of standing or sluggish water—now largely destroyed; marshy vegetation. | Madrona Marsh<br>Harbor Lake (Bixby Slough) | Tules *(Scirpus)*<br>Bulrushes and cattails *(Typha)*<br>Sedges *(Carex)* | Elliptic Water Scavenger *(Tropisternus ellipticus)*<br>Big Red Skimmer *(Libellula saturata)* |
| **Coastal Sage Scrub:** Clay slopes of high inland hills and coastal mountains. Open growth of small shrubs. | Verdugo Hills<br>Griffith Park | California Sagebrush *(Artemisia tridentata)*<br>Black Sage *(Salvia mellifera)*<br>California Buckwheat *(Eriogonum fasciculatum)*<br>Our Lord's Candle *(Yucca whipplei)* | Chalcedon Checker-spot *(Euphydryas chalcedona)*<br>California Mantid *(Stagmomantis californica)*<br>Yucca Moth *(Tegeticula maculata)* |

*The vegetation is well characterized by use of the system of plant communities devised by Munz and Keck in *A California flora* (University of California, 1959; pp. 10-18), which is largely followed here.

Table 4. Insect macrohabitats of the Los Angeles Basin

| | | Characteristic or Indicator Species | |
|---|---|---|---|
| Macrohabitat | Example | Plants | Insects |
| **Chaparral:** Mountain slopes, mostly above 1000-foot (300-meter) contour (the upper limit of Coastal Sage Scrub). Dense growth of tall thick-leaved shrubs. | Middle slopes of San Gabriel Mountains | Chamise *(Adenostoma)*<br>Scrub Oak *(Quercus dumosa)*<br>Wild lilacs *(Ceanothus)*<br>Manzanitas *(Arctostaphylos)* | Timemas *(Timema)*<br>Ceanothus Silk Moth *(Hyalophora euryalus)*<br>Pale Swallowtail *(Papilio eurymedon)* |
| **Southern Oak Woodland:** Inland valleys and canyons. Trees with grass beneath. | Woodland Hills | Coast Live Oak *(Quercus agrifolia)*<br>Valley Oak *(Quercus lobata)* | California Oak Moth *(Phryganidia californica)*<br>Ironclad Beetle *(Phloeodes pustulosus)* |
| **Walnut Woodland:** Dense growths of California Walnut on steep hillsides | Mount Washington<br>Rowland Heights | California Walnut *(Juglans californica)* | Walnut Underwing *(Catocala piatrix)* |
| **Grass-Herbland:** Intermittent patches between areas dominated by Coastal Sage Scrub: Native perennials outnumbered by introduced grasses and weeds. | Original vegetation of metropolitan Los Angeles; now totally extinct. | Native perennial herbs, introduced weedy annuals and grasses | California Trapdoor Spider *(Bothriocyrtum californicum)*<br>Red-winged Grass Cicada *(Tibicinoides cupreosparsus)* |
| **Riparian Woodland:** Along streams and around ponds. Trees with herbaceous or shrubby undergrowth | Big Tujunga Canyon<br>Chevy Chase Canyon<br>Rustic Canyon<br>Topanga Canyon | Sycamore, alders, maples, cottonwoods, bays<br>Nettle, *Baccharis*, and willow undergrowth | Lorquin's Admiral *(Limenitis lorquini)*<br>Velvety Tree Ant *(Liometopum occidentale)*<br>Western Tiger Swallowtail *(Papilio rutulus)*<br>California Glowworm *(Ellychnia californica)* |

| Macrohabitat | Example | Characteristic or Indicator Species: Plants | Insects |
|---|---|---|---|
| **Dry River Beds:** Washes, arroyos, and basins below mountains where water is seldom present; vegetation is partly riparian, partly Coastal Sage Scrub, but very sparse. During long dry periods, Coastal Sage may infringe and isolated Chaparral species may occur | San Gabriel Wash | (Partly Riparian, partly Coastal Sage; sparse) | Velvet-ants *(Dasymutilla)*<br>Harvester ants *(Pogonomyrmex)*<br>Checkered Metalmark butterfly *(Apodemia mormo)* |
| **Aquatic Areas:** Natural streams, ponds, and lakes and waterways as yet unpaved. | Arroyo Seco River above Devil's Gate Dam<br>Topanga Creek<br>San Gabriel Wash | Willows *(Salix),* mulefat *(Baccharix),* rushes *(Juncus)* | Backswimmers *(Notonecta)*<br>California Dobsonfly *(Neohermes californicus)*<br>Giant Crane Fly *(Holorusia hespera)* |

ARTIFICIAL

Areas disturbed, created, or modified to a considerable degree by people.

| Macrohabitat | Example | Plants | Insects |
|---|---|---|---|
| **Uncultivated Areas:** Ground that is unused or with only transient human occupancy. | | | |
| 1. *Untenanted Areas:* | Vacant lots; fallow fields | Weeds: wild radish, wild mustard, cheeseweed | Painted Lady *(Vanessa cardui)*<br>Harlequin Bug *(Murgantia histrionica)*<br>European Cabbage Butterfly *(Pieris rapae)* |
| 2. *Grass-covered hills:* Some in use as grazing areas. | Repetto Hills<br>Baldwin Hills | Wild oats *(Avena fatua)* and other nonnative grasses | Band-wing Grasshopper *(Dissosteira pictipennis)*<br>Alfalfa Looper *(Autographa californica)* |

| Macrohabitat | Example | Characteristic or Indicator Species: Plants | Characteristic or Indicator Species: Insects |
|---|---|---|---|
| **Cultivated Areas:** Agricultural, horticultural, and recreational areas; other planted places that are continually tended | | Various introduced weeds and cultivated plants and trees | |
| 1. *Agricultural areas.* | Planted fields<br>Fruit orchards<br>Nut groves | | Cutworm moths (Noctuidae)<br>Green Fruit Beetle *(Cotinus mutabilis)* |
| 2. *Livestock areas.* | Dairies, farms, ranches, stables, aviaries, zoos | | Stable Fly *(Stomoxys calcitrans)* |
| 3. *Gardens.* | County and State Arboretum<br>Backyards<br>Nurseries, greenhouses<br>Indoor plantings | | Broad-winged Katydid *(Microcentrum rhombifolium)*<br>Spider mites (Tetranychidae)<br>Honey Bee *(Apis mellifera)* |
| 4. *Turf areas.* | Forest Lawn Memorial Park<br>Brookside Golf Course | | Fiery Skipper *(Hylephila phyleus)*<br>Southern Chinch Bug *(Blissus insularis)* |
| 5. *Road and freeway embankments, airport runways, powerline and railroad easements.* | San Diego Freeway shoulders; LAX runway | | Blue-green Sharpshooter *(Graphocephala atropunctata)* |
| 6. *Planted woodlands.* | Windbreaks<br>Parks, poplar groves | | Eucalyptus Long-horn Borer *(Phoracantha semipunctata)*<br>Sow Bug Killer *(Dysdera crocota)*<br>Elm Leaf Beetle *(Xanthogaleruca luteola)* |

| Macrohabitat | Example | Characteristic or Indicator Species: Plants | Insects |
|---|---|---|---|
| **Storage Areas:** Large areas that are infrequently visited by people | Lumber yards, granaries, warehouses, movie backlots, dumps, landfills, compost piles | Weeds | Western Black Widow *(Latrodectus hesperus)*<br>Argentine Ant *(Iridomyrmex humilis)*<br>Green Bottle Fly *(Phaenicia sericata)* |
| **Structures** | Homes, commercial buildings, parking lots, amusement parks, pet houses | | American Cockroach *(Periplaneta americana)*<br>Tropical Rat Mite *(Ornithonyssus bacoti)*<br>House Centipede *(Scutigera coleoptrata)*<br>Cat Flea *(Ctenocephalides felis)* |
| **Aquatic Environments** | | | |
| 1. *Standing water:* Reservoirs, fish ponds, swimming and wading pools, fountains, sewage treatment ponds, settling basins, wells, golf course hazards | Silver Lake Reservoir<br>Johnson's Lake<br>Legg Lake | Algae, introduced water plants | Water midges *(Chironomus)*<br>Foul Water Mosquito *(Culex peus)* |
| 2. *Canals:* Tidal or stagnant brackish channels near the coast. | Venice canals | Sedge *(Scirpus)* | Marsh Boatman *(Trichocorixa reticulata)* |
| 3. *Flood control channels and basins:* Concrete canals (where pavement is broken and the water is still unpolluted, an essentially natural Riparian or native flora may return) | Los Angeles River bottom<br>Hansen Dam<br>Santa Fe Dam Recreational Area | Algal scum, *Potomageton, Salix, Cyperus junens* (umbrella plant) | Punkie *(Leptoconops foulki)*<br>Oregon Tiger Beetle *(Cicindela oregona)* |
| 4. *Water-collecting containers.* | cans, bottles, cemetery flower urns, cisterns, discarded automobile tires | | Cool Weather Mosquito *(Culiseta incidens)* |

6 (opposite). Yucca Moth (*Tegeticula maculata*) resting within the blossom of the Common Yucca (*Yucca whipplei*). The two species live in symbiosis: the female moth transports pollen from flower to flower, thus ensuring fertilization and development of the plant's seed pod, in which the moth's larvae will feed. The specialized relationship between yucca and yucca moth represents an extreme example of the value of insects as pollinators, in that the plant is rarely able to pollinate itself by any other means. Photograph by D. Frack.

sand roaches and other types more characteristic of southern California deserts. Insects frequently arrive in the basin on objects brought in from other environments (for example, firewood containing beetles).

REFERENCES. Anonymous. 1969. Report and general soil map of Los Angeles County. U.S. Department of Agriculture, Soil Conservation Service.

Editors of Sunset Books and Sunset Magazine. 1979. The West's 24 climate zones. Pages 8-27 in *Sunset New Western Garden Book,* 4th edition. Menlo Park, Calif.: Lane.

Jaeger, E., and A. Smith. 1966. *Introduction to the natural history of southern California.* Natural History Guide Series. Berkeley, Calif.: University of California.

Kevan, D. K. McE. 1962. *Soil animals.* New York: Philosophical Library.

Munz, P. A., and D. D. Keck. 1959. California plant communities. Pages 10-18 in *A California flora.* Berkeley, Calif.: University of California.

Ornduff, R. 1974. *Introduction to California plant life.* Berkeley, Calif.: University of California.

Yerkes, R. F., T. H. McCulloh, J. E. Schoellhamer, and J. G. Vedder. 1965. Geology of the Los Angeles Basin, California: An introduction. U.S. Geological Survey, Professional Papers, no. 420-A.

The few ecological or faunal compilations that are relevant to insect habitats of the Los Angeles Basin or nearby areas are listed in the general bibliography, under "Insect fauna of the Los Angeles Basin."

## ■ CONSERVATION

OF ALL ENVIRONMENTAL PRESSURES on the insect fauna of the Los Angeles Basin, human activity has had the most profound effect. Urban density is steadily and inexorably increasing, and natural habitats are disappearing under acres of asphalt and concrete. Intensive use of insecticides and the clean culture of plants, introduction of alien vegetation, conversion of land to monoculture, air and water pollution, and use of off-road vehicles are examples of people's activities that compromise living space for insects and other small terrestrial arthropods.

Habitat destruction is resulting in the elimination, one by one, of insect populations *(extirpation)* and even species *(extinction)*. Some well-known examples of extirpation in relatively recent times are the loss of the El Segundo Blue butterfly except at the Los Angeles International Airport dunes, and the probable disappearance of the Giant Water Scavenger Beetle *(Hydrophilus triangularis)* and Giant Water Bug *(Lethocerus americanus)*. The great decline in the but-

terflies in the city and its outskirts is very evident to collectors old enough to have lived here since the 1920s and 1930s. The obliteration of the very limited macrohabitat of the Palos Verdes Blue butterfly has just recently resulted in its complete extinction.

The negative impacts of human activities on native basin insects probably began thousands of years ago with overhunting of the Pleistocene mammal fauna. Dung beetles, which lay eggs in and feed as larvae on the droppings of large herbivorous mammals, are known through fossils at the La Brea Tar Pits (species of *Copris* and *Onthophagus)* but are no longer found locally.

Eventually, if the trend continues, our only insects will be those that live in our homes or are adapted to the conditions created as a result of human actions. Our only herbivorous insects will be those that feed on agricultural, ornamental, or weed plants; our only parasites and carnivores will be those that prey on the few herbivores, on our pets and domestic animals, or on us; we will have practically no aquatic insect fauna; and the species adapted to living in trash and garbage will be rampant. From time to time a native species may wander into the basin from the neighboring mountains or deserts. Perhaps a few of those indigenous and precariously adapted species of special beauty or interest, such as the Western Short-horned Walkingstick, the Yucca Moth (Figure 6), and the El Segundo Blue, will be preserved in small nature parks.

This scenario could be avoided. But at present the conservation of insect life is not a popular or widely understood cause. Because of an overemphasis on controlling injurious species, the general public tends to consider all insects bad and worthy only of eradication efforts. This, of course, is a patently ignorant and illogical viewpoint because of the tremendously important roles most insects play in our lives—as pollinators, as food for other life forms, as reducers of organic waste, and as regulators of the populations of other organisms. Many insects are simply aesthetically valuable—as ornamental as flowers or birds.

Fortunately, some appreciation for the worth and beauty of insects is slowly developing among California residents (see Monarch butterfly). Insects are now considered in Environmental Impact Reports, and many gardeners and landscapers plan "butterfly gardens." It remains to be seen, however, if the populace and politicians will expend sufficient effort at conservation to save the majority of species, the lesser known or humbler insects.

## LOS ANGELES BASIN INSECT FAUNA

THE INSECT FAUNA of the greater Los Angeles area, like that of any such region, is composed of a mixture of species with diverse origins. A detailed analysis of the geographic history of the local forms has never been made. It is also not known exactly how many species there are, but estimates of between 3,000 and 4,000 species are probably correct and in line with what would be expected in an area of similar size and ecological diversity in California. The majority of these species—at least 90 percent—are natives, and a very few of the native types—such as the El Segundo Blue butterfly—are *endemic* (found nowhere else).

Most of our insects belong to genera typical of North America and to some extent Eurasia. We do have some ancient intruders from the tropics to the south, such as the Monarch butterfly, but by and large our fauna is that of the Holarctic zoogeographic region and includes insects found in the northern parts of both the Old and New Worlds.

Although made notorious by publicity, the number of insect species in the basin that threaten human health or economic welfare is fortunately very small—certainly less than 1 percent of the total species present. The remainder should be considered beneficial or benign.

---

### ■ PREHISTORIC INSECTS

A PARTICULARLY FINE RECORD of the insects that were living in a portion of the Los Angeles Basin 10,000 to 40,000 years ago (in the upper Pleistocene geologic period) is preserved in the famous La Brea Tar Pits located at Rancho La Brea in Hancock Park on Wilshire Boulevard in Los Angeles. Thousands of specimens have been extracted from the asphalt deposits, in company with the bones of saber-toothed cats, giant ground sloths, and mammoths.

The following are the most common types of insects found in the tar pits.

■ Aquatic beetles (families Hydrophilidae and Dytiscidae) and bugs (Notonectidae, Belostomatidae). The La Brea site offered the riparian environment required by these species. Some of the aquatic insects found in the Rancho La Brea deposits may have mistaken an asphalt seep for water and become entrapped after landing on or crawling onto its sticky surface.

■ Carrion beetles (Silphidae) and flies (Sarcophagidae, Calliphoridae). These terrestrial insects fed upon the decaying corpses of birds and mammals that had become mired in the tar; their remains are often found with the bones of the vertebrate fossils.

■ Ground-dwelling beetles (Carabidae, Tenebrionidae). These insects may have become trapped in the sticky asphalt during the course of their terrestrial wanderings.

Well over 100 species of insects and related terrestrial arthropods have been identified from the La Brea deposits, and many more are certain to come to light as the material is studied by paleoentomologists. Most species are near relatives of modern insect residents of the basin or are still found in the area today. Only two species represent apparent extinctions: both are dung beetles (genera *Copris* and *Onthophagus*), which must have fed on the feces of the mammals that were abundant near the pools. Today, close relatives of these beetles occur no closer to Los Angeles than southern Arizona.

REFERENCES. Doyen, J. T., and S. E. Miller. 1980. Review of Pleistocene darkling beetles of the California asphalt deposits (Coleoptera: Tenebrionidae, Zopheridae). *Pan-Pacific Entomologist,* vol. 56, pp. 1-10.

Harris, J. M., and G. T. Jefferson, editors. 1985. *Rancho La Brea: Treasures of the tar pits.* Science Series, no. 31. Los Angeles: Natural History Museum of Los Angeles County.

Miller, S. F. 1983. Late Quaternary insects of Rancho La Brea and McKittrick, California. *Quaternary Research,* vol. 20, pp. 90-104.

Miller, S. E., and S. B. Peck. 1979. Fossil carrion beetles of Pleistocene California asphalt deposits, with a synopsis of Holocene California Silphidae (Insecta: Coleoptera: Silphidae). *Transactions of the San Diego Society of Natural History,* vol. 19, pp. 85-106.

Miller, S. E., R. D. Gordon, and H. F. Howden. 1981. Reevaluation of Pleistocene scarab beetles from Rancho La Brea, California (Coleoptera: Scarabaeidae). *Proceedings of the Entomological Society of Washington,* vol. 83, pp. 625-630.

Pierce, W. D. 1946. Fossil arthropods of California, part 10: Exploring the minute world of the California asphalt deposits. *Bulletin of the Southern California Academy of Sciences,* vol. 45, pp. 113-118.

_______.1946. Descriptions of the dung beetles (Scarabaeidae) of the Tar Pits. *Bulletin of the Southern California Academy of Sciences,* vol. 45, pp. 119-132.

_______. 1947. Fossil arthropods of California, part 13: A progress report on the Rancho La Brea asphaltum studies. *Bulletin of the Southern California Academy of Sciences,* vol. 46, pp. 136-143.

_______. 1949. Fossil arthropods of California: The silphid burying beetles in the asphalt deposits. *Bulletin of the Southern California Academy of Sciences,* vol. 48, pp. 55-70.

Stock, C. [1949] 1992. *Rancho La Brea: A record of Pleistocene life in California,* revised by J.M. Harris. Science Series, no. 37. Los Angeles: Natural History Museum of Los Angeles County. [See especially "Invertebrates."]

## ■ ALIEN INSECTS

QUITE A FEW of our species are relatively recent exotic immigrants. Some of these (the Green Fruit Beetle, for example) have arrived on their own and are here to stay. Others are seasonal or occasional migrants; good examples of these are the Black Witch moth and Senna Sulfur butterfly, which arrive from the south during our warm months but lack the ability to survive the cold winters.

A considerable number of species is known to have been introduced with the help of human beings, usually inadvertently but very rarely for a definite purpose. Many agricultural and public health pests were accidentally introduced by people; prime examples are the varied fruit flies that are detected by authorities almost every year, some after they have become established (see Appendix A). The circumstances surrounding these introductions are often historically interesting. For example, the Gulf Fritillary butterfly probably arrived in the Los Angeles area with early Mexican settlers, who brought it along unknowingly with its passion vine host, a plant that held special religious significance to these people. However, the exact time, place, and method of arrival of most of the alien insects introduced by people (for example, all of the domestic cockroaches) are usually obscure.

Purposeful introductions have been made primarily for biological control. Many beneficial alien species of parasitoid flies (Cottony Cushion Scale Killer, *Crytochaetum iceryae)* and wasps (genus *Aphytis)* and predatory beetles (genus *Vedalia)* now reside in the basin, helping to protect us from pest species. We have introduced some vegetarian insects with specific tastes for plants that we consider undesirable, and these immigrants are now doing considerable service in weed control. These insect helpers have usually been imported from lands originally native to their host species.

New arrivals are certain, and we can expect additions to the list of local insects. Some emigrants are not altogether welcome: the so-called Killer Bee is slowly expanding its range northward from Brazil, through Central America and Mexico, toward the United States and California.

REFERENCE. Dowell, R. V., and R. Gill. 1989. Exotic invertebrates and their effects on California. *Pan-Pacific Entomologist,* vol. 62, pp. 132-145.

## ■ INSECTS AND BRUSHFIRES

BRUSHFIRES ARE A WAY OF LIFE to Angelenos. Fire regularly consumes large acreages of grassland, coastal sage, and chaparral vegetation in and around the basin. Most insects living in fire-striken areas are destroyed in the blaze, although some may escape to neighboring terrain. As plants regenerate following a burn, the insects return as well, and they aid the reestablishment of the plant community by pollinating fire annuals and helping to decompose dead wood.

Like their plant counterparts, some chaparral insects seem to be fire-adapted. The "fire beetles" (genus *Melanophila,* family Buprestidae) are actually attracted to heat and smoke and may arrive in a burning plot even before the flames recede (the beetles sometimes land on fire fighters!). Smoke flies (family Platypezidae) swarm in the smoke of campfires and may also be among the first to arrive on postfire scenes. The larvae of the beetles bore in dead wood, and those of the flies feed on the fungi that grow on wood, which would explain the appearance of both types where numerous trees and shrubs have been killed by fire.

REFERENCES. Evans, W. G. 1966. Perception of infrared radiation from forest fires by *Melanophila acuminata* De Geer (Buprestidae, Coleoptera). *Ecology,* vol. 47, pp. 1061-1065.

_______. 1973. Fire beetles and forest fires. *Insect World Digest,* vol. 1, no. 3, pp. 14-18.

Force, D. C. 1981. Postfire insect succession in southern California chaparral. *American Naturalist,* vol. 117, pp. 575-582.

_______. 1982. Postburn insect fauna in southern California. Pages 234-240 in U.S. Forest Service, Pacific Southwest Forest and Range Experiment Station, General Technical Report, PSW-58.

Kessel, E. L. 1952. Another American fly attracted to smoke. *Pan-Pacific Entomologist,* vol. 28, pp. 56-58.

Linsley, E. G. 1943. Attraction of *Melanophila* beetles by fire and smoke. *Journal of Economic Entomology,* vol. 36, pp. 341-342.

## ■ INSECTS ON SANTA CATALINA ISLAND

ALTHOUGH CATALINA ISLAND is not a part of the Los Angeles Basin, its close proximity makes it deserving of some comment here. The insects there are mostly of the same species and found in like proportions as those of the mainland. These are mainly chaparral and coastal sage types, although the lengthy and varied shoreline supports a well-developed assemblage of marine insects.

Because of its isolation, Catalina like many islands has a number of endemic insect species, including the Avalon Hairstreak butterfly *(Strymon avalona)*, four scarab beetles *(Coenonycha clypeata* and *C. fulva, Phobetus ciliatus,* and *Serica catalina)*, the walkingstick *Pseudosermyle catalina,* and the Catalina Shield-back Cricket (*Neduba propsti*).

Most of the island is protected by its rugged nature and its present status as a recreational preserve, so its insect life can be expected to persevere. As is the case with all of southern California's Channel Islands, much study is still needed to fully understand the nature of Catalina's insect populations.

REFERENCES. Bennett, S. G. 1988. Medically important and other ectoparasitic acarines on vertebrates from Santa Catalina Island, California. *Bulletin of the Society of Vector Ecologists,* vol. 12, pp. 534-538.

Gorelick, G. A. 1987. Santa Catalina Island's endemic Lepidoptera, part 3: The Avalon Hairstreak, *Strymon avalona* (Lycaenidae)—An ecological study. *Atala,* vol. 14, pp. 1-12.

Miller, S. E. 1985. Entomological bibliography of the California Islands, supplement 1. Pages 137-169 in *Proceedings of the First Symposium on the Entomology of the California Channel Islands,* edited by A. S. Menke and D. R. Miller. Santa Barbara Calif.: Santa Barbara Museum of Natural History. [A second supplement by Miller is in press.]

______. 1985. The California Channel Islands—Past, present, and future: An entomological perspective. Pages 3-27 in *Proceedings of the First Symposium on the Entomology of the California Channel Islands,* edited by A. S. Menke and D. R. Miller. Santa Barbara, Calif.: Santa Barbara Museum of Natural History.

# Part II
# THE INSECTS

THE MOST AMAZING CHARACTERISTIC of the insect fauna of the Los Angeles Basin is that it exists at all. In such a large area that is so densely urbanized and so heavily populated it would seem that very little of nature could survive. But the result of the city's varied topography, fluctuating climate, mobile population, and uneven development is a surprisingly high diversity and intensity of insect life.

Los Angeles tolerates and shelters species tied to its original vegetation of desert, grassland, and chaparral species. Where water still runs and stands, aquatic insect forms abound. Homes and buildings—and even freeways—create abodes for other types. And people continue to bring new insects into the city.

Although no detailed data are available for comparison, I would speculate from my observations and experience that we have several times more insect species here in the Los Angeles Basin than exist in any other large metropolitan area in the United States.

Jerusalem cricket (*Stenopelmatus* species), views of the insect shown in Figure 48. These insects are nocturnal and live in the soil and so are not often seen by Los Angeles residents. When they are encountered, usually by a gardener turning the soil, they cause a great deal of curiosity because of their large size and humanoid heads. See pages 80-82. Photographs by C. Hogue.

# 3 PRIMITIVE WINGLESS INSECTS

7. Tidepool Springtail. Drawing by T. Ross after E. Essig, 1942, *College Entomology,* fig. 35 (New York: Macmillan).

THE MOST ANCIENT and primitive insects lack wings in all their life stages. They possess vestiges of legs on some abdominal segments, indicating their closer relationship to the centipede-like ancestors of insects than to winged insects. Metamorphosis is also absent in these species: the immatures resemble the adults except for their smaller size and sexual immaturity.

## SPRINGTAILS (Order Collembola)

THESE MINUTE INSECTS (most are less than 1/8 in., or 3 mm, in length) derive their name from the curious method of locomotion of many species, in which the *furcula,* a tail-like appendage on the underside of the abdomen (Figure 10), is extended and snaps against the substratum, propelling the insect upward.

Springtails are usually seen when a group of them is uncovered under a board or rock lying on damp soil or plant litter. When set in motion by the sudden exposure to light, they resemble fleas or a troupe of dancing, hopping particles. These insects are primarily vegetarians but seldom do any perceptible damage to gardens in our area.

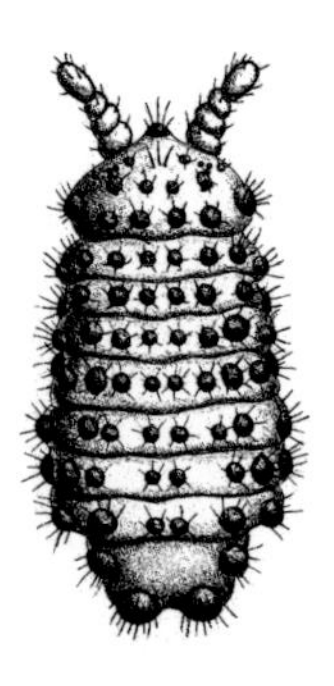

8. Obese Springtail. Drawing by T. Ross after various sources.

Springtails are among the most numerous of animals found in the soil and are also commonly encountered in compost piles and grass cuttings, in turf, under flower pots, in cellars, or among stored plant bulbs—wherever it is humid and dark. Masses of an unidentified small white species occasionally appear on the surface of the water in swimming pools.

Some of the more common species of springtails in the basin are the following:

9. Lawn Springtail. Drawing by T. Ross after E. Maynard, 1951, *A monograph on the Collembola of New York State,* fig. 568 (Ithaca, N.Y.:Comstock).

■ Tidepool Springtail *(Entomobrya laguna;* Figure 7). This moderate-sized species (1/4 in., or 3 mm) lives on rocks in the intertidal zone along the coast; it was discovered at Laguna Beach, hence its scientific name. Its body is covered with long spatulate hairs.

■ Obese Springtail *(Morulina multatuberculata;* Figure 8). This springtail is unusually large (almost 1/4 in., or 6 mm, long). It has a bulky body with rows of dark tubercles on a blue-white background, and it lacks the spring organ. It is fairly abundant under objects lying on oak leaf mold.

■ Lawn Springtail *(Bouletiella arvalis;* Figure 9). This species, which is often found in lawns, is very small (less than 1/16 in., or 1.5 mm), globular in shape, and pale yellow in color.

■ Varied Springtail *(Isotoma viridis;* Figure 10). This widespread species comes in many colors (dark brown or greenish to yellow, often mottled with purple). The body is elongate and up to 3/16 inch (5 mm) long; the spring organ is well developed.

Furcula

10. Varied Springtail. Drawing by T. Ross after various sources.

REFERENCES. Christiansen, K., and P. Bellinger. 1980. *The Collembola of North America, north of the Rio Grande,* 2 parts. Grinnell, Iowa: Grinnell College.

Coleman, T. C. 1941. The Poduridae of southern California. *Pomona College Journal of Entomology and Zoology,* vol. 33, pp. 1-12.

## SILVERFISH (Order Zygentoma)

SILVERFISH ARE SO CALLED because of their body covering of shiny slick scales and their wriggling motion when running, which makes them slippery and difficult to catch. In our warm climate, the local species may be seen indoors or outdoors. Indoors, they appear behind furniture and among books and other items around the house and garage. Out of doors, they live in wood piles, under rocks or boards on the ground, or in leaf litter and are sometimes seen scurrying over junk in storage areas.

Silverfish feed on dry organic debris and have a taste for paper, especially that which contains sizing such as starch, gum, or glue. For this reason they may do considerable damage to wallpaper and books, especially the bindings. Other domestic foods for silverfish include sugar, flour, breakfast cereals, fabric, and insulating materials.

The following are the species most likely to be encountered in the Los Angeles Basin. They are difficult to distinguish (the shape of the upper plate of the last abdominal segment is helpful in identifying the species). All reach about 1/2 inch (13 mm) in length on maturity. All are introduced, and their occurrence is spotty. In the first three species, the large bristles on the back are barbed or feathered.

■ Silverfish *(Lepisma saccharina;* Figure 5). This is primarily an indoor species. The body has a silvery sheen and is monocolorous slate gray.

■ Long-tailed Silverfish *(Ctenolepisma longicaudata;* Figure 11). This mostly gray species, which is characterized by very long terminal filaments, has prominent hair tufts along the sides of its body.

■ Lineated Silverfish *(Ctenolepisma lineata pilifera)*. The Lineated Silverfish is most commonly found outdoors. The pattern of its markings varies, although the back tends to be banded lengthwise, and the legs and terminal filaments ringed.

■ Spinulate Silverfish *(Allacrotelsa spinulata;* Figure 12). This usually free-living species has a robust body and relatively short terminal filaments. It is solid dark brown with finely banded terminal filaments.

■ Fire Brat *(Thermobia domestica)*. The smallest local silverfish, the Fire Brat is about 7/16 inch (11 mm) long and has a mottled pattern. The species is decidedly domestic, with a strong preference for extra warm—even hot—places; it frequents boiler rooms and bakeries, staying close to steam pipes, ovens, and autoclaves.


REFERENCES. Adams, J. A. 1933. Biological notes upon the fire brat, *Thermobia domestica* Packard. *Journal of the New York Entomological Society,* vol. 41, pp. 557-562.

Lindsay, E. 1940. The biology of the silverfish, *Ctenolepisma longicaudata* Esch., with particular reference to its feeding habits. *Proceedings of the Royal Society of Victoria* (new series), vol. 52, pp. 35-78.

Mallis, A. 1941. Preliminary experiments on the silverfish *Ctenolepisma urbana* Slabaugh. *Journal of Economic Entomology,* vol. 34, pp. 787-791.

_______. 1964. Silverfish. In *Handbook of pest control,* 4th edition, edited by A. Mallis. New York: MacNair Dorland.

McIver, S. B., and R. F. Harwood. 1966. Thermal discrimination in the Firebrat, *Thermobia domestica* (Thysanura: Lepismatidae). *Journal of the Kansas Entomological Society,* vol. 39, pp. 535-541.

Smith, E. L. 1970. Biology and structure of some California bristletails and silverfish, Apterygota: Microcoryphia, Thysanura. *Pan-Pacific Entomologist,* vol. 46, pp. 212-225.

Sweetman, H. L. 1938. Physical ecology of the firebrat, *Thermobia domestica* (Packard). *Ecological Monographs,* vol. 8, pp. 285-311.

Wygodzinsky, P. W. 1972. *A review of the silverfish (Lepismatidae, Thysanura) of the United States and the Caribbean area.* American Museum Novitates, no. 2481. New York: American Museum of Natural History.


## BRISTLETAILS (Order Microcoryphia)

THE BRISTLETAILS are closely related to silverfish and are sometimes lumped with them into one order. Bristletails are more cylindrical than the flattened silverfish and have mouthparts that project downward rather

11. Long-tailed Silverfish. Photograph by C. Hogue.

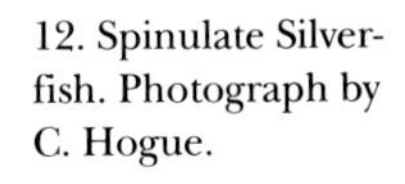

12. Spinulate Silverfish. Photograph by C. Hogue.

13. Bristletail (unidentified genus and species). Photograph by C. Hogue.

than toward the front of the head. The front part of the bristletail body tends to be arched (Figure 13) rather than straight as in silverfish. Like silverfish, most bristletail species are nocturnal, abhor light, and are secretive in their habits.

The several local species of bristletails are not easy to distinguish, nor are they even known with any certainty. They are typically found outdoors (beneath debris, under bark, in leaf litter, or on rocks), but some, including unidentified species in the genera *Neomachilis* and *Machilinus,* occasionally appear in houses.

# 4 PRIMITIVE WINGED INSECTS

THE DEVELOPMENT OF WINGS was a major step in the evolution of insects. At first these were unsubstantial organs that lacked a folding mechanism at their bases and gave the insect only feeble flight capabilities. Many groups reached this level long ago, only to then become extinct. The two orders comprising the mayflies and the dragonflies and damselflies are the sole surviving representatives of this bygone assemblage. The insects in these groups exhibit incomplete metamorphosis.

## MAYFLIES (Order Ephemeroptera)

ALTHOUGH THEY ARE NOT commonly seen in the Los Angeles Basin proper, mayflies belonging to several genera are plentiful enough near streams and ponds in the nearby foothills and mountains. These insects

14. Mayfly male *(Callibaetis pacificus)*. Photograph by C. Hogue.

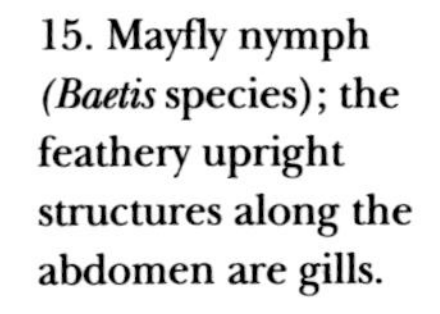

15. Mayfly nymph *(Baetis* species); the feathery upright structures along the abdomen are gills.

are completely aquatic as immatures, and the adults, which have only weak flight capabilities, never stray far from the water. Our commonest basin species are small and belong to the genus *Callibaetis* (Figure 14), in which the males have a brown-tinted front margin on the fore wings. Two other species, *Tricorythodes minutus* and *Baetis* sp., which are small and dark with clear wings, may also be found locally.

The nymphs (Figure 15) live in flowing or (more rarely) still water and are found on the bottom, on rocks, or in moss, trash, and other hiding places; nymphs of local species are herbivores. Adults emerge as dull-colored "subimagos" with opaque gray wings, but they soon molt and change into the familiar glassy-winged form. This is the only order of insects in which the molting process occurs in a winged stage.

Adults, which do not feed, live only a few hours or days at the most, hence the order name Ephemeroptera, from the Greek *ephemeros* (lasting but a short while).

Both mature and young mayflies may be considered beneficial: the adults are an important food item for birds and certain beneficial insects, and the nymphs are eaten by fish (freshwater anglers pattern many dry and wet "fly" lures after these insects).

REFERENCE. Edmunds, G.F., Jr., S. L. Jensen, and L. Berner. 1976. *The mayflies of North and Central America.* Minneapolis, Minn.: University of Minnesota.

## DRAGONFLIES (Order Odonata; Suborder Anisoptera)

LIKE MAYFLIES, DRAGONFLIES LIVE AS IMMATURES in water. But because they are strong fliers, dragonflies are frequently seen long distances from the ponds, lakes, and streams in which they breed.

Dragonflies are better known than their smaller, more timid and delicate close relatives, the damselflies (the common names of the two groups, which are both members of the order Odonata, refer to this difference). Dragonflies, or "dragons of the air," are also known colloquially as "devil's darning needles," "snake doctors," and "horse stingers." The first name is derived from their supposed ability to sew up the mouths of children who tell lies. Actually, of course, dragonflies cannot sew. Nor can they sting as the last name implies: a live specimen held between the fingers seems to be trying to sting as it coils its abdomen around and jabs with the tip, but it has no stinging apparatus.

The dragonfly is an extremely agile aeronaut, spending most of its waking hours in flight, patrolling pond or stream margins, darting low over the surface or high into the air. Most of this aerial energy is directed at catching the small flying insects that are its food. The dragonfly catches its prey in midflight in a "basket" made by folding its legs in front of and beneath its head (the legs are useless for walking; at rest the dragonfly uses them only for clasping its perch). While engrossed in snatching prey from the air, the hunter may be fooled, momentarily, by a small stone or wadded piece of paper tossed into the air in its path. With a rapid dive, however, the dragonfly will detect the ruse and veer away before actually catching the object.

The aquatic nymphs are as voraciously predatory as the adults, but their feeding method is very different. The prey, which consists of almost any small animals swimming closely enough, is captured by a swift strike with the jointed lower lip. The action is similar to that displayed by the toad or chameleon seizing insects with an extendable sticky tongue.

Human beings are latecomers in the use of jet propulsion. By forcibly expelling water from its rectum, the dragonfly nymph can drive its body forward through the water at great speed. This is an emergency method of locomotion that is employed principally to evade enemies.

The posture assumed by dragonflies and their relatives, damselflies, during copulation is unique (Figure 16). The genitalia of the male are located in a pocket on the underside of the second and third abdominal segments, not at the tip of the abdomen as in most other insects. Prior to mating, the male curls his abdomen forward and expels sperm from the genital orifice, which is at the apex of his abdomen,

16. Damselfly male (top) and female in copulating attitude. Photograph by R. Garrison.

into a copulatory organ in the genital pocket. During copulation, the male grasps the female behind the head with pinchers carried at the tip of his abdomen; the female then curls her abdomen forward to receive sperm from the male's genital pocket. This embrace, which in most dragonfly species is maintained during flight, has led casual observers to make reports of "Siamese dragonflies."

REFERENCES. Needham, J. G., and H. B. Heywood. 1929. *A handbook of the dragonflies of North America.* Springfield, Ill.: Charles C. Thomas.

Needham, J. G., and W. Westfall. 1955. *A manual of the dragonflies of North America (Anisoptera).* Berkeley, Calif.: University of California.

Paulson, D. R., and R. W. Garrison. 1977. A list and new distributional records of Pacific Coast Odonata. *Pan-Pacific Entomologist,* vol. 53, pp. 147-160.

Seemann, T. M. 1927. Dragonflies, mayflies, and stoneflies of southern California. *Journal of Entomology and Zoology,* vol. 19, pp. 1-69.

## ■ DARNERS (Family Aeshnidae)

■ COMMON GREEN DARNER *(Anax junius)* Figures 17, 18

This is one of our largest dragonflies (the body is 3 1/4 in., or over 8 cm, long). Its body is mostly green, and it has black legs, a yellow face, and a bluish abdomen. The wings are clear; however, all but the tips are often tinged with amber.

The nymphs are good swimmers, elongate and streamlined. The legs and body are more or less unicolorous, and the abdomen has small dark spots. They often hide in ambush, with their heads low, among submerged vegetation.

Although the adults emerge at night, they are diurnally active.

REFERENCE. Trottier, R. 1973. Influence of temperature and humidity on the emergence behavior of *Anax junius* (Odonata: Aeshnidae). *Canadian Entomologist,* vol. 105, pp. 975-984.

■ MULTICOLORED DARNER *(Aeshna multicolor)* Figures 19, 20

This is another fairly large dragonfly (its body length is 2 1/2 in., or 64 mm, or more). It is predominantly blue with a checkerboard pattern of blue and black on the abdomen; it has completely clear wings. The eyes in life are brilliant blue. The abdomen is very long and hairy on the back of the basal portion; segment three is contracted.

The nymph is elongate and a good swimmer, like that of the Common Green Darner. It is patterned with irregular longitudinal stripes on the body and

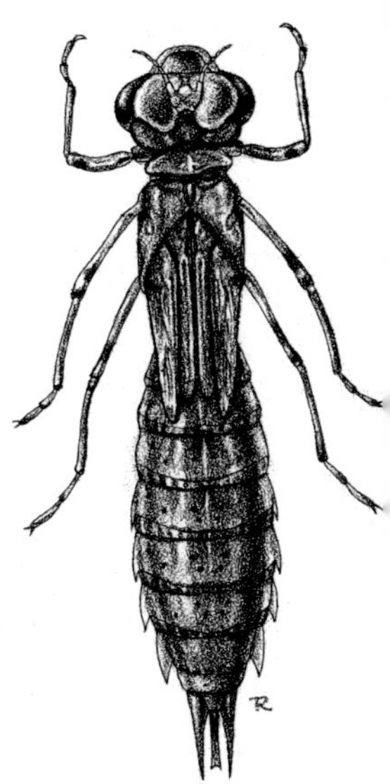

17. Common Green Darner. Photograph by C. Hogue.

18 (above right). Common Green Darner nymph. Drawing by T. Ross after J. G. Needham and W. Westfall, 1955, *A manual of the dragonflies of North America (Anisoptera),* fig. 165 (Berkeley, Calif.: University of California).

19 (below right). Multicolored Darner nymph. Drawing by T. Ross after various sources.

20. Multicolored Darner. Photograph by C. Hogue.

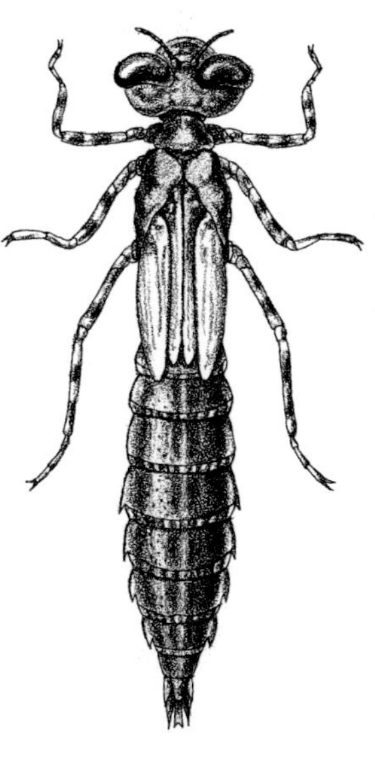

bands on the legs; there are no spots on its abdomen.

A strong flyer, the Multicolored Darner seldom comes to rest during the day and frequently forages far from water.

## ■ SKIMMERS (Family Libellulidae)

**■ Big Red Skimmer**
*(Libellula saturata)*
Figures 21, 22

This dragonfly is conspicuous because of its large size (its body length is 2 to 2 1/2 in., or 50 to 65 cm) and its orange-red color. The wings are usually bright orange-red over the entire basal half, although this color may be reduced to just the heavily veined area toward the leading edges of the wings. The species is usually seen near sluggish streams and ponds, flying strongly to and fro "hawking" small insect prey. The squat hairy nymph lives in the ooze on the bottom of stagnant pools.

21. Big Red Skimmer. Photograph of LACM specimen by C. Hogue.

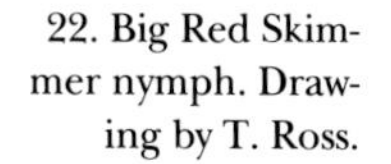

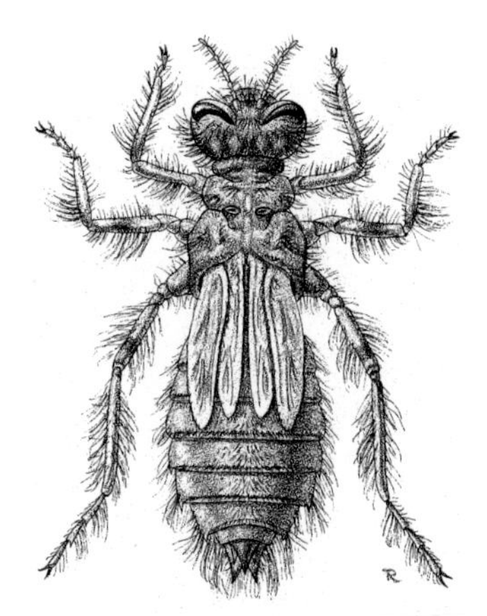

22. Big Red Skimmer nymph. Drawing by T. Ross.

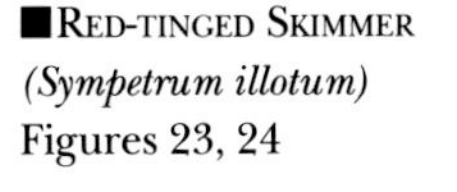

■ RED-TINGED SKIMMER
*(Sympetrum illotum)*
Figures 23, 24

This species is similar in color to the Big Red Skimmer, but the veins in its wings are redder (not orange) and it is considerably smaller (its body length is about $1\frac{1}{2}$ in., or 38 mm). Also, the red-orange of the wing membrane never consists of more than light streaks near the base surrounded by a yellowish hue. There are two faint oblique stripes on the sides of the thorax, and the red legs have black spines.

The nymph has a wide abdomen, which is flattened on the bottom, and a wide oval head. It is a bottom dweller.

■ PASTEL SKIMMER
*(Sympetrum corruptum)*
Figure 25

This dragonfly is a close relative of the Red-tinged Skimmer and is about the same size (body length of $1\frac{3}{4}$ in., or 45 mm). It is colored in soft pastel shades of yellow, red, and olive. The oblique stripes on the

23. Red-tinged Skimmer. Photograph by N. Baker.

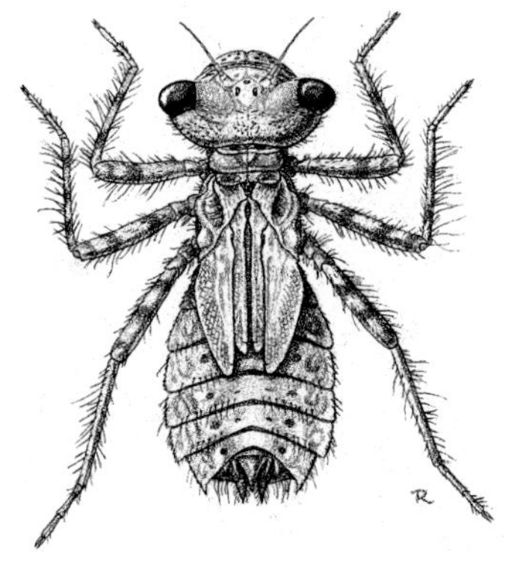

24. Red-tinged Skimmer nymph. Drawing by T. Ross after G. Roldan, 1988, *Guia para el Estudio de los Macroinvertebrados Acuaticos del Departamento de Antioquia,* p. 48, fig. 33

25. Pastel Skimmer. Photograph by R. Garrison.

side of the thorax are whitish in the middle and overlaid at the lower end by a yellow spot, the lower half of which is rimmed in black. The entire wing membrane is clear.

The nymph is a little larger than that of the Red-tinged Skimmer, *Sympetrum illotum*, but different otherwise only in the relative length of the terminal appendages (in *S. corruptum*, the outer lobe is half as long as the inner; in *S. illotum*, it is two-thirds as long).

Adults are often seen resting on vegetation in the midsummer and fall, even during the middle of the day. They migrate in the fall, when they may be found far from water. Their habitats are semipermanent ponds, marshes, and lake margins.

REFERENCE. Arnaud, P. H., Jr. 1973. Mass movement of *Sympetrum corruptum* (Hagen) at Pacifica, California (Odonata: Libellulidae). *Pan-Pacific Entomologist*, vol. 49, p. 84.

## DAMSELFLIES (Order Odonata; Suborder Zygoptera)

DAMSELFLIES HAVE MANY OF THE HABITS and characteristics of their larger and more agressive cousins, the dragonflies. The immature forms of both groups are aquatic, but adult damselflies are not strong fliers and are usually found near the bodies of water in which they breed. Damselflies employ the same method of catching their prey, small flying insects, and they copulate on the wing, in the same manner as dragonflies (the male damselfly clasps the female by the front of the thorax instead of behind the head). Most damselfly species fold their wings back over the body when resting (Figure 26) rather than leaving them outstretched, as do dragonflies.

26. Common Ruby Spot male. Photograph by R. Garrison.

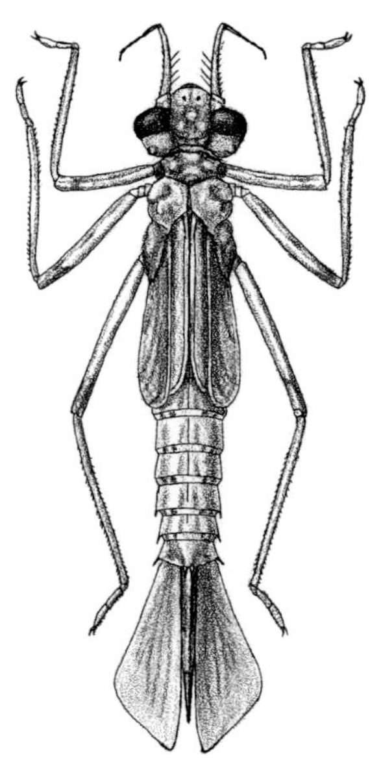

27. Common Ruby Spot nymph. Drawing by T. Ross after D. Geijskes, 1943, *Annals of the Entomological Society of America*, vol. 36, p. 169,

## ■ BROAD-WINGED DAMSELFLIES
(Family Calopterygidae)

**■ Common Ruby Spot**
*(Hetaerina americana)*
Figures 26, 27

The Common Ruby Spot is likely to be seen only near rapidly flowing streams in the mountain canyons surrounding the Los Angeles Basin. The body length in this species is 1½ to 2 inches (37 to 50 mm).

In contrast to the female's wings, which are amber or clear, the wings of the male are conspicuously marked at their bases with brilliant ruby or carmine (Figure 26). Males advertise their presence in flight by displaying these bright wing colors. The male's thorax is coppery-bronze with metallic green reflections.

The nymph is a long-legged sprawling creature that lives among bottom trash in streams of moderate current. The gills are large and elongatedly oblong.

REFERENCE. Johnson, C. 1961. Breeding behavior and oviposition in *Hetaerina americana* (Fabricius) and *H. titia* (Drury) (Odonata: Agriidae). *Canadian Entomologist,* vol. 93, pp. 260-266.

## ■ LESTIDS (Family Lestidae)

### ■ California Archilestes (*Archilestes californica*)

Figures 28, 29

It is possible to distinguish this species from the several other damselflies common in our area by its relatively large size (the body is nearly 2 in., or 50 mm, long) and the fact that it rests like a dragonfly, with its wings outstretched rather than folded back in normal damselfly fashion. Adults are on the wing in the late fall and sometimes rest in groups on vegetation.

The female inserts her eggs into willow twigs (six eggs per puncture) several feet over water. The nymph is slender with elongate terminal gills, which are dark at the tips and in bands near the midpoints.

28. California Archilestes. Photograph by C. Hogue.

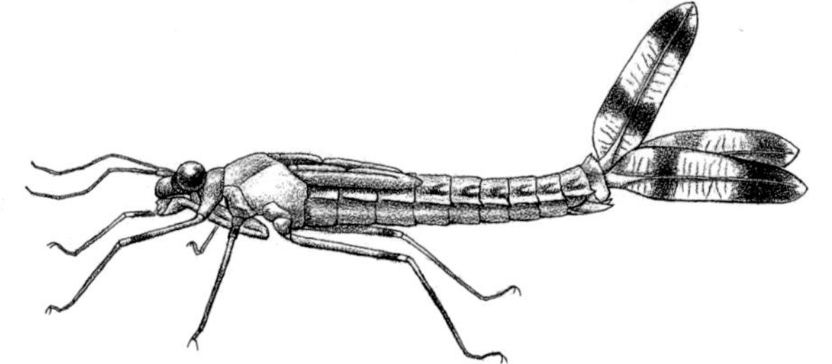

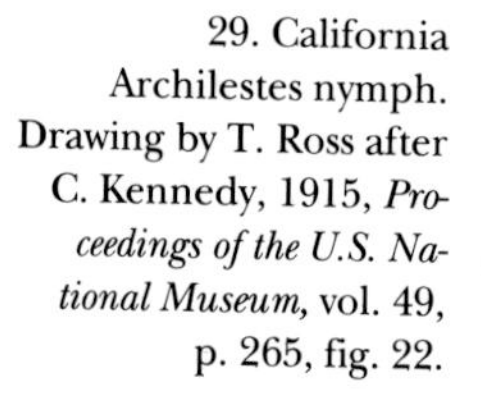

29. California Archilestes nymph. Drawing by T. Ross after C. Kennedy, 1915, *Proceedings of the U.S. National Museum,* vol. 49, p. 265, fig. 22.

REFERENCE. Kennedy, C. H. 1917. Notes on the life history and ecology of the dragonflies (Odonata) of central California and Nevada. *Proceedings of the U.S. National Museum,* vol. 52, pp. 482-635. [Contains a major section on the California Archilestes.]

## ■ DANCERS (Family Coenagrionidae)

### ■ Violet Dancer
*(Argia vivida)*
Figures 30, 31

This damselfly, which is normally seen around small streams in the mountains, is commonest in the San Gabriels and Santa Monicas. The body is 1 to 2 inches (25 to 50 mm) in length, and its predominant color is an intense medium-blue; the back third of each abdominal segment is black-banded. The wings are unmarked except for a spot near the forward margin at the tips.

The nymph is short and robust, with dark-colored legs and short elliptical gill plates.

REFERENCE. Pritchard, G. 1989. The roles of temperature and diapause in the life history of a temperate-zone dragonfly: *Argia vivida* (Odonata: Coenagrionidae). *Ecological Entomology,* vol. 14, pp. 99-108.

30. Violet Dancer. Photograph by R. Garrison.

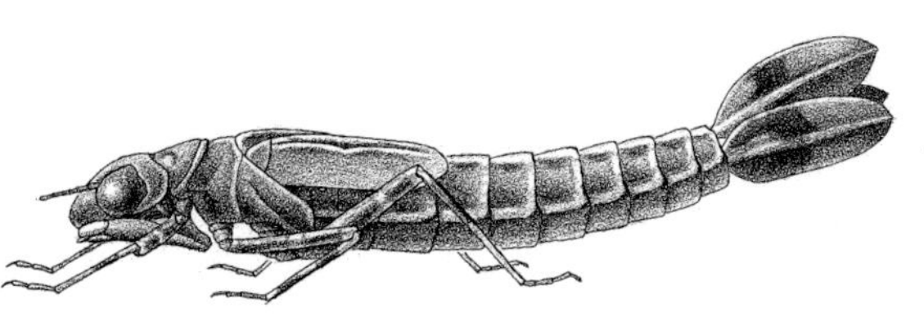

31. Violet Dancer nymph. Drawing by T. Ross after C. Kennedy, 1915, *Proceedings of the U.S. National Museum,* vol. 49, p. 278, fig. 37.

# 5 ORTHOPTEROIDS

THESE INSECTS—grasshoppers, cockroaches, termites, and their relatives—are considered to represent survivors of the first stage in the evolution of modern flying insects. They are primarily terrestrial and often have leathery or hardened fore wings that are almost rigid and frequently form protection for large, fanlike hind wings, which are the main flight organs. Their metamorphosis is incomplete.

## EARWIGS (Order Dermaptera)

EARWIGS ARE WELL KNOWN to those of us who have cleaned up lumber piles, rocks, or other debris from vacant lots or backyards. They are usually found hiding under a stone or board that is lying on slightly damp soil, and they are easily recognized by the dangerous-looking pincers, or "forceps," that they carry at the tip of the abdomen. Although these structures can give a mild nip, they are weak and otherwise harmless; they are used by the insect for catching and manipulating prey and sometimes for fending off enemies.

No one is quite sure how earwigs became so named. One guess is that the early Anglo-Saxons, who named them *earwicga* (ear beetle or worm) and who lived in sod huts, where these insects also lived, occasionally found them in their ears upon waking from a sound sleep on a straw mattress. The warm and tight ear opening of a slumbering person might well have been a snug hiding place for these crevice-loving creatures.

Although several species are occasionally found in the Los Angeles Basin, only the two described in the following sections are well established and common.

REFERENCE. Langston, R. L., and J. A. Powell. 1975. The earwigs of California. *Bulletin of the California Insect Survey*, vol. 20, pp. 1-25.

■ RING-LEGGED EARWIG
*(Euborellia annulipes)*
Figure 32

Although a long-time resident, this insect is not native to the basin, having immigrated prior to the 1880s from somewhere in Europe, where it is almost universally found. It is small (length including forceps 1/2 to 3/4 in., or 13 to 19 mm) and without flight wings and sometimes without fore wings as well (another basin resident, the European Earwig, has fully developed wings). The body is shiny and dark brown to black, the legs pale brown. There are one or two light segments near the ends of the antennae, and there are dark brown spots or rings on the larger leg segments. The Ring-legged Earwig is omnivorous and sometimes a pest, but it prefers an insect diet and in some cases may even be beneficial as a predator.

This earwig exhibits some maternal care. The female sits on her eggs like a brooding hen. If disturbed, she will either try to nip the intruder with her forceps or to run for cover, holding an egg protectively in her mouth. Care of the young in the species usually continues only until the nymphs are "second instars" (have undergone their first molt).

REFERENCES. Bharadwaj, R. K. 1966. Observations on the bionomics of *Euborellia annulipes* (Dermaptera: Labiduridae). *Annals of the Entomological Society of America,* vol. 59, pp. 441-450.

Langston, R. L., and J. A. Powell. 1975. The earwigs of California (*Euborellia annulipes,* pp. 8-10). *Bulletin of the California Insect Survey,* vol. 20, pp. 1-25.

32. Ring-legged Earwig. Photograph by C. Hogue.

33. European Earwig female guarding a cluster of nymphs. Photograph by C. Hogue.

■ EUROPEAN EARWIG
*(Forficula auricularia)*
Figure 33

This conspicuous earwig continues to be more and more common in the basin since its introduction from Europe sometime around 1930; it probably arrived in southern California from the northern parts of the state. It tends to displace the Ring-legged Earwig, although the two species may live together—the latter in subdued numbers, however. Under proper conditions, the numbers of European Earwigs in patios, backyards, and storage areas may build into the thousands. Large populations sometimes hide in the tubular framework of lawn furniture as well as under stones and in more normal habitats such as compost piles and vegetable trash. The species occasionally makes a nuisance of itself by coming indoors. It is active at night.

This earwig's food consists of other insects, which it may catch with its terminal forceps. But it is fond of tender young plants: it often chews seedlings, soft lettuce leaves, and corn silk and is rightly considered a garden and crop pest. According to Langston and Powell (1975), there were programs to eradicate the species until 1952, when a survey revealed extensive colonies established throughout the Los Angeles Basin.

The European Earwig is distinctive because of its large forceps (these are longer and more bowed in the male than in the female) and its well-developed wings, which are folded fanwise under squarish wing covers. Unlike the dark brown or black Ring-legged Earwig, this species is light brown, and its legs and antennae are not banded. It is also larger than the Ring-legged; its body length, including forceps, is 1/2 to 3/4 inch (12 to 22 mm). In spite of its well-developed hind wings, the insect seldom flies.

Maternal care is highly developed in this earwig. Females are often found in the spring in small hollows in the soil under stones or concrete, brooding over a cluster of eggs or young nymphs (Figure 33). This protection partly accounts for the success of the species.

The forceps are effective defensive weapons. The European Earwig is protected also by noxious secretions from two pairs of small sack-like glands that are situated on the back of the abdomen and open on the rear margins of the third and fourth abdominal segments.

REFERENCES. Eisner, T. 1960. Defense mechanisms of arthropods, part 2: The chemical and mechanical weapons of an earwig. *Psyche,* vol. 67, pp. 62-70.

Lamb, R. J. 1976. Parental behavior in the Dermaptera with

special reference to *Forficula auricularia* (Dermaptera: Forficulidae). *Canadian Entomologist,* vol. 108, pp. 609-619.

Lamb, R. J., and W. G. Wellington. 1975. Life history and population characteristics of the European earwig, *Forficula auricularia* (Dermaptera: Forficulidae), at Vancouver, British Columbia. *Canadian Entomologist,* vol. 107, pp. 819-824.

Langston, R. L., and J. A. Powell. 1975. The earwigs of California (*Forficula auricularia,* pp. 20-22). *Bulletin of the California Insect Survey,* vol. 20, pp. 1-25.

## GRASSHOPPERS, CRICKETS, AND KATYDIDS (Order Orthoptera)

A GOOD NUMBER of the insects in this group possess hind legs that are specialized for jumping or leaping. The femora are swollen to accommodate greatly enlarged muscles, which can powerfully kick the tibiae downward and backward to propel the insect forward. This action is used by wingless species when in a hurry to escape an enemy and by winged species as propulsion to become airborne.

In addition to their jumping abilities these insects are noted for their musical talents. Who hasn't heard a cricket or a katydid singing through a summer night? These noises are not for the insect's or our entertainment but serve a definite purpose in sexual behavior: males chirp to attract females.

The sounds made by insects in this group are complex auditory patterns generated by anatomical structures specialized for the purpose. There are three primary types of sound production: Orthopterans that "sing" have specially modified structures at the bases of the fore wings or on the legs—opposible rows of fine ridges, knobs, or pegs—which are rubbed together to make the sound, and membranous areas, which vibrate and amplify it. This is one form of "stridulation"; another is the making of high-pitched creaking noises by rubbing roughened places on the hind leg against the fore wings. "Crepitators" crack or buzz by rapidly opening and closing the fan-like hind wings. "Drumming" is accomplished by thumping some part of the body against the substratum.

It follows that, if the grasshoppers and their relatives make music with a purpose, they must also possess devices with which to hear it. The auditory organs ("ears") of these insects are well developed and consist of thin membranes connected to sensory nerves. They are located in various places on the body but never on the head. These ears hear most of the sounds in the insect's environment, but they are especially tuned to those generated by the species' own sound-producing organs.

REFERENCES. Helfer, J. R. 1963. *The grasshoppers, cockroaches and their allies.* Dubuque, Iowa: W. E. Brown.

Rentz, D. C. F., and D. B. Weissman. 1981. Faunal affinities, systematics, and bionomics of the Orthoptera of the California Channel Islands. *University of California Publications in Entomology,* vol. 94, pp. 1-240. [Contains much information on mainland species.]

## ■ GRASSHOPPERS (Family Acrididae)

GRASSHOPPERS ARE DISTINGUISHED from most other orthopteroids by their short, relatively stout antennae and by the location of their auditory organs, which are on either side of the first abdominal segment. These insects are stridulators: the resting grasshopper rubs tiny peglike projections on the inner side of the hind leg rapidly against the fore wing, causing the wing to vibrate and produce a high-pitched noise. Some species also make a crackling sound (crepitation) during flight.

The hopping activity of these insects prompted early Los Angeles residents to rename South Figueroa Street "Calle de las Chapules" (Spanish for "Street of the Grasshoppers"). Pedestrians used to leap about while sour-faced policemen whistled and chased the hordes from one street corner to the other.

There are many species of grasshoppers in the basin; only those seen most frequently are described here.

REFERENCES. Otte, D. 1970. *A comparative study of communicative behavior in grasshoppers.* Miscellaneous Publications of the University of Michigan Museum of Zoology, no. 141, 168 pp.

_______. 1981. *The North American grasshoppers,* vol. 1: Acrididae. Gomphocerinae and Acridinae. Cambridge, Mass.: Harvard University.

Strohecker, H. F., W. W. Middlekauff, and D. C. Rentz. 1968. The grasshoppers of California (Orthoptera: Acridoidea). *Bulletin of the California Insect Survey,* vol. 10, pp. 1-177.

Willey, R. B. 1975. Slowed motion analysis of sound production in the grasshopper *Arphia sulphurea* (Acrididae: Oedipodinae). *Psyche,* vol. 82, pp. 324-340.

■ PALLID BAND-WING
*(Trimerotropis pallidipennis)*
Figure 34

This moderate-sized insect (head to wing tips 1 1/4 to 1 5/8 in., or 31 to 42 mm) is the commonest of our so-called "band-wing" grasshoppers, all of which sport hind wings with vivid colors inside a black submarginal band. The Pallid Band-Wing's hind wing is translucent pale yellow, but other local species have wings that are bright yellow *(Lactista gibossus)*, orange *(Arphia conspersa)*, red *(Dissosteira pictipennis)*, or blue *(Leprus intermedius)*; the colors are used in sexual displays. The fore wings of all these species are mottled, giving the hopper camouflage when it is resting on gravelly or sandy ground (Figure 34).

Male band-wings make a crackling sound in flight by snapping their plaited wings (crepitation). They also make a noise by rubbing the wings against a row of pegs along the inside of the hind leg (stridulation). Both forms of sound are produced to attract females.

REFERENCES. Chappell, M.A. 1983. Metabolism and thermoregulation in desert and montane grasshoppers. *Oecologia*, vol. 56, pp. 126-131.

_______. 1983. Thermal limitations to escape responses in desert grasshoppers. *Animal Behavior*, vol. 31, pp. 1088-1093.

Otte, D. 1984. *The North American grasshoppers*, vol. 2: Acrididae. Oedipodinae. Cambridge, Mass.: Harvard University.

---

■ **Gray Bird Grasshopper** *(Schistocerca nitens)* Figure 35

This is a large species (from head to wing tips, males are 1½ in., or 38 mm; and females are 2½ in., or 6 cm). It is fairly common and is often found feeding on garden shrubbery.

I have noticed adults only in the spring; they are

34. Pallid Band-wing. Photograph by C. Hogue.

35. Gray Bird Grasshopper female. Photograph by C. Goodpasture.

gray or brownish in general color, and the hind wings are uniformly transparent olive-green. The light green nymphs attain noticeable size in the late summer. Both stages feed on various garden crops and ornamentals.

This species has some notorious relatives, for the genus *Schistocerca* also contains the infamous Desert Locust *(Schistocerca gregaria),* which was responsible for the Biblical plagues and is a major pest in Africa and the Middle East today. In our state, the Devastating Grasshopper *(Melanoplus devastator)* is the injurious, massing species.

OTHER SCIENTIFIC NAMES. *Schistocerca vaga.*

## ■ CRICKETS (Family Gryllidae)

TRUE CRICKETS ARE KNOWN to most of us only by their nightly song. The most common local garden and residential species, a tree cricket, is seldom seen or found even when searched for because its voice, like that of the ventriloquist, is difficult to trace to its source.

Many folk beliefs exist about crickets. Their presence in the home is an omen of good fortune in many parts of the world, and in China they are kept in captivity. The Chinese also match crickets for combat; the sport has been as popular there as cockfighting is in other countries, and extravagant wagers are made on the outcome of championship fights.

Like katydids, crickets "sing" by making frictional movements of long roughened veins at the bases of the fore wings. However, in crickets, only the males possess the necessary structures. When chirping, the cricket elevates its wings above its abdomen.

■ TREE CRICKETS
*(Oecanthus* species)
Figure 36

This name actually refers to a series of closely related species, which are distinguishable only by minute differences in anatomy and song. They are all about ½ inch (13 mm) long and transparent greenish, brown, or white in color. The males have flattened bodies and broad wings; the females are cylindrical with wings folded around the body.

Tree crickets live among the leaves of trees, shrubs, and herbaceous vegetation from which their familiar calls punctuate the warm nights of summer and fall. The song consists of a pulsed, whistled "treet-treet-treet" *(Oecanthus fultoni* and *O. rileyi),* or a continuous trill *(O. californicus* and *O. argentinus).* The last is our most abundant species.

The chirping is a prelude to courtship and mating, which are in themselves complicated processes. The sequence of behavior is described by Fulton (1915) as follows:

> While the male is vigorously singing, the female lingers near and repeatedly nudges him, until finally the male ceases his song and holds his wings in a raised position; thereupon, the female promptly climbs on top of the male. In this position, she eats from the metanotal gland just behind the wings of the male (from whence an alluring substance is secreted). The insects are then in a situation proper for mating, which takes place with the female on top of the male.

It is common knowledge that the chirp rate varies with changes in the surrounding (ambient) temperature, increasing at higher degrees and decreasing at lower. This fact has inspired formulas for calculating the temperature from the number of chirps per minute in different cricket species. The Snowy Tree Cricket *(Oecanthus fultoni),* also called the "thermometer cricket," indicates the temperature in degrees Fahrenheit if one counts the number of chirps in 13 seconds and adds 40. But the species is rare in the Los Angeles Basin. The much more common species with a pulsing call, *O. rileyi,* chirps about twice as fast as *O. fultoni* at the same temperature.

These insects do eat plant matter but are generally considered beneficial because of their greater inclination to feed on plant pests such as aphids, scale insects, leaf hoppers, and the like.

Tree crickets, especially males, are occasionally attracted to light.

REFERENCES. Block, B. C. 1966. The relation of temperature to the chirp-rate of male Snowy Tree Crickets, *Oecanthus fultoni* Orthoptera: Gryllidae). *Annals of the Entomological Society of America,* vol. 59, pp. 56-59.

Fulton, B. B. 1915. The tree crickets of New York: Life history and bionomics. New York Agricultural Experiment Station Technical Bulletin, vol. 42, 47 pages.

36. Tree cricket male *(Oecanthus* species). Photograph by C. Hogue.

Funk, D. H. 1989. The mating of tree crickets. *Scientific American,* vol. 261, no. 2, pp. 50-55, 58-59.

Walker, T. J. 1962. The taxonomy and calling songs of United States tree crickets (Orthoptera: Gryllidae: Oecanthinae), part 1: The genus *Neoxabea* and the *niveus* and *varicornis* groups of the genus *Oecanthus. Annals of the Entomological Society of America,* vol. 55, pp. 303-322.

_______. 1969. Acoustic synchrony: Two mechanisms in the Snowy Tree Cricket. *Science,* vol. 166, pp. 891-894.

---

■ **Field Crickets** (*Gryllus* species) Figure 37

There are several species of this genus locally. They are not nearly as common in residential areas in Los Angeles as tree crickets, although in some field situations they predominate. With body lengths of 1/2 to 1 1/4 inches (20 to 30 mm), they are considerably larger and more robust than tree crickets. They are largely black, brown, or reddish in color (often only the wings are brown, the remainder of the body being black).

Field crickets live on the ground in fissures and under litter, vegetation, and stones. They sometimes sing in the morning or late afternoon but more usually at night when they come out to feed on all sorts of organic matter. They may constitute a minor agri-

37. Field cricket *(Gryllus* species). Photograph by L. Brown.

38. European House Cricket female. Photograph by C. Hogue.

cultural pest when invading gardens and fields to eat cultivated plants. They also occasionally enter homes and become a nuisance by their unwelcome presence and incessant chirping. In general, our species are heard and seen in spring to early summer.

The calling songs of members of the genus *Gryllus* are varied and species specific: major differences in chirp rate and other call features exist between species (these are best measured with an oscilloscope from sound recordings).

REFERENCE. Weissman, D. B., D. C. F. Rentz, R. D. Alexander, and W. Loher. 1980. Field crickets *(Gryllus* and *Acheta)* of California and Baja California, Mexico (Orthoptera: Gryllidae: Gryllinae). *Transactions of the American Entomological Society,* vol. 106, pp. 327-356.

---

■ EUROPEAN HOUSE CRICKET
*(Acheta domesticus)*
Figure 38

This cricket is generally straw-brown, with an irregular dark transverse bar extending between the eyes. Its body length is about 3/4 inch (20 mm), making it somewhat smaller than most species of field crickets.

The species was apparently introduced into the eastern United States from Europe, although its original home may have been Africa. It has since become widespread in southern California, where it is usually associated with human habitations. Lacking a dormancy period and hence being easy to raise, it is sold as fish bait and animal food in pet stores. Its chirp is frail and attracts less attention than that of its field cricket relatives and *Grylloides suplicans,* another immigrant that appears to be becoming increasingly common in our area.

REFERENCE. Weissman, D. B., and D. C. F. Rentz. 1977. Feral house crickets *Acheta domesticus* (L.) (Orthoptera: Gryllidae) in southern California. *Entomological News,* vol. 88, pp. 246-248.

## ■ KATYDIDS (Family Tettigoniidae)

KATYDIDS ARE SO NAMED because of their supposed participation in a legendary love affair. Involved were two maidens: one was fair, the other (whose name was Kate) was more on the stately side. The masculine corner of the triangle, an anonymous lad, fell in love with the fair one and scorned the passions of Kate. When he mysteriously died, the question was: "Did or didn't the proud Kate do him in?" The insect lives today as the deceased's spirit, continually proclaiming the answer each summer night—"Kate-she-did," or the variation "Katy-did."

Only the males of the common eastern katydid

*Pterophylla camellifolia* make this novel sound. In the two relatives of this species that live in the basin, both sexes can make sounds. The male has a song repertoire, and the female answers with a single click, which draws the male to her. Both sexes produce sound like crickets do, by rubbing together roughened veins at the bases of the fore wings. The ears of katydids are located on the fore legs, in small depressions near the base of the tibia.

Katydids are easily distinguished from true grasshoppers by their antennae, which are much longer (almost as long or longer than the body) and more slender than those of grasshoppers.

REFERENCE. Spooner, J. 1968. Pair-forming acoustic systems of phaneropterine katydids (Orthoptera, Tettigoniidae). *Animal Behavior,* vol. 16, pp. 197-212.

---

■ BROAD-WINGED KATYDID *(Microcentrum rhombifolium)* Figures 39, 40

This species usually reveals its presence only by its song, which can be heard at dusk and in the evening in late summer and fall (August to October). The male emits two distinct and inconsistently related sounds: loud zips (in series of two or three, separated by about 3 seconds) and series of ticks (fifteen to thirty-five, at the rate of eight or nine per second). The female responds to the ticks after about an eighth of a second with a low-intensity ticking, which attracts the male to her.

The male is 2 inches (5 cm) in body length, smaller than the female, which is 2 1/2 inches (63 mm) from head to tip of wings. The adult insects are leaf-green and difficult to see high in the foliage of deciduous trees and shrubs where they spend most of their time. The nymph is green with tiny black flecks. The front of the nymph's head is smoothly rounded, and its body is compressed with a convex crested back.

Nymphs are of interest for the gymnastics they perform. They do "push-ups" with the front four legs or sway from side to side in a manner suggestive of the movements made in some Oriental dances, where the body is maintained level and is moved from side to side as the knees alternately bend and straighten.

Females line up their eggs on the surfaces of twigs, in single or double overlapping files.

---

■ FORK-TAILED BUSH KATYDID *(Scudderia mexicana)* Figure 41, 42

This katydid is common on all kinds of vegetation in the late summer and fall. It is easily distinguished from the Broad-winged Katydid by its elongate wings and slightly smaller size (its length from head to wing tips is 1 1/2 in., or 38 mm). Its song is also slightly different:

39. Broad-winged Katydid. Photograph by C. Hogue.

40. Broad-winged Katydid nymph. Photograph by C. Hogue.

41. Fork-tailed Bush Katydid female. Photograph by C. Hogue.

42. Fork-tailed Bush Katydid nymph. Photograph by C. Hogue.

males usually lisp ("zeek") in series of three or four a few seconds apart, although a shorter-pulsed sound ("zip") is also made; ticking seems to be rare. Females respond after a little over a second with ticking, which attracts the males.

The females insert their eggs lengthwise into the edges of the leaves of shrubs and trees.

Although similar to that of the Broad-winged Katydid, the nymph of the Fork-tailed Bush Katydid possesses a "unicorn" horn on the front of the head. The body is cylindrical, strongly arched, and multicolored, with small orange spots and blue patches on the front part of the thoracic shield.

---

■ CHAPARRAL KATYDID
*(Platylyra californica)*
Figure 43

This katydid is small (length from head to wing tips 3/4 to 7/8 in., or 20 to 23 mm) with fully developed leaf-green wings. In males the specialized sound-producing area at the wing bases is mottled brown, and the front and side margin of the wing is thinly brown-tinged as well. The female has a broad, curved ovipositor with very fine teeth at its tip.

The species is found in the chaparral, coastal sage, and oak woodland plant communities in the basin's fringing foothills. Females lay their eggs on the leaves of deciduous trees, especially oaks, in the late summer and fall. They are nocturnally active and sing with a distinctive strident chirp.

REFERENCE. Grant, H. J., Jr., and D. C. Rentz. 1966. The katydid genus *Platylyra. Pan-Pacific Entomologist,* vol. 42, pp. 81-88.

---

■ CHAPARRAL SHIELD-BACKED KATYDID
*(Idiostatus aequalis)*
Figure 44

This katydid is moderate-sized (females are 1 5/8 in., or 41 mm, from head to tip of the long ovipositor; males are about 1 1/2 in., or 38 mm, long) and elongate, with long slender legs. The wings in both sexes are missing and are represented only by small pads. In our populations, the body is reddish-brown; all males and some females have a dark spot at the tip of the wing pads.

The species occurs in the Santa Monica Mountains in chaparral, riparian woodland, and grassy areas. Adults are active from late June to mid-September at night. Their song is a low continuous buzz. They are nocturnal, venturing to the tops of bushes to feed on blossoms, foliage, or other insects.

REFERENCE. Rentz, D. C. 1973. The shield-backed katydids of the genus *Idiostatus. Memoirs of the American Entomological Society,* vol. 29, pp. 1-211.

## ■ CAMEL CRICKETS
(Family Rhaphidophoridae)

MEMBERS OF THIS FAMILY are brown or gray in color and lack wings and auditory organs. Many species possess strongly curved bodies, giving them a hump-backed appearance—hence the name "camel cricket."

■ CALIFORNIA CAMEL CRICKET
*(Ceuthophilus californianus)*
Figure 45

This species is secretive and may only occasionally be seen, creeping on the ground at night or, more commonly, hiding by day in cavities under stones, logs, or boards lying on the soil. Rodent burrows may also serve as shelters. In these habitats the cricket feeds on other living or dead insects.

43. Chaparral Katydid male. Photograph by C. Hogue.

44. Chaparral Shield-backed Katydid female. Photograph by C. Hogue.

45. California Camel Cricket. Photograph by C. Hogue.

It is medium-sized (length 1/2 to 3/4 in., or 13 to 20 mm), with a body that is more or less uniformly pale yellowish to reddish-brown in color and often appears to be banded transversely with a slightly darker shade. The hind legs of both sexes are very long, with straight tibiae.

REFERENCE. Hubbell, T. H. 1936. *A monographic revision of the genus* Ceuthophilus *(Orthoptera: Gryllacrididae: Rhaphidophorinae).* University of Florida Publications in Biology Series, vol. 2, 551 pp.

■ MUSHROOM CAMEL CRICKETS *(Pristoceuthophilus* species) Figure 46

Although very similar to the California Camel Cricket, members of this genus are generally smaller (1/2 in., or 13 mm, long) and more mottled. There are two local species. In both, males have large bowed hind tibiae.

Little is known about the habits of either species, except that specimens are usually found under rocks, boards, and similar cover on the ground in wooded or otherwise protected areas. They are common in foothill grassy areas under stones and concrete rubble partly buried in the ground. Their food probably consists of the many other small arthropods that also live in this same habitat. They sometimes infest commercial mushroom farms.

Adults are seen in the fall and winter months, and young nymphs appear in the spring. Older nymphs are found throughout the summer.

■ SOUTHERN CALIFORNIA CHAPARRAL CAMEL CRICKET *(Gammarotettix genitalis)* Figure 47

Proportionately shorter hind legs (the tibiae of which are straight) and a blunt rather than tapering abdomen distinguish this species from the mushroom camel crickets. This species generally lives not on the ground but in brush and small trees, where it apparently feeds on leaves. It is approximately 1/2 inch (13 mm) in length.

## ■ STENOPELMATIDS
(Family Stenopelmatidae)

■ JERUSALEM CRICKETS *(Stenopelmatus* species) Figure 48

There are several species of Jerusalem crickets in the basin, none apparently named. The insects are immediately recognizable by their large size, bald round heads, winglessness, heavily spined hind legs, and fat abdomens ringed in black. They are known by several other common names, including "sand crickets," "children of the Earth" (Niñas de la Tierra), and "Chacos." (Locally, they have also been called "potato bugs," probably because of the damage they occasionally do to potatoes in the ground. But Jerusalem crickets are not to be confused with the Colorado Potato Beetle

*(Leptinotarsa decimlineata)*, a distinctive striped beetle that feeds on the foliage of this crop in field and gardens in the eastern United States.)

The large size of Jerusalem crickets (they are up to 2 in., or 50 mm, long), and their amber-colored humanoid heads caused them to be the object of superstition and fear by some southwestern and Mexican Indians. The Navajo thought them deadly poisonous and called them "wó see ts´inii," which means "skull insect" or "bone neck beetle." Although their

46. Mushroom Camel Cricket male. Photograph by C. Hogue.

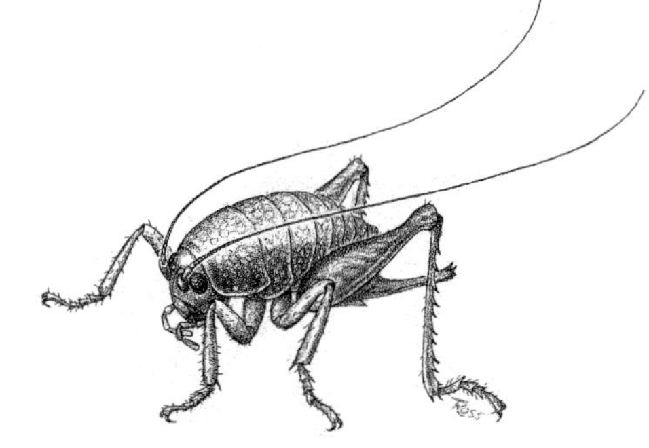

47. Southern California Chaparral Camel Cricket. Drawing by T. Ross.

48. Jerusalem cricket. Photograph by C. Hogue.

strong jaws can bite with considerable force, Jerusalem crickets are not poisonous.

These insects are nocturnal and live in the soil. Individuals are usually seen when they become stranded after a nightly sojourn above ground or when they are uncovered by the gardener's spade. They require high humidity and are most active in the spring, after the winter rains have loosened the soil; in the dry summer they burrow deeply to escape the heat of the day but may wander on the soil surface during the night. Their food consists mainly of roots and tubers, although they sometimes eat dead animal matter and may even be cannibalistic at times.

To attract each other, males and females drum the abdomen against the bottoms of their burrows and on the ground. Mating begins with vigorous "wrestling matches" between the sexes, followed by copulation. The eggs are large and are laid in groups in soil pockets. The typical life cycle extends over two years, and the developing insect may molt up to ten times.

Jerusalem crickets are not easily kept in captivity. It is important to provide them with a humid environment: fresh apple or potato slices provide moisture as well as food.

In my position as entomologist at the Natural History Museum of Los Angeles County, I am asked by the curious public to identify these local insects more than any other.

REFERENCES. Baker, N. W. 1971. Jerusalem cricket. *Pacific Discovery,* vol. 24, no. 2, pp. 12-13.

Tinkham, E. R., and D. C. Rentz. 1969. Notes on the bionomics and distribution of the genus *Stenopelmatus* in central California with the description of a new species (Orthoptera: Gryllacrididae). *Pan-Pacific Entomologist,* vol. 45, pp. 4-14.

Weissman, D. B. 1986. Revisionary studies of Jerusalem crickets *(Stenopelmatus)* (Orthoptera: Stenopelmatidae). *Proceedings of the 4th Triennial Meeting of the Pan American Acridological Society,* vol. 4, p. 150.

## COCKROACHES (Order Blattodea)

COCKROACHES are much maligned insects. A few pesky species have given a bad name to the whole order of thousands of species, including more than fifty in North America. The few "bad" cockroaches are common household pests in most warm parts of the world. By far the majority of kinds, however, are very interesting "wild" cockroaches that inhabit caves, burrow in sand dunes, live in ant nests, or exhibit other unusual life histories.

Cockroaches are also not all drably colored like the familiar household varieties. Many tropical species sport yellow, red, green, and other colors on their bodies and wings and are quite beautiful.

Our all-too-common household cockroaches are active primarily at night. During the day they hide in dark crevices—in or behind kitchen cabinets, drawers, stoves, and refrigerators. The female lays her eggs in a hard dark brown purse-shaped capsule, which she may carry about for several days, protruding from the end of the abdomen. The capsule is eventually dropped, and later the young cockroaches hatch and scatter. Both young and adults are general feeders. Almost any crumb of food or other organic material left exposed around the house will serve them as a meal. For this reason, good housekeeping is the key to cockroach control.

The importance of cockroaches in transmission of human diseases seems overrated, although most of the domiciliary species have been found capable of mechanically transmitting some disease organisms, especially dysentery bacteria.

Local residents sometimes report "albino roaches." These are nothing more than freshly molted specimens whose white integument has not yet become pigmented by the tanning process that follows the development of the new skin.

The major kinds of cockroaches found in our area may be grouped into three categories on the basis of their place of origin:

- Cosmopolitan household species long associated with mankind: American Cockroach, Oriental Cockroach, German Cockroach, Brown-banded Cockroach.
- Occasional imports (usually from the American tropics), which become temporarily or weakly established, often only in port areas: green cockroaches.
- Natives: Sand cockroaches, Western Wood Cockroach, Field Cockroach.

REFERENCES. Roth, L. M., and E. R. Willis. 1957. *The medical and veterinary importance of cockroaches.* Smithsonian Miscellaneous Collections, vol. 134, 147 pp.

Wood, F. E., W. H. Robinson, S. A. Kraft, and P. A. Zungoli. 1981. Survey of attitudes and knowledge of public housing residents toward cockroaches. *Bulletin of the Entomological Society of America,* vol. 27, pp. 9-13.

■ American Cockroach
*(Periplaneta americana)*
Figure 49

This is the species that Angelenos are referring to when they brag about some huge roach of record-breaking size that they have seen on their premises. This cockroach seems to inspire more exaggeration than any other; in spite of what anyone says, however, adult American Cockroaches in our area rarely exceed 2 inches (50 mm) in length from the front of the head shield to the wing tips. (In Florida there is a much larger cockroach, known as the Palmetto Cockroach *(Blaberus craniifer),* which measures up to 3 inches, or 75 mm, in length.)

The wings completely cover the abdomen of mature individuals of both sexes and are glossy reddish-brown. The head shield is paler reddish-brown, with indistinct central markings.

This is the local species that most often attracts attention, but it occurs only in small numbers. Two close relatives of this cockroach sometimes appear in the basin, but they are not ubiquitous like the American Cockroach. These are the Brown Cockroach *(Periplaneta brunnea)* and the Smoky Brown Cockroach *(P. fuliginosa)*. The latter species is established locally in isolated spots.

REFERENCES. Bell, W. J., and K. G. Adiyodi, editors. 1981. *The American Cockroach.* London: Chapman Hall.

Powell, P. K., and W. H. Robinson. 1980. Descriptions and keys to the first-instar nymphs of five *Periplaneta* species (Dictyoptera: Blattidae). *Proceedings of the Entomological Society of Washington,* vol. 82, pp. 212-228.

Waldron, W. G., and F. Hall. 1972. Mode of entry into Los Angeles County, California, of the Brown Cockroach, *Periplaneta brunnea* Burmeister. *California Vector Views,* vol. 19, pp. 1-2.

Wharton, D. R. A., J. E. Lola, and M. L. Wharton. 1967. Population density, survival, growth, and development of the American Cockroach. *Journal of Insect Physiology,* vol. 13, pp. 699-716.

■ Oriental Cockroach
*(Blatta orientalis)*
Figure 50

Next in size to the American Cockroach is this shiny black to brownish species, which is about 1¼ inches (32 mm) long. The male has short wings (their tips do not quite reach the middle of the abdomen), and the female has almost no wings at all, only small pads emerging from beneath the head shield.

Because of their black color, Oriental Cockroaches are sometimes confused with large ground beetles. They are sometimes called "water bugs" because of their habit of living in wet areas around toilet bowls, bath tubs, water pipes, and faucets.

The Oriental Cockroach seldom enters houses and is usually found outdoors near the house or in the garage. It is reported to be very common in the sewers of Los Angeles.

The species has a long life cycle, up to two years. The egg cases of the female contain about sixteen eggs.

■ GERMAN COCKROACH
*(Blattella germanica)*
Figure 51

This species is only about 1/2 inch (13 mm) long and is light brown, with two darker longitudinal stripes on the head shield. The wings extend beyond the tip of the abdomen in the adults of both sexes. A wary and

49. American Cockroach male. Photograph by C. Hogue.

50. Oriental Cockroach female. Photograph by C. Hogue.

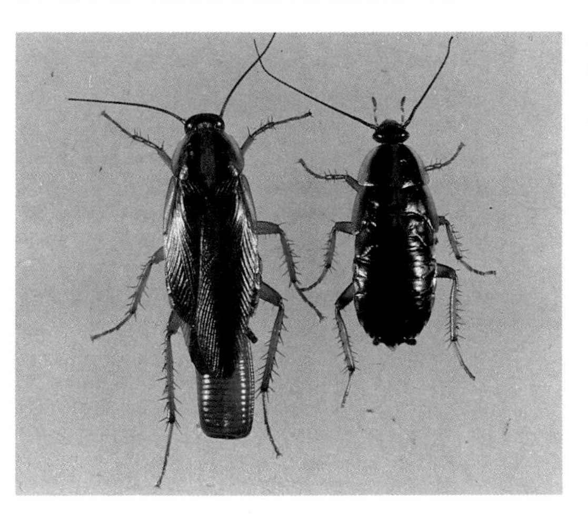

51. German Cockroach female with egg case (left) and nymph. Photograph by R. Pence.

active cockroach, it tends to reach higher population levels than the American Cockroach. Furthermore, it is gregarious, so that tremendously dense infestations may develop. This is by far the most prolific cockroach of those mentioned here—it has about forty-eight young per egg case. It can complete its life cycle in as little as six weeks.

The closely related Field Cockroach *(Blattella vaga),* which is distinguished by the contrasting dark pigment band between the eyes, is an outdoor species found primarily at the eastern and southern fringes of the basin.

Both *Blattella* species are introduced.

REFERENCES. Buxton, G. M., and T. J. Freeman. 1968. Positive separation of *Blattella vaga* and *Blattella germanica* (Orthoptera: Blattidae). *Pan-Pacific Entomologist,* vol. 44, pp. 168-169.

Roth, L. M. 1985. A taxonomic revision of the genus *Blattella* Caudell (Dictyoptera: Blattaria: Blattellidae). *Entomologica Scandinavica,* supplement 22, pp. 1-221.

Twomey, N. 1966. A review of the biology and control of the German Cockroach, *Blattella germanica* (L.), in California. *California Vector Views,* vol. 13, pp. 27-37.

---

■ **Brown-banded Cockroach** *(Supella longipalpa)* Figure 52

This insect is only about $^{3}/_{8}$ inch (10 mm) long and is our smallest domestic cockroach. It is unusual in its markings, with two pale bands crossing the fully developed pale brown wings. The least prevalent of the local cockroach species, it tends to be solitary in its habits or to live in small groups. The females form cases with about eighteen eggs.

---

■ **Green Cockroaches** *(Panchlora* species) Figure 53

These cockroaches, which are immediately recognizable by their pale translucent green color, are not permanently established in the basin but instead come in occasionally on produce shipped from Central or South America. Consequently, they will be found mostly around the harbor area, but they may spread inland at times. They do not seem to be as common now as in the past, probably because of the recently adopted practice of packaging bananas in cardboard cartons. The various species are all about 1 inch (25 mm) long.

REFERENCE. Roth, L. M., and E. R. Willis. 1958. The biology of *Panchlora nivea* with observations on the eggs of other Blattaria. *Transactions of the American Entomological Society,* vol. 83, pp. 195-207.

■ Sand Cockroaches
*(Arenivaga* species)
Figure 54

The males and females of these cockroaches look vastly different (Figure 54). Males are just under 1 inch (25 mm) long and have the normal cockroach form with fully developed, mottled, pale-brown wings. The females, however, are oval wingless creatures that look something like trilobites, the extinct aquatic arthropods of the Paleozoic Era.

These native cockroaches are found near the coast on sand dunes or inland in our neighboring

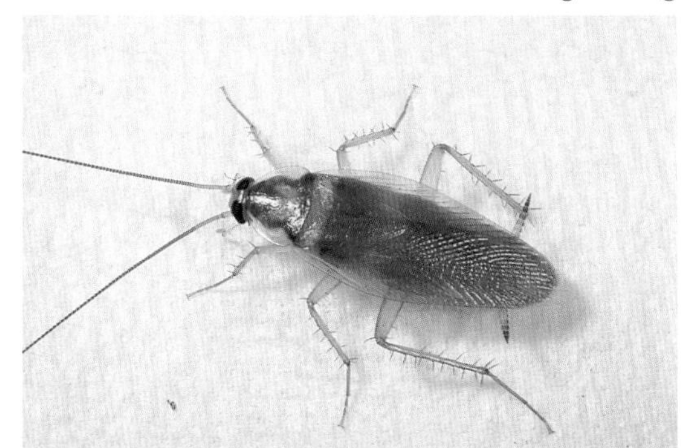

52. Brown-banded Cockroach. Photograph by C. Hogue.

53. Green cockroach male (left) and female *(Panchlora* species). Photograph from L. Roth and E. Willis, 1960, *The biotic associations of cockroaches,* Smithsonian Miscellaneous Collections, vol. 141, pl. 16 (Washington, D.C.: Smithsonian Institution).

54. Sand cockroach male (left) and female *(Arenivaga* species). Photographs by M. Badgley (left) and L. Reynolds.

deserts and foothills. Their presence is indicated by the small serpentine ridges of their burrows, which are formed near the surface like miniature mole runs. The ridges are most evident at night when the roaches are active and when the light of a lantern casts strong shadows over the sand surface. Males are attracted to light.

REFERENCE. Appel, A. G., A. M. Van Dyke, and M. K. Rust. 1983. A technique for rearing and some notes on the biology of a desert sand cockroach, *Arenivaga investigata* (Dictyoptera: Polyphagidae). *Proceedings of the Entomological Society of Washington,* vol. 85, pp. 598-600.

---

■ WESTERN WOOD COCKROACH *(Parcoblatta americana)* Figure 55

The males of this native species are a glossy dark brown and fully winged. The females resemble small Oriental Cockroaches (their body length is about 1 in., or 25 mm) and have only small wing pads rather than wings. The species is found locally in the spring under rocks, in loose bark and litter, and on grassy slopes in Coastal Sage and chaparral plant communities. The males are often attracted to light. Nothing else is known of their biology.

55. Western Wood Cockroach male. Photograph by C. Hogue.

## PRAYING MANTIDS (Order Mantodea)

MOST OF THE ATTENTION paid these curious insects centers around their feeding habits. Exclusively carnivorous, they catch small animals with quick thrusts of their front legs, which are structurally modified for grasping.

The mantid does not actively seek its food but usually waits for it, remaining motionless for long periods at a time, with the head upraised and the fore legs together in front as if the "hands" are held in prayer. This attitude justifiably led to the common name, although "preying" mantid is probably more realistic: to quote Gurney (1951) "the only thing mantids would seem to pray for is a square meal."

Because they feed voraciously on flies, grasshoppers, caterpillars, and other insects, many of which are garden and crop pests, mantids are considered beneficial. There is even a record of a mantid consuming a mouse! Nevertheless, many people are afraid of these insects and even hold superstition beliefs about them that probably stem from the creatures' pugnacious temperament and human-like actions. Adding to the mystique is the mantid's relatively narrow neck, which like the owl's allows the head to be rotated freely almost completely around. The mantid appears to be staring agressively as it follows one's movements with its head and eyes.

These peculiarities have given rise to a variety of common names for praying mantids, including "nuns," "saints," "preachers," "rear horses," and "devil horses." The word "mantid" in Greek *(mantikos)* means "prophet."

Mantid females enclose their eggs in a frothy mass, which hardens and forms a protective capsule. Females have been observed decapitating the male during mating, but this behavior seems to be only occasionally displayed by laboratory-reared individuals.

Local residents rarely see mantids in spite of the fact that these insects are reasonably common. Their occurrence is limited to native shrubbery on the few relatively undisturbed hilly areas remaining in the basin. In the fall, males may occasionally be seen flying among vegetation or near lights.

There are three permanent basin mantid residents (the California Mantid and Minor Ground Mantid are natives), although two or three exotics (the Carolina Mantid, Chinese Mantid, and European Mantid) occasionally show up as escapees from imported egg masses, which are sold to gardeners for pest control by biocontrol insectaries and seed companies.

REFERENCES. Gurney, A. B. 1951. Praying mantids of the United States, native and introduced. Annual report for 1950, Smithsonian Institution, pp 339-362.

Liske, E., and W. J. Davis. 1984. Sexual behavior of the Chinese Mantid. *Animal Behavior,* vol. 32, pp. 916-917.

Nickle, D. A. 1981. Predation on a mouse by the Chinese Mantid *Tenodera aridifolia sinensis* Saussure (Dictyoptera: Mantoidea). *Proceedings of the Entomological Society of Washington,* vol. 83, pp. 801-802.

Roeder, K. D. 1935. An experimental analysis of the sexual behavior of the praying mantis *(Mantis religiosa* L.). *Biological Bulletin,* vol. 69, pp. 203-220.

Wood, S. F. 1978. Notes on mantids *(Stagmomantis, Iris)* as possible predators of conenose bugs *(Triatoma, Paratriatoma). Pan-Pacific Entomologist,* vol. 54, pp. 17-18.

---

## ■ KEEPING MANTIDS IN CAPTIVITY

MANY PEOPLE like to keep adult mantids alive as pets, and they are easily maintained in a glass jar (a small potted house plant placed in an aquarium or large screen cage makes an even better home).

Because mantids are carnivorous, they must be kept supplied with fresh animal food. Domestic flies, grasshoppers, moths, and other garden insects may be introduced alive into the cage. Small pieces of uncooked meat or dead insects held with forceps may be fed to the mantid by hand, but it will probably ignore meat or immobile bodies of insects placed on the cage floor.

One may acquire mantids by searching for the adults or by rearing them from purchased egg capsules. The latter, usually of the Chinese Mantid, are available from plant nurseries and biological supply dealers in the fall and winter. Eggs often hatch unpredictably in southern California: merely bringing the case indoors or into our warm winter climate may cause the eggs to hatch almost immediately.

The young are cannibalistic, so they should be kept in individual containers. They may be fed fruit flies, aphids, or other small insects until they are larger.

REFERENCE. Fye, R. E., and R. L. Carranza. 1979. A simplified method of feeding large numbers of individual mantids. *Journal of Economic Entomology,* vol. 72, pp.83-84.

---

■ CALIFORNIA MANTID
*(Stagmomantis californica)*
Figure 56

This native species is moderately large (2 to 2 1/2 in., or 50 to 64 mm, long) and mottled brown or greenish in color. The male has fully developed wings that, when folded, extend beyond the tip of the abdomen; the female's wings are very short, reaching only to about the middle of the abdomen. The species prefers an arboreal habitat and is primarily found on shrubs of the Coastal Sage plant community. Males are often attracted to lights.

---

■ MINOR GROUND MANTID
*(Litaneutria minor)*
Figure 57

This is a small, slender mantid (about 1 1/4 in., or 32 mm, long) that is pale tan or grayish. The male often has a dark spot near the base of its fully developed hind wings; the female has practically no wings, only short stubs. The species is very active and is usually not found on vegetation as it prefers to run on the ground among rocks and short grass in search of prey.

As with other mantids, there is one generation per year. Eggs are laid in the fall; they hatch the following spring; and the nymphs grow during the summer to mature as adults in the fall.

REFERENCE. Roberts, R. A. 1937. Biology of the minor mantid, *Litaneutria minor* Scudder (Orthoptera, Mantidae). *Annals of the Entomological Society of America*, vol. 30, pp. 111-121.

56. California Mantid male. Photograph by P. Bryant.

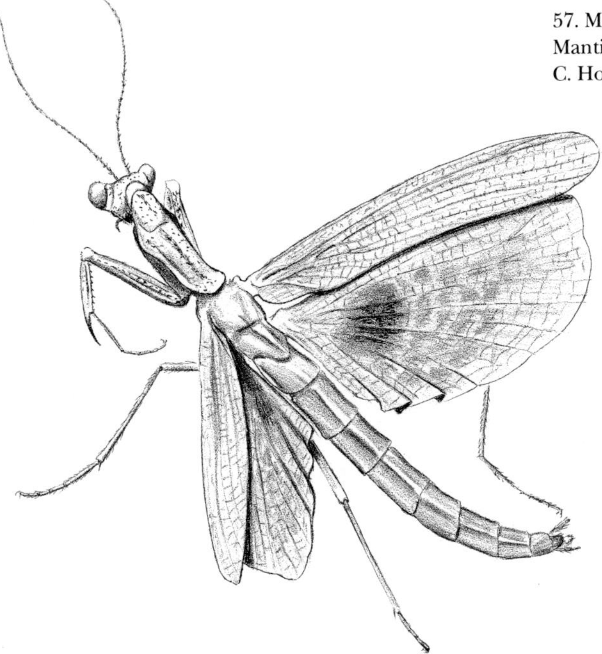

57. Minor Ground Mantid. Drawing by C. Hogue.

**■ Mediterranean Mantid** *(Iris oratoria)* Figure 58

This exotic mantid, which is commonly found in the San Joaquin, Coachella, and Imperial Valleys, has made its way into the Los Angeles Basin and appears to have established itself here. The species is native to the Mediterranean region but somehow was introduced into the United States, probably in the 1930s or 1940s.

This mantid resembles the greenish variety of the California Mantid and is about the same size (2 in., or 50 mm, long). It is most easily distinguished by the color pattern of its hind wing, which has a shiny bluish-black blotch near the base and an area of dull brick-red toward the tip (the hind wing of the California Mantid is generally brown).

REFERENCE. Gurney, A. B. 1955. Further notes on *Iris oratoria* in California. *Pan-Pacific Entomologist,* vol. 31, pp. 67-72.

## WALKINGSTICKS (Order Phasmatodea)

Like mantids, walkingsticks are seldom seen in the Los Angeles Basin, and probably for the same reasons—cryptic form and color rather than rarity. Walkingsticks are virtually impossible to distinguish from the twigs of the shrubby plants in which they live. Their cautious behavior adds to the camouflage effect.

Unlike mantids, walkingsticks are herbivorous, feeding on various species of native vegetation. Their tastes are fairly specific, and it is difficult to keep caged specimens without fresh leaves of their proper host plant.

We have only three main types locally. All are wingless and more or less stick-like in form and thus are by no means typical of their order: many of the hundreds of tropical species have fully developed wings or leaf-like forms.

REFERENCE. Bedford, G. O. 1978. Biology and ecology of the Phasmatodea. *Annual Review of Entomology,* vol. 23, pp. 125-149.

**■ Gray Walkingstick** *(Pseudosermyle straminea)* Figure 59

Its long antennae (longer than the femur of the front leg) and the roughened longitudinally ridged or spined thorax distinguish this species from the Western Short-horned Walkingstick. It is generally light gray or whitish in color, with a body $1^1/_2$ to 3 inches (40 to 75 mm) long. A desert species, it occurs on tall grasses and dry scrub, such as Rabbitbrush *(Chrysothamnus nauseosus),* Burro-brush *(Ambrosia dumosa),* and sagebrush *(Artemisia).* It is known from the Santa Monica Mountains but seems to be more prevalent in the drier valleys. Adults are active in the fall.

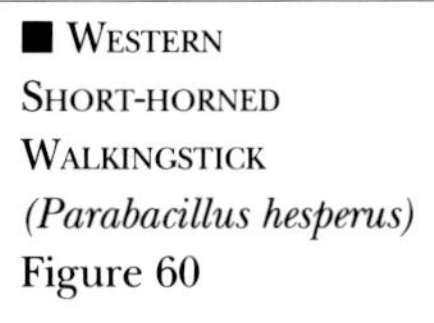
■ Western Short-horned Walkingstick (*Parabacillus hesperus*) Figure 60

This is the species most likely to be encountered in the basin. It is readily recognized by its short antennae (which, unlike those of the Gray Walkingstick, are much shorter than the femur of the front leg) and its smooth body. It is light to dark brown, sometimes pinkish, and the body is 2 1/2 to 3 1/2 inches (65 to 90 mm) long. It can be found in the chaparral and on tall grasses and native herbaceous plants in hilly areas. Specific hosts are burroweed *(Haplopappus)*, globemallow *(Sphaeralcea)*, mountain mahogany *(Cercocarpus)*, and buckwheat *(Eriogonum)*. Adults are found in early summer to early fall.

58. Mediterranean Mantid male. Photograph by C. Hogue.

59. Gray Walkingstick. Photograph by P. Bryant.

60. Western Short-horned Walkingstick. Photograph by C. Hogue.

**■ Timemas (*Timema* species) Figure 61**

Timemas are not typical walkingsticks. They are only 1/2 to 1 inch (13 to 25 mm) long and are considerably shorter and stouter than the other local species. Although usually mottled green in color, brown or pinkish individuals frequently occur.

There are several local species; *Timema californica* seems to be dominant. It lives on chaparral shrubs and oaks in our surrounding mountains, especially the Santa Monicas and San Gabriels. Individuals may be common in the spring on Scrub Oak *(Quercus dumosa)* and wild lilac *(Ceanothus)*. When disturbed this insect can produce a disagreeable odor.

REFERENCE. Gustafson, J. 1966. Biological observations on *Timema californica* (Phasmoidea: Phasmidae). *Annals of the Entomological Society of America,* vol. 59, pp. 59-61.

61. Timemas *(Timema* species) copulating. Photograph by C. Hogue.

## TERMITES (Order Isoptera)

Of all the insects that live around us in the city, termites probably cause the most concern. The resident desiring to sell a house is made painfully aware of them when told by an inspector that there is an infestation that must be corrected before transfer of ownership. Most loan companies make termite inspection one of the requisites of a clear title.

Termites are like cockroaches in that not all species are despicable. Most of the many hundreds of species of termites in the world actually benefit humanity by cleaning up cellulose debris from the earth's surface. Symbiotic microorganisms in the intestine of the worker caste enable termites to obtain nutrition from this otherwise indigestible material. When it reaches the termite's intestine, the wood swallowed by the insect is broken down into simple digestible sugars. "Defaunated" termites (those that have had their protozoans and bacteria killed by special treatment) die of starvation even though they may still ingest wood.

Termites are social insects. The colony is founded by a mated ("royal") pair of winged adults. Soon after the nymphs of the first brood hatch, they are able to take care of themselves as well as the royal pair, freeing the latter of all tasks save the essential one of producing new offspring. Subsequently the colony grows in size and complexity, and different forms develop, each with a specialized anatomy and particular job to do in the colony, which now is really a large diversified family. The different types (Figures 64, 65) are called "castes" and typically consist of (1) the "queen," the original female member of the royal pair, (2) the "king," the original male member of the royal pair, (3) the "workers," sexually undeveloped individuals that tend and feed the royal pair in addition to building and maintaining the nest, and (4) the "soldiers," which are modified workers with massive heads and enlarged mandibles that they employ in fighting off intruders to the colony.

The mature colony periodically produces generations of adult broods. Swarms of these reproductive forms (Figure 62) are often seen issuing from frame houses or other wooden structures on warm days following a rain. These forms are winged and are sometimes confused with carpenter ants; termites are readily distinguished, however, by their four similarly shaped wings (in ants, the hind wing is much smaller than the fore wing) and their lack of the narrowed waist region that is characteristic of ants.

The formation of adults is correlated with the rainy season in the Western Subterranean Termite; swarming is stimulated by precipitation and is most pronounced locally following rains in the fall and to a lesser extent in the spring. The drywood and dampwood species are on an independent schedule and swarm regularly in the fall (September or October), regardless of rainfall.

Males and females pair off during the swarming phase and then drop their wings. Pairs of the Western Subterranean Termite burrow into the ground in search of wood. Drywood and Dampwood types enter bare wood surfaces through cracks and crevices, although they usually do not chew through well painted or finished surfaces. After making a chamber for themselves in the substrate, the pairs mate and begin to rear a new brood.

Contrary to popular belief, redwood, cedar, and eucalyptus woods are not immune to termite attack. When newly cut, these woods contain oils that naturally repel termites, but they become susceptible after moisture and age have modified their original composition.

REFERENCES. Ebeling, W., and R. J. Pence. 1965. Termite control. California Agricultural Experiment Station Extension Service, Circular 469, revised edition. Berkeley, Calif.: University of California.

Krishna, K., and F. M. Weesner, editors. 1969-70. *Biology of Termites,* 2 vols. New York: Academic Press.

## ■ TRUE TERMITES (Family Rhinotermitidae)

■ WESTERN SUBTERRANEAN TERMITE *(Reticulitermes hesperus)* Figures 62-64

On warm sunny days following the first autumn rains, swarms of the winged adult forms of this termite are commonly noticed emerging from frame houses, fence posts, and other wooden structures that touch soil. The species has a high humidity requirement, which forces it to maintain contact with the ground, traveling up and down between its subterranean galleries and the wood through protected cracks in mortar or concrete foundations, or through earthen tubes that it constructs from soil, saliva, and chewed bits of wood.

In Los Angeles and much of the West, this is the species that causes the greatest damage. It is probably safe to say that the majority of older houses in the Los Angeles area are infested to some degree with this termite. In general, however, damage is not noticeable until tunneling activity has proceeded to the point of weakening structural members in stressed areas, such as flooring and stairways. Severe damage requires a period of years to develop: our termites do not reduce a house to a pile of sawdust overnight! Homeowners are urged to have periodic inspections to determine the presence of termites. This is simply good insurance and should be done regardless of how many preventative methods were employed in the original construction.

This species is distinguished from others that are prevalent in the basin by the black heads of its sexual forms, its earthen tubes, and the fact that it does not make pellet piles. Its tunneling pattern is also different: the workers attack wood only in the soft spring growth region of the annual rings. Thus a cross-section of an infested timber shows a characteristic pattern of concentric circles or arcs.

62. Western Subterranean Termite winged males and females swarming. Photograph by C. Hogue.

63. Western Subterranean Termite queen (center) among soldiers (the two insects with dark mandibles) and workers. Photograph by R. Pence.

64. Western Subterranean Termite workers. Photograph by C. Hogue.

## ■ DRYWOOD TERMITES
(Family Kalotermitidae)

■ **Western Drywood Termite**
*(Incisitermes minor)*
Figure 65

This species is about as common in southern California buildings as the Western Subterranean Termite, but it ordinarily causes less serious damage. It looks like the Subterranean Termite but is a little larger, and the head of its sexual form is reddish rather than black. The Western Drywood Termite's tunneling action in wood is also distinctive: the workers burrow across as well as with the wood grain, excavating large pockets in all directions.

Drywood termites can become established anywhere and are especially common in attics; they require no connection with the ground and make no

65. Western Drywood Termite winged adult (a), queen after shedding wings (b), soldier (c), and fecal pellets (d). Photographs by R. Pence.

earthen tubes. Their presence is often indicated by "sawdust" piles of dry fecal pellets. This material has a grainy texture; under magnification (Figure 65d), the roughly hexagonal cross-section of the individual pellet reveals it to be of termite origin, rather than from other wood-boring insects.

Colonies contain the usual castes except true workers. The nymphs perform the worker function until they mature into reproductive or soldier forms.

## ■ ROTTENWOOD TERMITES
(Family Hodotermitidae)

■ PACIFIC DAMPWOOD TERMITE
*(Zootermopsis angusticollis)*
Figures 66, 67

This is a much larger species than either of the foregoing. The nymphs are about 1/2 inch (13 mm) long, the soldiers 3/4 inch (20 mm). All types of wood are attacked, but wood that is buried in the ground and subject to moisture and decay is the most vulnerable. Although of relatively occasional occurrence in the basin, this termite is extremely common in the San Gabriel Mountains in conifer logs and stumps.

Late afternoon or early evening swarming may take place throughout the year; but July to October is the typical swarming season. The winged forms are strong fliers and are attracted to lights at night. There are only two well-defined castes, the soldier and reproductive; true workers are absent, and—as in the Western Drywood Termite—the worker function is performed by nymphs.

By vibrating their heads on the walls of their chambers, soldiers and nymphs produce an audible ticking sound that is believed to function as an alarm to the colony.

REFERENCES. Castle, G. B. 1934. The damp-wood termites of the western United States, genus *Zootermopsis* (formerly,

66. Pacific Dampwood Termite soldier; Argentine Ant in the background. Photograph by C. Hogue.

67. Pacific Dampwood Termite winged adult. Photograph by C. Hogue.

*Termopsis).* Pages 273-310 in *Termites and termite control,* 2nd edition, edited by C. A. Kofoid. Berkeley, Calif.: University of California.

Heaton, S. S. 1966. Life of a damp-wood termite. *Pest Control,* vol. 34, no. 2, pp. 28a-30.

Howse, P. E. 1964. The significance of the sound produced by the termite *Zootermopsis angusticollis* (Hagen). *Animal Behavior,* vol. 13, pp. 284-300.

# 6 PSOCIDS AND LICE

BECAUSE THE PARASITIC LICE are believed to be descended from the free-living psocids, the orders to which these insects belong form a natural group. Both psocids (pronounced "so-sids") and lice are small and often without wings and exhibit a variety of mouthpart structures and feeding habits. The more primitive psocids, which eat plant materials, are often winged; the always wingless lice are specialized for feeding on animal products. Metamorphosis is incomplete.

## PSOCIDS (Order Psocoptera)

IN SPITE OF THE FACT that many species of psocids live in our area, they are seldom noticed because of their small size and secretive habits. Most are found on foliage or tree bark and have been given the name "bark lice." Although the wingless forms resemble true lice, psocids are nonparasitic insects that feed principally on various organic materials of plant origin (for example, pollen and mold spores).

### ■ COMMON BARK LICE (Family Liposcelididae)

■ BOOK LICE
*(Trogium* and *Liposcelis* species and others)
Figure 68

These wingless species of psocids are called "book lice" or "paper lice" because they are so commonly found scurrying over books and newspapers, especially those stored in damp cellars and garages. They are very small (less than 1/8 in., or 2 mm, in length) and light brown or gray in color.

There are probably several closely related domiciliary basin species. The most common is the true Book Louse *(Liposcelis bostrychophila),* a cosmopolitan pest for the food industry, households, museums, and libraries. Although it is a contributor to the allergens found in house dust and its feeding may do minor damage to book bindings and paper, the presence of the Book Louse is usually no more than an annoyance. Out of doors this species lives on woody vegetation, on fungi in ground litter, in soil, or in animal nests.

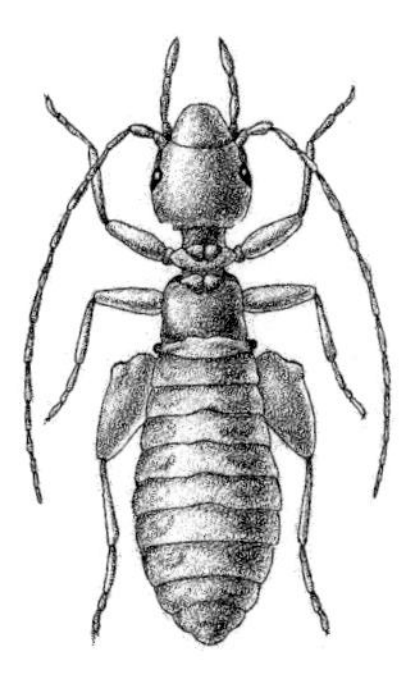

68. Book louse *(Liposcelis* species). Drawing by T. Ross after various sources.

OTHER SCIENTIFIC NAMES. The name *Liposcelis divinitorius,* commonly used for *Liposcelis bostrychophila,* has been declared invalid by entomologists studying the order.

REFERENCES. Broadhead, E. 1946. The book louse and other library pests. *British Book News,* vol. 68, pp. 77-81.

Lienhard, C. 1990. Revision of the western Palaearctic species of *Liposcelis* Motschulsky (Psochoptera: Liposcelidae). *Zoologische Jahrbucher, Abteilung für Systematik der Tiere,* vol. 117, pp. 117-174.

## TRUE LICE (Orders Anoplura and Mallophaga)

TRUE LICE BELONG TO TWO ORDERS, the sucking lice (Anoplura) and the chewing lice (Mallophaga). The groups are also called, respectively, Mammal Lice and Bird Lice because of their usual hosts; however, some bird lice are also found on mammals. The principal distinction, as the former names imply, is in the type of mouthparts and method of feeding employed. Sucking lice use a set of long hypodermic-like stylets to pierce the skin and withdraw blood; chewing lice typically gnaw on fragments of feathers, hair, and skin with a simple pair of mandibles.

All true lice are wingless ectoparasites, which means that they live closely attached to their warm-blooded hosts and cannot pursue a free life as do most other insects. Like other parasites in general, they show many interesting structural modifications that suit their demanding way of life and their unusual environment. Of special note are their claw-like legs, which enable them to cling tenaciously to feathers and hairs in spite of their host's preening and scratching.

In addition to feeding on a host's tissues, many species of true lice transmit diseases or harm their hosts in other ways.

### ■ PRIMATE LICE
(Order Anoplura; Family Pediculidae)

■ HUMAN LOUSE *(Pediculus humanus)* Figure 69

Like the bedbug, this menace to human beings is not as prevalent today as in the past because of improved public and personal hygiene. Yet it still pops up here and there, most often among school children and indigents, and it remains the lone true companion of the hobo.

This is a sucking louse found only on humans, to whom it causes much discomfort in exchange for its meal of blood. Two forms are known: the head louse infests the hair of the scalp, and the body louse lives in clothing near the body surface. Both are small ($^1/_{16}$ to $^3/_{16}$ in., or 1.5 to 4 mm, long) and oval, with pointed legs. Unfed individuals are flat and yellowish to medium brown in color; after ingesting blood they are

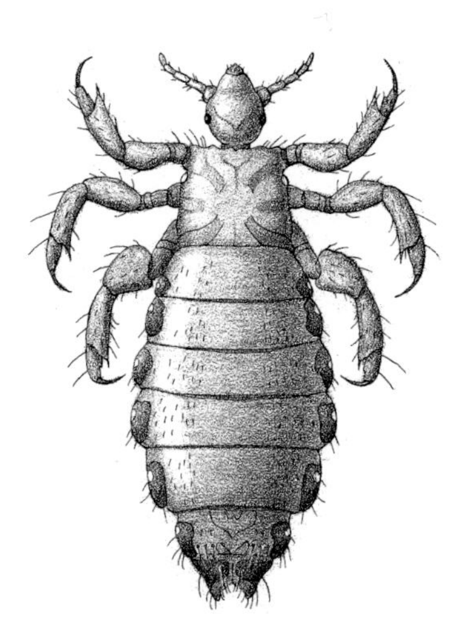

69. Human Louse female. Drawing by T. Ross after various sources.

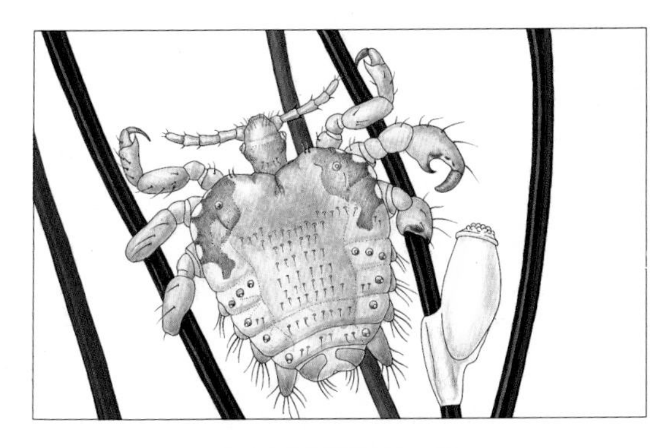

70. Crab Louse and egg attached to human hair. Drawing by T. Ross after K. Smith, editor, 1973, *Insects and other arthropods of medical importance,* fig. 180 (London: British Museum of Natural History).

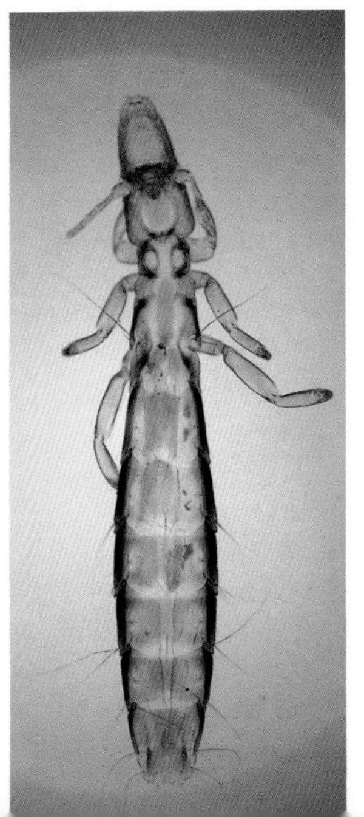

71. Slender Pigeon Louse. Photograph by C. Hogue.

swollen and show a dark clot of blood in the abdomen.

Bites of the Human Louse cause a slight local reaction accompanied by itching. Prolonged infestation may result in a general thickening and increase in pigmentation of the skin ("Vagabond's Disease").

The direct effects of the louse's bite are minor compared to the misery it is capable of causing by transmitting the microorganism *(Rickettsia prowazeki)* that causes Epidemic Typhus, which is a very serious disease. The disease does not occur in the basin; but it has few rivals as a devastator of populations in Europe, where it has actually turned the tide of history.

In the vernacular, the Human Louse is known as the "cootie." Its eggs, which are firmly attached to the hairs of the head and body, are the familiar "nits."

REFERENCES. Buxton, P. A. 1947. *The louse.* London: Edward Arnold.

Keh, B. 1979. Answers to some questions frequently asked about pediculosis. *California Vector Views,* vol. 26, pp. 51-62.

Zinsser, H. 1935. *Rats, lice and history.* Boston: Little, Brown.

■ CRAB LOUSE *(Pthirus pubis)* Figure 70

The Crab Louse is so named because of its broad flat body and crab-claw legs. It is 1/16 inch (1.5 mm) long—much smaller than the body louse, and it normally lives only among the hairs of the pubic regions of the human body. The most common form of transference from person to person is through sexual intercourse, although it may be picked up casually from infested bed linen, clothing, and other sources.

This louse is responsible for local irritation and itching, but it is not a vector of any human disease.

REFERENCES. Keh, B., and J. Poorbaugh. 1971. Understanding and treating infestations of lice on humans. *California Vector Views,* vol. 18, pp. 24-31.

Nuttall, G. H. F. 1918. The biology of *Phthirius pubis. Parasitology,* vol. 10, pp. 383-405.

## ■ NARROW-HEADED BIRD LICE (Order Mallophaga; Family Philopteridae)

■ SLENDER PIGEON LOUSE *(Columbicola columbae)* Figure 71

This chewing louse is 3/32 inch (2.5 mm) long and is characterized by its very slender shape. It is quite abundant on domestic pigeons, especially those making up the feral flocks around the center of Los Angeles. It feeds on the barbules of the feathers. The eggs are laid in groups on the underside of the pigeon's wing, principally on the long flight feathers. Infestations apparently do little if any harm to the bird.

REFERENCE. Martin, M. 1934. Life history and habits of the pigeon louse. *Canadian Entomologist,* vol. 66, pp. 6-16.

## ■ BROAD-HEADED BIRD LICE
(Order Mallophaga; Family Menoponidae)

■ CHICKEN BODY LOUSE
*(Menacanthus stramineus)*
Figure 72

This chewing species of louse lives directly on the skin of chickens and turkeys (it also occurs on guinea fowl, peafowl, and game birds raised with chickens). It is a small species ($^{1}/_{16}$ to $^{3}/_{32}$ in., or 2 to 2.5 mm, long).

The constant action of the louse's mouthparts and sharp leg claws on its host's skin causes irritation, and heavy infestations can seriously affect the vigor of poultry.

The eggs are fastened to the basal barbs on the feather shafts, especially those below the vent. The Chicken Body Louse is most abundant about the vent and under the wings.

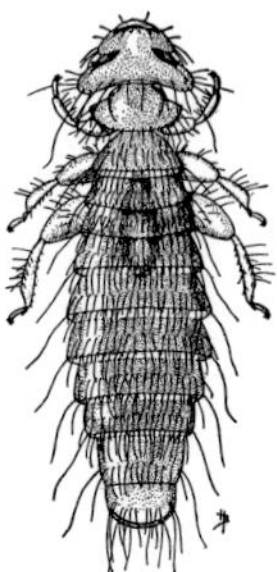

72. Chicken Body Louse. Drawing from F. Bishop and H. Wood, 1921,

# 7 TRUE BUGS (Order Heteroptera)

As a group, the true bugs are characterized by few consistent features except the universal possession of sucking mouthparts. These structures form a jointed beak, which contains long slender stylets that act like a hypodermic syringe to pierce plant tissues (or animal tissues, in some cases) and withdraw vital fluids, weakening the host or transmitting disease. Because of this habit many species have become agricultural or medical pests.

These are the only insects which, according to the philologist, rightfully deserve to be called "bugs." The word probably comes from the Old English word *bwg,* meaning hobgoblin or apparition, or from *buk,* a long-standing Arabic term for the Common Bedbug.

## STINK BUGS (Family Pentatomidae)

The bugs in this family are the real "stinkers" of the insect world. Glands opening near the coxae of the hind legs produce a strong acrid-smelling vapor, which serves to discourage or repulse enemies. Several species and genera are found locally, but the Harlequin Bug and Uhler's and Say's Stink Bugs are those most likely to be recognized.

■ **Uhler's and Say's Stink Bugs** (*Chlorochroa uhleri* and *sayi*) Figure 73

These are two closely related species. Both are about 1/2 inch (13 mm) long and bright green, with light spots on the scutellum (the triangular area in the middle of the back). On Uhler's (the more slender of the two species), these spots are more distinct, and the spot at the tip of the scutellum is orange, as are both edges of the body. Say's species (Figure 73) has yellow mottling on the wings and purple flecks on the wing membrane; the wings in Uhler's are solid green.

Both species are common in the summer on weedy vegetation in vacant lots and on some garden plants. Adults are sometimes attracted to lights.

REFERENCE. Buxton, G. M., D. B. Thomas, and R. C. Froeschner. 1983. Revision of the species of the *sayi*-group of *Chlorochroa* Stål (Hemiptera: Pentatomidae). Occasional Papers in Entomology of the California Department of Food and Agriculture (Division of Plant Industry–Laboratory Services), no. 29, 23 pp.

■ HARLEQUIN BUG
*(Murgantia histrionica)*
Figures 4, 74

The common name of this small stink bug (body length 1/4 in., or 6 mm) is derived from its multicolored pattern (it is also called Calico Back). It is variegated black, red, and white, with a reddish or light colored "+" on the red-tipped scutellum and black on the tips of the wings. The white eggs are also colorfully marked with black rings and a spot—they resemble miniature barrels, the rings emulating hoops and the spot the bunghole. The coloration of the nearly circular nymphs is also characteristic: the front third of the body is shiny black, and the abdomen is bright red except for three transverse black plates on the center of the back.

The Harlequin Bug is usually seen in the garden on cabbage, sweet alyssum, and related plants of the family Brassicacaea, or in vacant lots and on hillsides on wild mustard. The species is a pest of Brassicacaea crops in all parts of the United States.

Mating pairs are often present. The male illicits copulation by tapping the female's antennae and body with his antennae. Females lay several sets of from five to twelve eggs each.

In the laboratory the total developmental time

73. Say's Stink Bug. Photograph by P. Bryant.

74. Harlequin Bug. Photograph by C. Hogue.

for the egg stage (six to eleven days) and the five nymphal stages is fifty-three days. Adults live for seventy to eighty days.

REFERENCES. Lanigan, P. J., and E. M. Burrows. 1977. Sexual behavior of *Murgantia histrionica* (Hemiptera: Pentatomidae). *Psyche,* vol. 84, pp. 191-197.

Streams, F. A., and D. Pimentel. 1963. Biology of the harlequin bug, *Murgantia histrionica. Journal of Economic Entomology,* vol. 56, pp. 108-109.

## BIG-LEGGED BUGS (Family Coreidae)

■ SQUASH BUG
*(Anasa tristis)*
Figure 75

The Squash Bug may be found in vegetable gardens and in areas where field crops are still grown. It feeds on the sap of a variety of viney cucurbit plants, especially squashes and pumpkins.

Adults survive through the winter to mate and begin laying eggs in the spring. Nymphs and newly matured adults may be evident all summer. The adults are "grainy" or speckled dark grayish-brown and are approximately 5/8 to 3/4 inch (13 to 17 mm) long. The lateral margins of the abdomen, which protrude beyond the folded wings, are orange and brown.

■ WESTERN LEAF-FOOTED BUG
*(Leptoglossus clypealus)*
Figure 76

This bug takes its name from the leaf-like enlargements of the hind tibiae. The body is generally brown with varied yellow markings, the most characteristic of which is an irregular narrow transverse band across the middle of the wings.

This species is also known in the west as the Chincha, a word in Spanish meaning "bug." It is usually found on junipers in the more arid eastern portions of the basin.

REFERENCE. Allen, R. C. 1969. A revision of the genus *Leptoglossus* Guerín (Hemiptera: Coreidae). *Entomologica Americana,* vol. 45, pp. 35-140.

## SEED BUGS (Family Lygaeidae)

■ SMALL MILKWEED BUG
*(Lygaeus kalmii)*
Figure 77

This is a pretty bug approximately 3/8 inch (10 mm) long, with paired black dots and an irregular lyre-shaped red pattern boldly emblazoned on its gray back. The twin white spots in the middle of the black terminal membranous region of the wing are also characteristic.

The Small Milkweed Bug is occasionally seen around the home garden where it has strayed from nearby milkweed. Although it may feed on other plants, especially those of the family Asteraceae (sunflower, asters, and ragweeds), it normally feeds on the pods, stems, and seeds of the milkweed. As this plant seems to be declining locally in the face of human progress, the insect will no doubt become increasingly rare.

REFERENCES. Simanton, W. A, and F. Andre. 1936. A biological study of *Lygaeus kalmii* Stål (Hemiptera–Lygaeidae). *Bulletin of the Brooklyn Entomological Society*, vol. 31, pp. 99-107.

Wheeler, A. G., Jr. 1983. The small milkweed bug, *Lygaeus kalmii* (Hemiptera: Lygaeidae): Milkweed specialist or opportunist? *Journal of the New York Entomological Society*, vol. 91, pp. 57-62.

75. Squash Bug. Photograph by M. Badgley.

76. Western Leaf-footed Bug. Photograph by C. Hogue.

77. Small Milkweed Bug. Photograph by C. Hogue.

■ SOUTHERN CHINCH BUG *(Blissus insularis)* Figure 78

This is a slender small bug about 1/8 inch (3 mm) long. In the adult, wings are either well developed (extending to the tip of the abdomen; Figure 78) or short (half the body length). In either case, the adults are black with white wings and reddish antennal bases and legs. The nymphs are brilliant red when young, changing to black as they mature.

The Southern Chinch Bug feeds on several grasses, but Saint Augustine is by far the preferred host plant. The insect's feeding may cause considerable damage: the grass becomes dwarfed, turns yellow and then brown, and dies. Because of the tendency of the species to form aggregations, the symptoms of attack are usually visible in scattered patches.

The species is not a native. It first appeared in the Los Angeles area in the late 1960s, having come from the southeastern states. It produces two generations per year and is most abundant in midsummer.

REFERENCE. Morishita, F. S., R. N. Jefferson, and L. Johnston. 1969. Southern chinch bug, a new pest of St. Augustine grass in southern California. *California Turfgrass Culture,* vol. 19, no. 2, pp. 9-10.

■ FALSE CHINCH BUG *(Nyssius raphanus)* Figure 79

This native species is small (1/8 in., or 3 mm), elongate, and light to dark gray; it bears a superficial resemblance to the Southern Chinch Bug.

The False Chinch Bug occasionally breaks out in large numbers locally. In the summer swarms may been seen crawling on the ground or over low vegetation. The species also takes over garden and field crops and can do considerable damage by its feeding.

## SCENTLESS PLANT BUGS (Family Rhopalidae)

■ WESTERN BOX-ELDER BUG *(Boisea rubrolineata)* Figure 80

This species is superficially similar to the Small Milkweed Bug in that it is about the same size (1/2 in., or 13 mm) and is gaily marked. The Western Box-elder Bug's distinguishing features are the bright red abdomen, three red lines running lengthwise on the pronotum, and reddish margins to the wings.

The bug feeds primarily on box-elder, ash, and maple and usually is found only where these trees grow; the insect seems to avoid male or staminate trees and nearly always remains only on female or seed-bearing trees. It is especially abundant in the canyons of the mountains bordering the basin. Adults are frequently seen in these canyons on warm spring days, swarming in the air and crawling over vegetation in large numbers. The species overwinters in the adult stage, aggregating and hiding in dry sheltered places, including our homes.

OTHER SCIENTIFIC NAMES. *Leptocoris rubrolineatus.*

REFERENCE. Wollerman, E. 1966. The boxelder bug. *Pest Control Operator's News,* vol. 26, pp. 18-20.

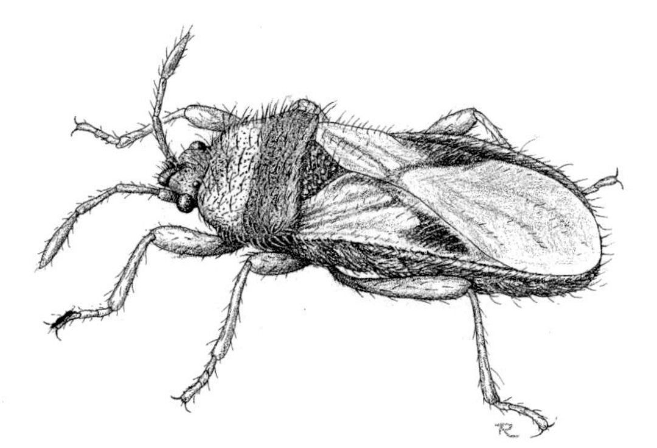

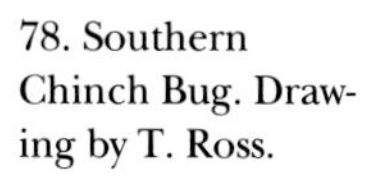

78. Southern Chinch Bug. Drawing by T. Ross.

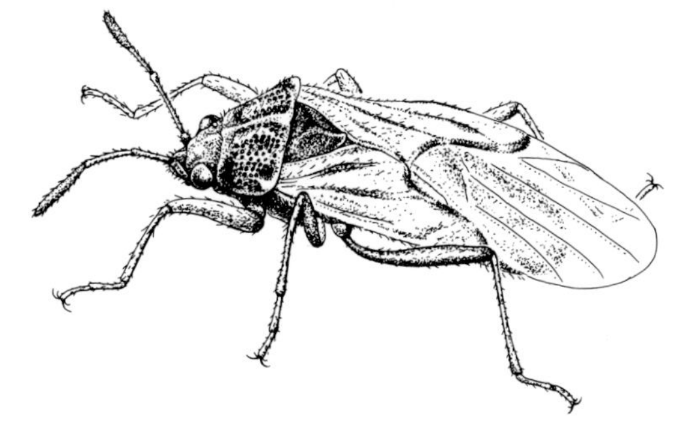

79. False Chinch Bug. Drawing by T. Ross.

80. Western Box-elder Bug. Photograph by C. Hogue.

## LARGID BUGS (Family Largidae)

■ BORDERED PLANT BUG *(Largus cinctus californicus)* Figure 81

This bug is conspicuous at times because of its habit of congregating in very large numbers on certain plants, especially herbaceous weedy shrubs. These aggregations are found only sporadically in the Los Angeles Basin proper but are quite frequently seen in the nearby desert areas.

The bug is about $^1/_2$ inch (13 mm) long as an adult, and its most characteristic markings are the orange margins and fine orange speckles on the basal portions of the wings.

## ASSASSIN BUGS (Family Reduviidae)

■ WESTERN CONE-NOSE BUG *(Triatoma protracta)* Figure 82

The Western Cone-nose Bug can be readily recognized by its medium size ($^5/_8$ to $^3/_4$ in., or 15 to 20 mm, long) and solid blackish or dark brown color. The abdomen has flared sides and is compressed in the center.

This bug has a bad reputation, rightfully earned. It belongs to a group of bugs called "kissing bugs" (from their habit of biting sleeping persons about the lips; they are also known as Bellows Bugs, Walpai Tigers (in Arizona), Cross Bugs, Big Bedbugs, China Bedbugs, or Sacred Bugs). The normal food of kissing bugs is the blood of vertebrate animals, including humans: among the many species in the American tropics are some that act as vectors of Chagas' Disease, a serious malady caused by a trypanosome protozoan similar to that which causes African Sleeping Sickness. Chagas' Disease does not appear in the basin; although a trypanosome parasite that seems identical to the one involved in the disease is harbored by the Western Cone-nose Bug, it is apparently a nonvirulent strain.

The bug is nonetheless a medical problem in Los Angeles because of its venomous bite. The bug's saliva contains substances foreign to the human system and capable of causing a serious allergic reaction. The symptoms range from simple itching, severe swelling, joint pain, nausea, chills, and dizziness to anaphylactic shock. Persons exhibiting severe allergic symptoms after a bite by one of these bugs are advised to consult a physician immediately and also to capture the bug and keep it alive for diagnosis. It should be emphasized, however, that the bug's bite causes little or no reaction in most individuals; like the sting of the Honey Bee, it is not to be unduly feared except by a few especially sensitive individuals.

The Western Cone-nose Bug does not occur in the heavily urbanized part of the basin because it normally lives in the dens of wood rats *(Neotoma* species), where it feeds on the blood of the rodent. Its only occasional contact with humans usually occurs when housing developments invade formerly uninhabited areas, especially brush-covered hillsides and canyons where the wood rats and their bug parasites have long been isolated from mankind. The bugs discover a new source of blood when they accidentally enter homes through cracks and vents or when, because of their attraction to lights, they fly into patios or through open doors.

REFERENCES. Marshall, N., M. Liebhaber, Z. Dyer, and A. Saxon. 1986. The prevalence of allergic sensitization to *Triatoma protracta* (Heteroptera: Reduviidae) in a southern California, USA, Community. *Journal of Medical Entomology,* vol. 23, pp. 117-124.

Swezey, R. L. 1963. "Kissing bug" bite in Los Angeles. *Archives of Internal Medicine,* vol. 112, pp. 977-980.

Wood, S. F. 1975. Home invasions of conenose bugs (Hemiptera: Reduviidae) and their control. *National Pest Control Operators News,* March 1975, pp. 16-18.

Wood, S. F., and F. D. Wood. 1967. Ecological relationships

81. Bordered Plant Bug. Photograph by C. Hogue.

82. Western Cone-nose Bug. Photograph by C. Hogue.

of *Triatoma p. protracta* (Uhler) in Griffith Park, Los Angeles, Calif. *Pacific Insects,* vol. 9, pp. 537-550.

_______. 1964. Nocturnal aggregation and invasion of homes in southern California by insect vectors of Chagas' Disease. *Journal of Economic Entomology,* vol. 57, pp. 775-776.

■ BEE ASSASSIN
*(Apiomerus crassipes)*
Figure 83

This robust black bug is 1/2 to 5/8 inch (13 to 16 mm) long and has ruby red markings. The legs are hairy and the tarsi diminutive. The tibiae of the fore legs have many long sticky hairs that aid in the capture of prey.

This bug preys on a variety of insects, including ants, bees (especially Honey Bees), and beetles. Adults and nymphs prowl vegetation or perch on flowers from late spring to fall, waiting to ambush unsuspecting insect visitors.

REFERENCES. Bouseman, J. K. 1976. Biological observations on *Apiomerus crassipes* (F.) (Hemiptera: Reduviidae). *Pan-Pacific Entomologist,* vol. 52, pp. 178-179.

Swadener, S. O., and T. R. Yonke. 1973. Immature stages and biology of *Apiomerus crassipes. Annals of the Entomological Society of America,* vol. 66, pp. 188-196.

■ WESTERN CORSAIR
*(Rasahus thoracicus)*
Figure 84

This bug, like the Assassin, has a fearsome bite—only more so. People who have received a bite say it gives a sharp burning sensation, more acutely painful than a Honey Bee's sting. The bug normally uses its beak to suck the blood of other insects and bites humans only in self-defense.

The Western Corsair is about 3/4 inch (20 mm) long. It has a harlequin pattern on an amber background on its back; a conspicuous circular amber spot in the center of the fore wing membrane marks it for easy identification. The edge of the abdomen is colored with alternating light and dark dashes.

Fortunately, the species is scarce in the basin. It is nocturnal and usually is encountered on warm summer evenings around house lights, to which it is readily attracted. It squeaks when alarmed.

## LACE BUGS (Family Tingidae)

LACE BUGS ARE ALL SMALL (1/8 to 3/16 in., or 3 to 5 mm, long). They are flat insects, and the surface of the wings and thorax is distinctively broken up into tiny transparent or translucent compartments outlined by ridges that give a reticulate or lacelike appearance. The prothorax is also frequently inflated, and the fore wings laterally expanded, hiding the abdomen completely.

Lace bugs are sap suckers, and some—like the Sycamore Lace Bug—occasionally cause damage to valuable plants when present in abundance.

■ Sycamore Lace Bug (*Corythuca ciliata confraternus*) Figure 85

Adults and nymphs of the Sycamore Lace Bug, a common local species, are found in groups on the undersides of the leaves of native and introduced sycamores. The insect is a typical lace bug in size (1/8 in., or 3 mm, long) and color—it is mostly white above with numerous light-brown markings.

83. Bee Assassin. Photograph by J. Levy.

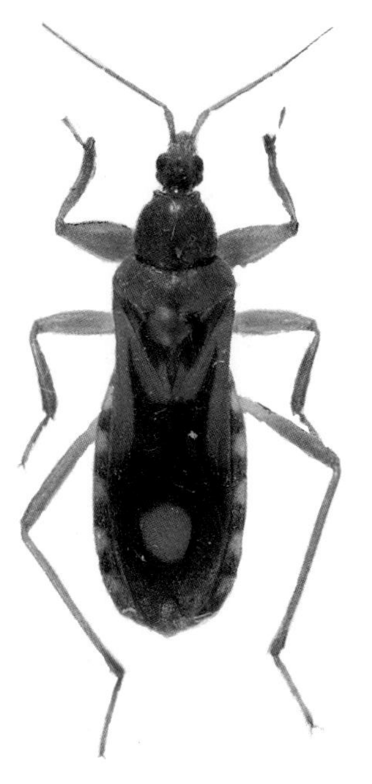

84. Western Corsair. Photograph of LACM specimen by L. Reynolds.

85. Sycamore Lace Bug. Photograph by M. Badgley.

■ COMMON BEDBUG
*(Cimex lectularius)*
Figure 86

Fortunately bedbugs are not as common locally as they are in other parts of the country. Our climate is probably too dry for them. Nevertheless, they persist in some places, especially unkept hotels, motels, rooming houses, and residences. The species may be recognized by its small size (its length is about 3/16 to 1/4 in., or 5 to 6 mm), dusky red color (which changes to bright red when the bug is ingesting blood), and flatness. It is entirely wingless, even as an adult.

When indoors, the Common Bedbug feeds exclusively on human blood, invading the bed at night for its meals. Although the bite may cause immediate pain in some individuals, the first indication of its presence is often only dark stains on the bed sheets from the bug's excrement or the itching of bites the next day. Heavy infestations of bedbugs also are accompanied by a characteristic disagreeable musky odor that comes from the bugs' scent glands, which are similar to those possessed by stink bugs.

Some people assume that the source of infestation is dirt or old clothing, and these mistaken ideas probably stem from the bug's ability to withstand long periods without food. Infestation always begins, of course, by introduction from other preexisting infestations, and the bug easily finds transportation on clothing, bedding, or overstuffed furniture. During the day bedbugs hide in crevices in walls and floors, behind wall decorations, and in furniture.

Although its primary host is man, the Common Bedbug also occurs on a wide variety of other mammals and birds, especially chickens and bats. In spite of its indiscriminate bloodsucking habits, the species is not known to be a vector of any human disease.

REFERENCES. Anonymous. 1973. *The bed-bug,* 8th edition. London: British Museum of Natural History.

Johnson, C. G. 1952. The bed bug. *New Biologist,* vol. 13, pp. 80-97.

■ SWALLOW BUG
*(Oeciacus vicarius)*
Figure 87

This parasitic bug normally lives in the mud nests of California Swallows. But at times it may invade homes upon which the swallows have built their nests and even bite the human occupants.

The bug somewhat resembles the Common Bedbug; it is about the same length (body length 1/4 in., or 6 mm) but is slightly more elongate in form and has longer, more numerous hairs.

When the hosts are away on their winter migratory travels Swallow Bugs are forced to endure long

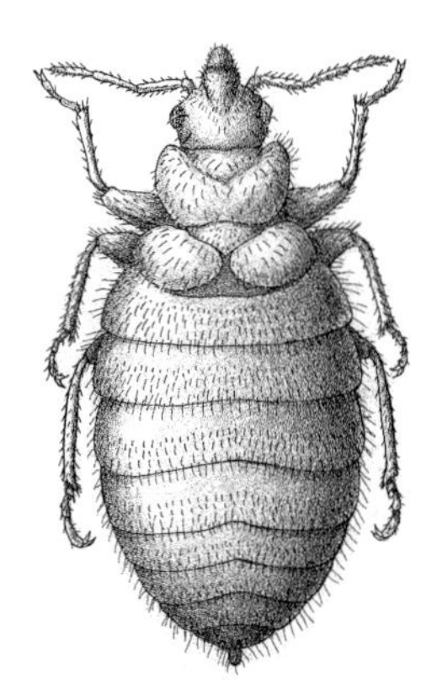

86 (left). Common Bedbug. Photograph by R. Pence.

87 (right). Swallow Bug. Drawing by T. Ross.

periods of fasting. It is at these times that they are most likely to attack humans.

REFERENCE. Eads, R. B., D. B. Francy, and G. C. Smith. 1980. The Swallow Bug, *Oeciacus vicarius* Horvath (Hemiptera: Cimicidae), a human household pest. *Proceedings of the Entomological Society of Washington,* vol. 82, pp. 81-85.

## WATER BUGS

THERE ARE SEVERAL TRUE BUGS that are ubiquitous in the basin's waters but are usually noticed only by fishermen, hikers, and owners of fish ponds and swimming pools. Winter is the best time to find the adults of the water bugs because most species are in this stage during the cold months.

These insects are of two types. The truly aquatic bugs live beneath the surface of water for an extended period of time; these include giant water bugs, backswimmers, and water boatmen. The semiaquatic bugs, such as water striders and toad bugs, are found on the surface or only near the water.

REFERENCE. Hungerford, H. B. 1920. The biology and ecology of aquatic and semi-aquatic Hemiptera. *University of Kansas Science Bulletin,* vol. 21, pp. 1-341.

## ■ GIANT WATER BUGS (Family Belostomatidae)

■ TOE BITER
*(Abedus indentatus)*
Figure 88

This is a fairly large ($1^1/_4$ in., or 32 mm, long) and completely aquatic bug found in running water in small streams. It clings motionless to underwater objects, moving only to catch and feed on other aquatic insects, tadpoles, and small fish or to come to the surface to replenish its air supply, which it carries in a cavity beneath its wings.

In the winter, males are frequently found with masses of large pale brown eggs attached to their backs (Figure 88; this seems to be the females' preferred place of egg deposition). Brooding males provide for the needs of the eggs by exposing them to the air and maintaining an intermittent flow of water over them by rocking up and down under water. Males also tend to live in protected places that foster the development of embryos and their successful hatching.

The generic name *Abedus* means "an eater" and refers to the voracious appetite of these bugs. The etymology of the common name "toe biter," which was given to this insect by swimmers, needs no explanation.

REFERENCES. Kraus, B. 1985. Oviposition on the backs of female giant water bugs, *Abedus indentatus:* The consequence of a shortage in male back space? (Hemiptera: Belostomatidae). *Pan-Pacific Entomologist,* vol. 61, pp. 54-57.

Smith, R. L. 1976. Male brooding behavior of the water bug *Abedus herberti* (Hemiptera: Belostomatidae). *Annals of the Entomological Society of America,* vol. 69, pp. 740-747.

## ■ BACKSWIMMERS (Family Notonectidae)

THESE AQUATIC BUGS are easily recognized in their stillwater habitat for they swim upside down, the hind legs moving to and fro like oars; they rise frequently to the surface to replenish the air supply they maintain at the tip of the abdomen. Like giant water bugs, backswimmers are predators, and their piercing

88. Toe Biter male carrying eggs. Photograph by C. Hogue.

mouthparts—which are normally employed in killing and sucking the blood of small aquatic invertebrates—can inflect a painful wound to the finger of the careless handler. Females usually stick their eggs onto or in the leaves and stems of aquatic plants.

There are several local species in two genera, *Notonecta* and *Buenoa*. *Buenoa* species are small (1/5 to 1/3 in., or 5 to 8 mm) and translucent whitish in color. They are graceful swimmers that remain in easy equilibrium with the water. Species of *Buenoa* are capable of making chirping noises.

In contrast, *Notonecta* species are larger (3/8 to 5/8 in., or 10 to 15 mm, long) and opaquely pigmented; they are awkward swimmers and must use jerky motions to move through the water and even just to stay submerged. Species of *Notonecta* are silent.

The backswimmers most often encountered are *Notonecta kirbyi, unifasciata,* and *shooteri* (Figure 89). There is a great deal of variation in color pattern among individuals of these species, which consequently are superficially difficult to distinguish (individuals of the latter two species range in color from almost all black to all white). Most specimens of *N. kirbyi* are 1/2 to 5/8 inch (13 to 15 mm) long and tan or reddish black with splotchy black markings across the middle of the body. Individuals of *N. unifasciata* are smaller (3/8 in., or 10 mm, long) and tend to be white with a distinct transverse black band across the body. *N. shooteri* is always monocolored white or black and is about the size of *N. unifasciata*.

REFERENCES. Hungerford, H. B. 1933. The genus *Notonecta* of the world. *University of Kansas Science Bulletin,* vol. 21, pp. 5-195.

Truxal, F. S. 1953. A revision of the genus *Buenoa*. *University of Kansas Science Bulletin,* vol. 35, pp. 1351-1523.

Voigt, W. G., and R. Garcia. 1976. Keys to the *Notonecta* nymphs of the West Coast United States. *Pan-Pacific Entomologist,* vol. 52, pp. 172-176.

89. Backswimmer *(Notonecta shooteri)*. Photograph by C. Hogue.

## ■ WATER BOATMEN (Family Corixidae)

In general, water boatmen resemble backswimmers, but they are much flatter, and they swim in a "normal" fashion, right side up. Also, unlike backswimmers, water boatmen are herbivorous; their mouthparts are modified away from the piercing beak that is common among the other true bugs, and they feed on plant microorganisms, which they ingest whole.

Several species of these aquatic bugs are found in the basin's freshwaters. The Marsh Boatman *(Trichocorixa reticulata;* Figure 90) is notable for its preference for salty waters. It is small (length about 3/16 in., or 5 mm) and light brown in color, with irregular transverse lines. It is often found in large numbers near the coast in salt marshes and estuaries, such as those at the mouths of Malibu and Ballona Creeks, and in the harbor area. It tolerates a wide range in salinity, from slightly brackish water to pure seawater.

## ■ WATER STRIDERS (Family Gerridae)

These semiaquatic bugs are sometimes called "water spiders," fittingly because of their long spider-like legs, which extend out from the body and support it delicately on the water's surface. Many people have noticed the strange shadows cast on the bottom of a clear still pool by the contact points (menisci) encircling the tips of the legs of these insects as they literally skate across the surface. The insects' light weight and pad-like feet tipped with water-repellent hairs permit them to remain on top of the surface film without breaking through it.

Water striders prey on all kinds of small insects—other aquatic and semiaquatic insects (including their own kind) as well as terrestrial insects that happen to fall upon the water surface.

The Common Water Strider *(Gerris remigis;* Figure 91) is an elongate bug of moderate size (the body length of adults is about 5/8 in., or 16 mm). The mid and hind pairs of legs are very long and are used for locomotion; the fore pair, which are much shorter, are used for feeding. Adults may be winged or wingless.

REFERENCES. Bowdan, E. 1976. The functional anatomy of the mesothoracic leg of the water strider, *Gerris remigis* Say (Heteroptera). *Psyche,* vol. 83, pp. 289-303.

Calabrese, D. 1974. Population and subspecific variation in *Gerris remigis* Say. *Entomological News,* vol. 85, pp. 27-28.

Caponigro, M. A., and C. H. Eriksen. 1976. Surface film locomotion by the water strider, *Gerris remigis* Say. *American Midland Naturalist,* vol. 95, pp. 268-278.

Galbraith, D. F., and C. H. Fernando. 1977. The life history of *Gerris remigis* (Heteroptera: Gerridae) in a small stream in southern Ontario. *Canadian Entomologist,* vol. 109, pp. 221-228.

Torre Bueno, J. F. 1917. Life-history and habits of the larger water strider, *Gerris remigis* Say (Hem.). *Entomological News,* vol. 28, pp. 201-208.

## ■ TOAD BUGS (Family Gelastocoridae)

■ COMMON TOAD BUG *(Gelastocoris oculatus)* Figure 92

These squat little semiaquatic bugs (body length 1/4 in., or 5 mm) are found on sandy or muddy shores of ponds and streams where their mottled color pattern blends with the background. Their hopping locomotion and habit of sitting on their haunches with the tip of the abdomen lowered suggest a toad in miniature in every way. They do not normally venture onto or below the surface of the water but prefer to remain close by on the shore, catching and feeding on small arthropods.

REFERENCE. Hungerford, H. B. 1922. The life history of the toad bug *Gelastocoris oculatus* Fabr. *University of Kansas Science Bulletin,* vol. 14, pp. 145-167.

90 (above, left). Marsh Boatman. Photograph by P. Bryant.

91 (above, right). Common Water Strider. Photograph by C. Hogue.

92 (left). Common Toad Bug. Photograph by P. Bryant.

# 8 HOMOPTERANS (Order Homoptera)

THIS ORDER is sometimes combined with the Heteroptera—the true bugs—into an order called Hemiptera because insects in the two groups share similar mouthparts and undergo incomplete metamorphosis. But important differences are in the wings, which in homopterans are generally of the same texture throughout (leathery and opaque or membranous and diaphanous) and are folded roof-like over the body when the insect is at rest. Virtually all homopterans have wax-producing glands in the integument, and many excrete honeydew, a sugary sticky solution that may attract symbiotic associates (especially ants). A great number are plant pests because of their great fecundity and ability to bleed their hosts of life-giving sap. Some also injure plants by transmitting pathogenic organisms, especially viruses. None are aquatic.

REFERENCE. Gill, G. J. [No Date.] Photo recognition key to the whiteflies and scale insect families of California. California Department of Food and Agriculture, Environmental Monitoring and Pest Management, Sacramento (Scale and whitefly key #1).

## CICADAS (Family Cicadidae)

THE DEAFENING SHRILL HIGH-PITCHED VOICE of the cicada, so familiar to Easterners and Midwesterners, is seldom heard locally. Only the males produce this sound: special organs at the base of the abdomen are composed of a thin membrane to which strong muscles are attached, and these muscles vibrate the membrane at a high frequency, setting up sound waves like those generated by a tuning fork or drumhead. The song is usually characteristic for a particular species and serves as a mating call to the female. Both sexes have a hearing organ (tympanum) located at the base of the abdomen; in the male it is close to the sound-producing organ.

Following courtship and mating, the female lays eggs in slits she makes with the ovipositor in plant stems and leaves. Upon hatching, the nymphs drop to the ground, burrow into the soil, and subsist by sucking the juices of roots. A considerable length of time, ordinarily two to five years, is required for nymphal development (one eastern and midwestern species, the Seventeen Year Cicada or Periodical Cicada, actually requires seventeen years to mature).

Mature nymphs emerge from the ground and transform into adults on some vertical or inclined object, frequently a tree trunk or rock. A cast skin may at times be found still clinging to the spot where a transformation occurred.

Cicadas are often called "locusts" by Easterners and Midwesterners. Doubtless the name was applied because some species that appear in swarms reminded early observers of migratory grasshoppers or true locusts.

Only a few species of cicadas occur in Los Angeles, and they are to be found only in our remaining undeveloped and vacant hill areas. The following three species are most common.

■ Red-winged Grass Cicada *(Tibicinoides cupreosparsus;* Figure 93). This species deserves note for its small size (it is only 3/4 in., or 20 mm, long, with a 1 1/4-in., or 32-mm, wing span) and the coloration of its wings, which have a broad brilliant red zone at the base. It is found on wild oats and other large grasses; the females insert their eggs into the host plants.

93. Red-winged Grass Cicada. Photograph of LACM specimen by C. Hogue.

94. Wide-headed Cicada. Photograph of LACM specimen by C. Hogue.

■ Wide-headed Cicada *(Platypedia laticapitata;* Figure 94). In this species the sound-producing organs are so reduced in size that they no longer function. Instead the insect makes a short sharp clicking noise by rubbing its wings together. It is found only in hilly or mountainous areas on shrubs and trees. It is characterized by medium size (a body length of 1/2 in., or 13 mm, and a wing span of 1 3/8 to 1 5/8 in., or 35 to 42 mm) and very broad fore wings that are strongly bowed along the front margins.

■ Vanduzee's Cicada *(Okanagana vanduzeei).* This cicada calls from trees or shrubs and is probably the commonest local singing species. It is medium-sized (its body length is 3/4 in., or 20 mm; its wing span is 2 in., or 50 mm) and has a shiny black body with irregular orange markings.

REFERENCE. Simons, J. N. 1954. The cicadas of California. *Bulletin of the California Insect Survey,* vol. 2, no. 3, pp. 1-192.

## LEAFHOPPERS (Family Cicadellidae)

LEAFHOPPERS ARE VERY COMMON INSECTS found on all types of plants throughout the basin. There are many species, all small and having generally the same long slender shape. Although most are greenish or brownish, some sport bright blue, red, and other colors. Both the nymphs and adults have the peculiar habit of running sideways. The Blue-green Sharpshooter and Smoke Tree Leafhopper are conspicuous basin species.

■ BLUE-GREEN SHARPSHOOTER *(Graphocephala atropunctata)* Figure 95

This is a beautiful blue or bluish-green iridescent species that is medium-sized (length 1/4 in., or 6 mm). The head, prothorax, legs, and underside of the body are lighter and yellow-green.

The Blue-green Sharpshooter is common on many native and cultivated plants and is usually seen on the leaves. The species is a vector of virus disease of grape vines (Pierce's Disease). It uses over 150 other plants as hosts, including many ornamentals and crops in addition to native plants.

OTHER SCIENTIFIC NAMES. *Hordnia circellata.*

REFERENCE. Purcell, A. H. 1976. Seasonal changes in host plant preference of the Blue-green sharpshooter *Hordnia circellata* (Homoptera: Cicadellidae). *Pan-Pacific Entomologist,* vol. 52, pp. 33-37.

■ SMOKE TREE LEAFHOPPER *(Homalodisca lacerta)* Figure 96

This is an unusually large leafhopper (it is 1/2 in., or 13 mm, long). Some females have a white globule in the middle of the fore wings.

Its hosts are numerous and varied through its

95. Blue-green Sharpshooter. Photograph by C. Hogue.

96. Smoke Tree Leafhopper. Photograph by M. Badgley.

broad range. In our area it prefers cultivated leguminous trees, mainly acacias; it is well known in the neighboring deserts on Smoke Tree *(Psorothamnus spinosa)*.

## SPITTLE BUGS (Family Cercopidae)

LOCAL RESIDENTS are often puzzled by masses of frothy white liquid resembling foam or spittle on parts of plants (Figure 97): they are surprised to learn that the froth is produced by the nymphs of spittle bugs and serves as a place of concealment and protection for the developing insects.

Adult spittle bugs (Figure 98) resemble leafhoppers. The common local species is *Aphrophora permutata,* which is primarily found as an adult on conifers. It is light brown and about 7/16 inch (11 mm) long.

97. Froth nest of spittle bug *(Aphrophora* species); nymph is visible at lower right. LACM photograph.

98. Spittle bugs *(Aphrophora permutata)*. Photograph by L. Brown.

## TREEHOPPERS (Family Membracidae)

TREEHOPPERS RESEMBLE SPITTLE BUGS and leafhoppers in their anatomy and habits but possess a uniquely developed prothorax. This region of the body is greatly enlarged and projects upward and backward over the abdomen. The profile of this structure gives the bug the look of a thorn or a roughening on a twig or branch and is a remarkable example of protective form.

Several species of treehoppers are found locally on many kinds of plants (not just on trees as the name implies); the Keelbacked Treehopper is our most common species.

■ KEELBACKED TREEHOPPER *(Antianthe expansa)* Figure 99

This moderate-size species (the adult is 3/16 in., or 5 mm, long) has a deep green back with minute yellowish flecks; the wings are transparent. The back is sharply keeled, and a short sharp spine extends out on each side behind the head.

The species feeds primarily on a variety of solanaceous garden plants, including peppers and

99. Keelbacked Treehopper. Photograph by C. Hogue.

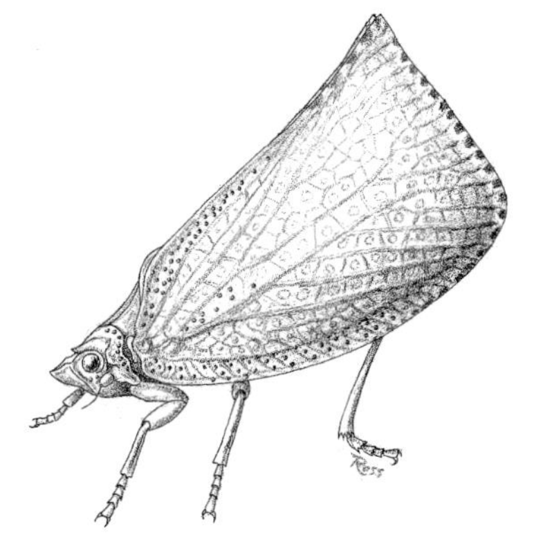


100. Torpedo Bug. Drawing by T. Ross.

tomatoes. The black spiny nymphs are usually the stage first noticed in the garden; they appear in groups, sometimes accompanied by one or more adults.

## PLANTHOPPERS (Family Flatidae)

■ Torpedo Bug
*(Siphanta acuta)*
Figure 100

This recent immigrant, first detected in 1982, is now established in both Los Angeles and Orange Counties. It is apparently native to Australia and reached our area through an accidental introduction; it may have arrived as an egg mass on one of its hosts, which include oleander, citrus, and acacia.

The adult is about 1/4 inch (6 mm) long, with a strongly compressed shape. The wings are roughly rectangular and squared off at the ends; the head is pointed in front. Its color is light green, except for the rear margins of the wings, which are pink. This bug is a great leaper.

REFERENCE. Myers, J. G. 1922. Life-history of *Siphanta acuta,* (Walk.), the large green plant-hopper. *New Zealand Journal of Science and Technology,* vol. 5, pp. 256-263.

## APHIDS (Family Aphididae)

APHIDS ARE NOTORIOUS PESTS of cultivated plants. Prolific breeders, they swiftly spread over the tender growing tips of prize roses and other plants, from which they withdraw large quantities of sap. The result is a wilted, curled, and unsightly mass of leaves or a dead plant. The aphid's harm is increased by its habit of copiously excreting from the anus a sugary solution called "honeydew," which covers the host plant with a sticky unsightly residue that often becomes blackened with a growth of sooty mold. Aphids also transmit viral diseases to plants.

Honeydew is a food greedily sought after by other insects. Ants are particularly fond of this substance, and certain species (especially the Argentine Ant) actually care for the aphids, moving them from place to place to ensure their safety from parasites and predators. The ants induce the aphids to give up the honeydew by "milking" them with strokes of the antennae. This practice is responsible for the name "ant cattle," which is often applied to aphids.

Aphids are remarkable for their peculiar modes of reproduction and development, which involve polymorphism (the capability of assuming different body forms). They display life cycles so complicated and varied that they are impossible to summarize here. Parthenogenesis (the development of unfertilized eggs), viviparity (the bearing of live young), and winged and wingless generations (Figure 105) are common reproductive phenomena.

Aphids have many enemies, particularly ladybird beetles and lacewings. Small parasitic wasps also kill aphids, leaving the latter's withered bodies (called "mummies"; Figure 413) as hollowed-out dry shells on leaves in evidence of their work. These predatory insects should be encouraged to live and multiply in the home garden.

Practically every kind of native and cultivated plant is susceptible to attack by aphids. Local gardeners and growers will most likely encounter the following species.

■ Rose Aphid *(Microsiphum rosae;* Figure 101). This is the small (3/32 in., or 3 mm), light green to pink species with black legs that is so numerous on roses in the spring.

■ Deodar Aphid *(Cinara curvipes;* Figure 102). This aphid is a giant over 1/8 inch (4 mm) long and gray in color, with black spots on tubercles on its abdomen. It

101. Rose Aphids. Photograph by P. Bryant.

102. Deodar Aphids. Photograph by J. Hogue.

103. Giant Willow Aphids. Photograph by C. Hogue.

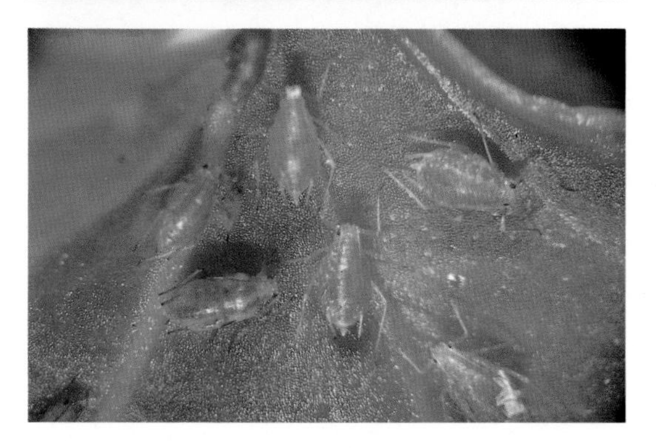

104. Green Peach Aphids. Photograph by M. Badgley.

105. Melon Aphids, winged and wingless forms. Photograph by M. Badgley.

forms dense colonies, which hang on the undersides of the horizontal branches of deodar trees.

■ Giant Willow Aphid *(Pterochlorus viminalis;* Figure 103). This is also a large species ($^{3}/_{16}$ in., or 5 mm, long); it is gray, with black spots, short black horns on the abdomen, and a large tubercle in the middle of the abdomen. Its host is willow *(Salix)*.

■ Green Peach Aphid *(Myzus persicae;* Figure 104). One of the commonest aphids in North America, this species is a general feeder on many cultivated plants. The immatures are yellow, pinkish, or pale green; adults are green with a dark blotch on the top of the abdomen.

■ Melon Aphid *(Aphis gossypii;* Figure 105). This aphid is usually a very dark green or brown but has paler forms as well. It is a cosmopolitan pest that feeds on numerous kinds of plants; in southern California, it is especially injurious to citrus.

REFERENCES. Leonard, M. D. 1972. Host plants of aphids collected at the Los Angeles State and County arboretum during 1966 and 1967 (Homoptera: Aphididae). *Proceedings of the Entomological Society of Washington,* vol. 74, pp. 95-120.

Walker, H. G., M. B. Stoltzel, and L. Enari. 1978. Additional aphid-host relationships at the Los Angeles State and County arboretum. *Proceedings of the Entomological Society of Washington,* vol. 80, pp. 575-605.

## SCALE INSECTS (Superfamily Coccoidea)

MALE SCALE INSECTS are rarely seen; they are tiny ($^{1}/_{32}$ to $^{1}/_{16}$ in., or 1 to 1.5 mm) two-winged, gnat-like creatures with a pair of long white tails. The females, on the other hand, are all too common but are seldom recognized as insects by the layman because of their amorphous body form. There are two general types, depending on whether or not the body is covered with a scale.

Those with shell or scale coverings (the so-called "armored scales," which are all in the Family Diaspididae) are soft skinned and usually pale colored. They live through their entire adult stage attached to one spot on the host plant by their tremendously long sucking mouthparts, which penetrate deeply into the plant tissue. The mouthparts can be seen by carefully catching the edge of the waxy scale with a pin and flipping it up (Figure 106): the female insect usually comes away in the scale, her mouthparts pulling out of the plant and appearing as microscopically fine curled hair-like filaments projecting from the body.

Unprotected by a shell are the "unarmored scales," which have been assigned to several families. The outer wall of the bodies of some species is thin ("soft scales"), but in most it is thickened and pigmented for protection; in other features, they are like the armored scales. Often in place of a shell they are covered with a powdery white waxy secretion. They normally attack woody plants, secreting copious quantities of honeydew in the process.

In all groups of scale insects the first-stage nymphs are active and free-living creatures called "crawlers." Only after a molt do they attach to the host and become stationary. There are many types of scale insects in the basin; the most conspicuous species are described below.

REFERENCES. Gill, R. J. [No date.] Color-photo and host keys to the soft scales of California. California Department of Food and Agriculture, Environmental Monitoring and Pest Management, Sacramento (Scale and white fly key #4).

_______. [No date.] Color-photo and host keys to the armored scales of California. California Department of Food and Agriculture, Environmental Monitoring and Pest Management, Sacramento (Scale and white fly key #5).

McKenzie, H. L. 1956. The armored scale insects of California. *Bulletin of the California Insect Survey,* vol. 5, pp. 1-209.

## ■ ARMORED SCALES (Family Diaspididae)

■ CALIFORNIA RED SCALE
*(Aonidiella aurantii)*
Figure 106

This scale is probably the most serious pest of citrus fruit in the world. It establishes itself readily on lemon and orange trees grown in orchards or in the garden. Although the scales cover leaves, twigs, and branches, they are easiest to see against the light background color of the ripe fruit. The reddish-colored scale is roughly circular or oval in shape with a diameter of slightly more than 1/16 inch (1.5 mm).

California Red Scale was apparently introduced into the state by 1877. Early records indicate that it was a severe pest throughout the Los Angeles and Orange

County areas. It is believed to have displaced another, closely related species, the Yellow Scale *(Aonidiella citrina),* which had been introduced sometime earlier.

REFERENCE. DeBach, P., R. M. Hendrickson, Jr., and M. Rose. 1978. Competitive displacement: Extinction of the Yellow Scale, *Aonidiella citrina* (Coq.) (Homoptera: Diaspididae), by its ecological homologue, the California Red Scale, *Aonidiella aurantii* (Mask.) in southern California. *Hilgardia,* vol. 46, pp. 1-35.

■ OLEANDER SCALE
*(Aspidiotus nerii)*
Figure 107

This is probably the most common and widespread armored scale in California. It has a very wide host spectrum and is a major pest on oleander, citrus, and ivy.

The female scale coverings are circular (1/16 to 3/32 in., or 1.5 to 2 mm, in diameter), flat, and light gray to light brown in color.

OTHER SCIENTIFIC NAMES. *Aspidiotus hederae.*

REFERENCE. DeBach, P., and T. W. Fisher. 1956. Experimental evidence for sibling species in the Oleander Scale, *Aspidiotus hederae* (Vallot). *Annals of the Entomological Society of America,* vol. 49, pp. 235-239

■ LATANIA SCALE
*(Hemiberlesia lataniae)*
Figure 108

This is an omnivorous armored scale with an endless list of woody plant hosts. Some common plants upon which it feeds in the Los Angeles area are *Acacia, Fuchsia,* avocado, willow, and yucca. The scale is white or tan and roughly circular, with a central node; it measures 1/16 to 3/32 inch (1.5 to 2 mm) in diameter.

■ FLUTED SCALES (Family Margarodidae)

■ COTTONY CUSHION SCALE
*(Icerya purchasi)*
Figure 109

The fame of this insect stems from the fact that it nearly destroyed the citrus industry in southern California at the end of the nineteenth century. It is native to Australia and appears to have been accidentally brought into the state at Menlo Park in 1868 or 1869. In 1888, the Vedalia Beetle *(Rodolia cardinalis),* an Australian member of the Ladybird family with a voracious appetite for Cottony Cushion Scales, was intentionally introduced, and it quickly reduced the intruder scale's numbers to harmless proportions. The case is cited today as one of the earliest and most successful examples of biological control of an insect pest.

Males are rare, and the females are hermaphroditic. They are reddish brown, with black legs and antennae; tufts of short black hairs ring the body in parallel rows. Most helpful for recognition, however, is the attached elongated white egg sac (Figure 109), which has fluted or grooved edges and usually forms

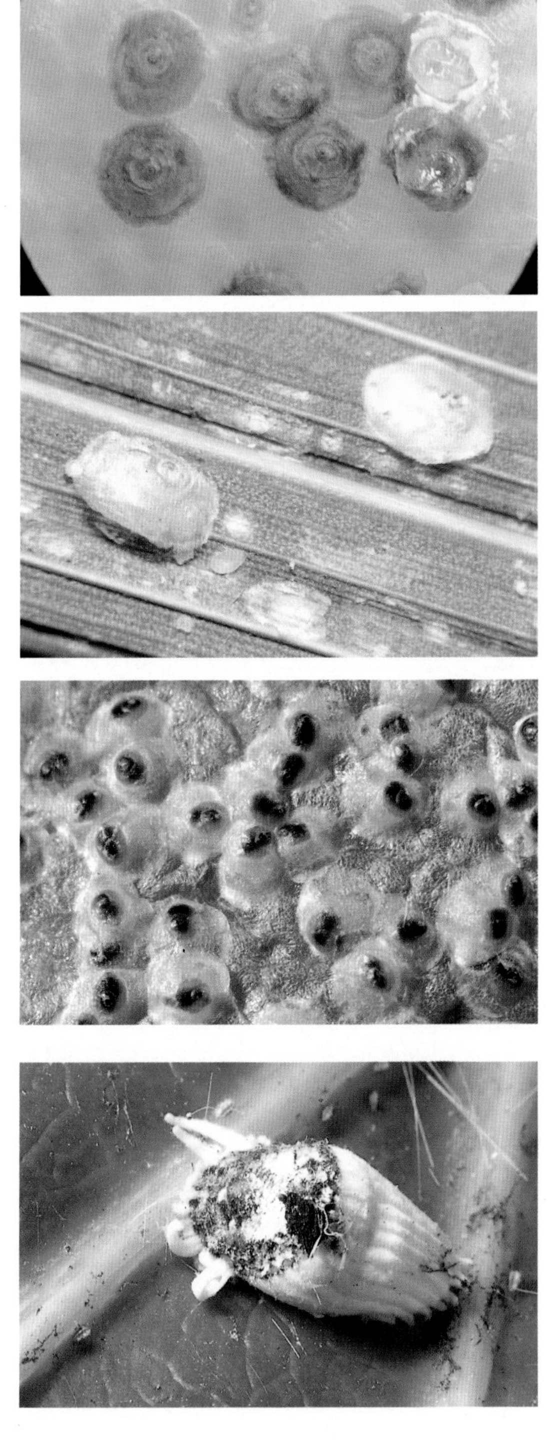

106. California Red Scales on surface of citrus fruit; one scale at right has been turned over to reveal female. Photograph by M. Badgley.

107. Oleander Scales. Photograph courtesy of Los Angeles County Agricultural Commissioner's Office.

108. Latania Scales. Photograph by M. Badgley.

109. Cottony Cushion Scale hermaphroditic female with attached white fluted egg sac; nymphs in background. Photograph by C. Hogue.

a mass larger than the insect itself. The scale is large, and the combined length of insect and sac is 3/8 to 1/2 inch (9 to 13 mm). The nymphs, which hatch from the eggs and leave the sac, are bright red.

Cottony Cushion Scale may be found on a variety of plants in addition to citrus.

REFERENCES. Doutt, R. L. 1958. Vice, virtue, and the Vedalia. *Bulletin of the Entomological Society of America,* vol. 4, pp. 119-123.

Gossard, H. A. 1901. The cottony-cushion scale. *Florida Agricultural Experiment Station Bulletin,* vol. 56, pp. 309-356.

Grossman, J. 1990. L.A.'s the place for biological pest control: How a little lady beetle saved California's citrus industry. *Terra,* vol. 28, no. 3, pp. 38-43.

### ■ SOFT SCALES (Family Coccidae)

■ Brown Soft Scale *(Coccus hesperidum)* Figure 110

This soft scale is small (1/8 to 3/16 in., or 3 to 5 mm, long) and very flat. It is oval in outline and pale greenish to brown in color; the surface of the body is covered with many minute brown dots, which may coalesce to form mottled patterns.

The Brown Soft Scale is a general feeder on all types of greenhouse and ornamental shrubs and trees as well as citrus. The insects nearly always cluster in great numbers in isolated spots on a tree, usually adjoining twigs; they may be so abundant that their bodies overlap.

### ■ WOOLY SAC SCALES (Family Eriococcidae)

■ European Elm Scale *(Glossyparia spuria)* Figure 111

In southern California, Chinese Elms in particular are affected by this scale. The insects may cluster in enormous numbers on the undersides of limbs, especially in the forks, weakening and even killing the trees by sucking the sap. They also secrete copious quantities of honeydew that covers anything beneath the tree and fosters the growth of sooty fungi.

The conspicuous stage is the female, which is medium-sized (1/6 to 3/8 in., or 1.5 to 9 mm) and oval; the body is reddish brown and coated at the rim with a fringe of thick white cottony wax. The insect is most abundant in late spring (May to June).

## MEALYBUGS (Family Pseudococcidae)

Mealybugs are related to the unarmored scale insects. The females are 1/8 to 1/4 inch (3 to 6 mm) long and have flattened oval bodies encrusted with a powdery white wax that readily rubs off, giving a "mealy" appearance to colonies of these insects. The males resemble male scale insects. Like scale insects and aphids, mealybugs are injurious to cultivated plants. There are several species of mealybugs in the basin, but the most generally injurious is probably the Long-tailed Mealybug.

REFERENCE. Gill, R. J. [No date.] Color-photo and host keys to the mealybugs of California. California Department of Food and Agriculture, Environmental Monitoring and Pest Management, Sacramento (Scale and white fly key #3).

110. Brown Soft Scales tended by Argentine Ants. Photograph by C. Hogue.

111. European Elm Scales. Photograph by C. Hogue.

112. Long-tailed Mealybugs on a lemon. Photograph by M. Badgley.

**■ LONG-TAILED MEALYBUG**
*(Pseudococcus longispinus)*
Figure 112

This species is distinguished by its very long terminal filaments; these are usually as long as or longer than the body, which is about 1/8 inch (3 mm) in length. The Long-tailed Mealybug also has conspicuous filaments around the edge of the body and a diffuse stripe down the middle of the back of the abdomen. Its hosts include citrus and many greenhouse plants; a favored species is the dracaena palm.

## COCHINEAL INSECTS (Family Dactylopiidae)

**■ COCHINEAL SCALES**
*(Dactylopius* species)
Figures 113, 114

Cochineal scales are medium-sized insects (females are 1/8 in., or 3 mm, long) that live in a protective mass of filamentous sticky white wax. These formations are commonly seen on the pads of the "beavertail" or *Opuntia* cactus, the host (Figure 113); the scales show a preference for the Burbank Spineless Cactus *(Opuntia ficus-indica)* and may become so abundant on this plant as to do it serious harm. The commonest local species is *Dactylopius opuntiae,* although *Dactylopius confusus* is also present in the basin.

Externally, the insect itself (Figure 114) is almost nothing more than a bag; it has vestigial nonfunctional legs and small mouthparts. The color of the scale's body in all stages is bright carmine red; however, the color is normally obscured by the waxy covering and is not visible unless the insect's body is broken, when the color seeps out and may stain the wax. The substance that gives these bugs their color can be extracted by crushing dry specimens to a powder and then boiling it in water. This cochineal, or "Spanish Red" as it is known, has been used for centuries as a dye by American Indians. It was discovered by the Spaniards during their conquest of the New World, and it quickly gained considerable commercial value as a crimson dye for textiles. Only with the advent of synthetic dyes has its importance subsided, although in recent years its use has been revived in the search for natural food colorings.

REFERENCES. Baranyovits, F. L. C. 1978. Cochineal carmine: An ancient dye with a modern role. *Endeavor* (new series), vol. 2, pp. 85-92.

Donkin, R. A. 1977. Spanish Red: An ethnogeographical study of Cochineal and the *Opuntia* cactus. *Transactions of the American Philosophical Society,* vol. 67, pp. 1-84.

Rodee, M. 1980. Bayeta and the history of cochineal dye. *Four Wings,* vol. 1, pp. 26-29.

Ross, G. N. 1986. The bug in the rug. *Natural History,* vol. 95, no. 3, pp. 66-73.

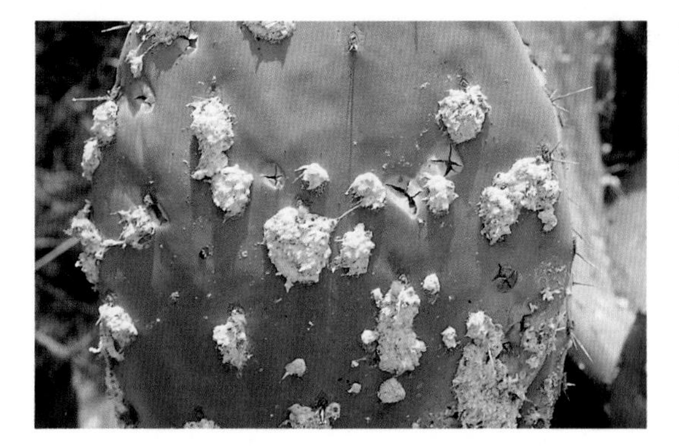

113. Wax formations of cochineal scale *(Dactylopius* species) on beavertail cactus. Photograph by J. Levy.

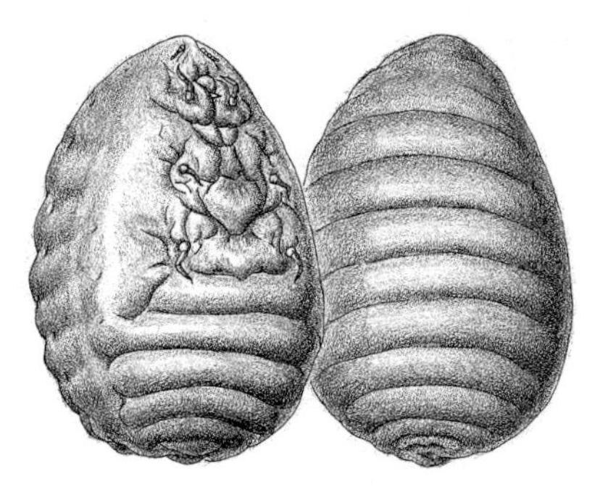

114. Cochineal scale *(Dactylopius coccus),* bottom (left) and top views. Drawing by T. Ross.

## WHITEFLIES (Family Aleyrodidae)

COLONIES OF TINY WHITE MOTH-LIKE INSECTS approximately 1/16 inch (1.5 mm) long frequently infest ornamental plants such as oak, iris, and morning glory. These are adult whiteflies, so called because of a fine white powder covering the body and wings. When disturbed, they fly from their perches in a flaky cloud.

The last-stage nymphs ("pupae") of some of the several local species, especially those found on the undersides of the leaves of oaks, secrete white wax plumes, ribbons (Figure 115), or filaments from the margin of the body; this secretion forms a fringe, giving the pupae a flower-like appearance (there may also be wax on the back in some species). Although the adults feed, the nymphs do most of the damage to the host by withdrawing quantities of sap in a manner similar to that of aphids, scale insects, and mealybugs. The nymphs are usually restricted to the lower surfaces of leaves.

The species described below are the whiteflies of primary local importance (the descriptions apply to the pupae; at other stages in their life cycles, these species are very difficult to distinguish from each other).

■ Crown Whitefly *(Aleuroplatus coronata;* Figure 115). The pupa has a large amount of snow-white wax, in wide ribbons, arranged in a crown-like pattern. It is the commonest species on oaks.

■ Mulberry Whitefly *(Tetraleurodes mori).* The pupa's marginal fringe of wax is long (about half the body width), and the back is without wax. This species infests mainly plants of the genera *Ceanothus* and *Rhamnus* (buckthorn family, Rhamnaceae).

■ Greenhouse Whitefly *(Trialeurodes vaporariorum;* Figure 116). This species occurs on citrus and many dooryard plants. The pupa has no fringe of wax and instead has a sparse cover of thin filaments. The body is translucent green.

■ Wooly Whitefly *(Aleurothrixus floccosus).* Infestation by this species is evident as a fluffy mass of waxy white material on the undersides of citrus leaves. The wax is produced in copious quantities by the nymphs and serves as their protection from parasitoids and predators. The nymph also freely secretes honeydew, which fosters growth of black sooty mold.

■ Ash Whitefly *(Siphoninus phillyreae;* Figure 117). This species, which first appeared in 1988, was apparently introduced from the Mediterranean area. It infests many trees, including citrus, ash, pear, apple, and apricot. In the absence of its natural enemies, the species has spread rapidly and has been so abundant at times that it formed swarms that looked like a fine ash blowing in the wind. Because it has resisted attempts at control by insecticides, predators and parasitoids have been introduced to help counter its spread.

REFERENCES. Bellows, T. S., and others. 1990. Biological control sought for ash whitefly. *California Agriculture,* vol. 44, no. 1, pp. 4-6.

Gill, R. J. [No date.] Color-photo and host keys to California white files. California Department of Food and Agriculture, Environmental Monitoring and Pest Management, Sacramento (Scale and whitefly key #2).

## JUMPING PLANT LICE (Family Psyllidae)

ADULT PSYLLIDS (pronounced "sill-ids") are like aphids in size and general shape but have shorter antennae (less than the body length) and more contracted abdomens; they look somewhat like miniature cicadas. Psyllids lack cornicles (the tube-like projections from the rear of the abdomen that are present on aphids), and they are missing the dark spot on the

115. Crown Whitefly nymphs. Photograph by L. Brown.

116. Greenhouse Whitefly adults. LACM photograph.

117. Ash Whitefly adults and immatures. Photograph by C. Hogue.

118. Pepper Tree Psyllid. Photograph by M. Badgley.

119. Eugenia Psyllid. Photograph by M. Badgley.

leading edge near the tip of the fore wing. They are also active jumpers, unlike the sedentary aphids, and they wag their bodies characteristically from side to side.

They feed on plant sap, and several species are plant pests. The following are two notable local examples:

■ Pepper Tree Psyllid *(Calophya schini;* Figure 118). This species is a native of Peru, which also is the home of its host, the Peruvian Pepper Tree *(Schinus molle).* (A relative of *S. molle,* the Brazilian Pepper Tree *(Schinus tenebrinthefolius),* is apparently not susceptible to this pest.) The Pepper Tree Psyllid was first discovered in Long Beach in 1984 and has recently become established in Los Angeles. Feeding by the nymphs causes the formation of crater-like depressions on the leaflets, usually on the lower surfaces, and these depressions are the most visible sign of the presence of the species. When it occurs in large numbers, the species can cause deformation of young leaves, and heavy infestations cause the trees to take on a grayish appearance and drop their leaves.

■ Eugenia Psyllid *(Trioza eugenii;* Figure 119). This psyllid is a new pest that appeared in 1988 on eugenia shrubs near the Los Angeles International Airport. That year and the next it spread rapidly throughout the basin to infest virtually every available host. It is similar to the Pepper Tree Psyllid and damages its host in much the same way. It is native to Australia.

REFERENCE. Downer, T. A., P. Svihra, R. H. Molinar, J. B. Fraser, and C. S. Koehler. 1988. New psyllid pest of California pepper tree. *California Agriculture,* vol. 42, no. 2, pp. 30-31.

# 9 THRIPS (Order Thysanoptera)

BECAUSE THEY ARE EXTREMELY TINY (less than 1/16 in., or 1.5 mm, long), thrips are hardly noticed until they reveal their presence by injuring garden plants by their feeding. A common symptom of their attack is leaf curl, a result of loss of the leaf's sap to the insects' piercing-sucking mouthparts. The thrips' feeding on fruit causes unsightly scars on the skin. Some species are predatory on plant pests and thus beneficial to the gardener and farmer.

When greatly magnified, thrips can be seen to possess distinctive wings that are very long and narrow with few or no veins and a fringe of long hairs.

Thrips live on plants and are especially common in flowers, particularly daisies and dandelions. If they accidentally fall onto human skin, they will prod about with their strong mouthparts and may even bite with a surprisingly sharp effect.

The significant local species are as follows:

- Western Flower Thrips *(Frankliniella occidentalis;* Figure 120). This species is variable in color, from light yellow to dark brown. It causes injury to a wide variety of crops and ornamentals.
- Greenhouse Thrips *(Heliothrips haemorrhoidalis;* Figure 121). This insect is dark brown except for some of its rear abdominal segments, which are yellowish-brown. It is a pest in greenhouses and on subtropical trees, especially citrus and avocados.
- Citrus Thrips *(Scirtothrips citri).* The body of this species is yellowish orange, the fore wings clear with orange veins. It is a serious citrus pest.

120. Western Flower Thrips. Photograph courtesy of Los Angeles County Agricultural Commissioner's Office, J. Davidson.

121. Greenhouse Thrips. Photograph by M. Badgley.

122. Cuban Laurel Thrips adults, nymphs, and eggs. Photograph courtesy of Los Angeles County Agricultural Commissioner's Office.

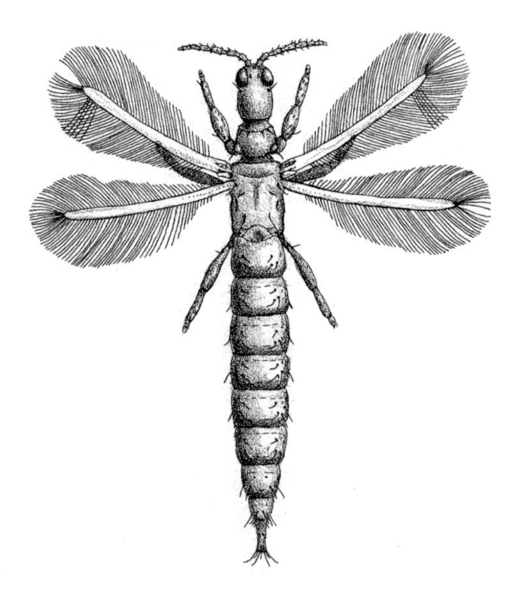

123. Black Hunter. Drawing by T. Ross after mounted specimen in S. Bailey, 1940, *Journal of Economic Entomology,* vol. 32, p. 542, fig. 1.

■ Cuban Laurel Thrips *(Gynaikothrips ficorum;* Figure 122). This species is larger than most thrips (1/8 in., or 3 mm, long) and dark brown to black in color; some segments of the antennae are yellow. It is found primarily on the outdoor ornamental fig, *Ficus nitida.* Colonies live under the rolled edges of infested leaves.

■ Black Hunter *(Haplothrips mali;* Figure 123). This is a beneficial species that feeds on the eggs of mites and insects and on small soft-bodied insects such as aphids and other thrips. It is black except for the intermediate antennal segments, which are yellow.

REFERENCES. Bailey, S. F. 1936. Thrips attacking man. *Canadian Entomologist,* vol. 68, pp. 95-98.

_______. 1957. The thrips of California, part 1: Suborder Terebrantia. *Bulletin of the California Insect Survey,* vol. 4, pp. 143-220.

Borrer, D. J., C. A. Triplehorn, and N. F. Johnson. 1989. *An introduction to the study of insects,* 6th edition. Philadelphia, Pa.: Saunders College.

Cott, H. E. 1956. Systematics of the suborder Tubulifera (Thysanoptera) in California. *University of California Publications in Entomology,* vol. 13, pp. 1-216.

Mound, L. A. 1976. The identity of the greenhouse thrips *Heliothrips haemorrhoidalis* (Bouche) (Thysanoptera) and the taxonomic significance of spanandric males. *Bulletin of Entomological Research,* vol. 66, pp. 179-180.

Paine, T. D. 1992. Cuban Laurel Thrips (Thysanoptera: Phlaeothripidae) biology in southern California: Seasonal abundance, temperature dependent development, leaf suitability, and predation. *Annals of the Entomological Society of America,* vol. 85, no. 2, pp. 164-172.

# 10 NERVE-WINGED INSECTS AND CADDISFLIES

Although they share common ancestry with the beetles and the flies, entomologists consider the nerve-winged insects and the caddisflies to be among the most primitive living insects that undergo complete metamorphosis. The body structure in the two groups is very similar, but their biologies diverge in that the nerve-winged insects are primarily terrestrial throughout their life cycles while the caddisflies are entirely aquatic in the immature states.

## NERVE-WINGED INSECTS (Order Neuroptera)

These insects are soft-bodied and generally poor fliers, in spite of the well-developed twin pairs of wings. The common name of the group refers to the complex nerve-like pattern of the veins in the wings. Because neuropterans are primarily predaceous in both the larval and adult phases, they often benefit humanity by eating insect pests.

### ■ DOBSONFLIES (Family Corydalidae)

■ California Dobsonfly *(Neohermes californicus)* Figure 124

This is a stream-loving species found sporadically in the Santa Monica Mountains and the northern foothills (it is common at higher elevations in the San Gabriels). It is conspicuous because of its large size (the wing expanse is 3 to 3 1/2 in., or 75 to 90 mm) and its habit of coming to light at night.

Eggs are deposited in oval masses on exposed flat surfaces of rocks and wood. The larvae, which are commonly called "hellgrammites," are aquatic and hide among leaf and twig debris jams in small rapidly

124. California Dobsonfly. Photograph by J. Hogue.

flowing streams. They prey on other aquatic insects, chiefly mayfly and caddisfly immatures. Young larvae may pass dry periods in dormancy under rocks and debris in contact with the moist stream bed. The pupae are very active and crawl onto sticks and leaves that accumulate as the streams dry up. Adults are active in the summer.

REFERENCE. Smith, E. L. 1970. Biology and structure of the dobsonfly, *Neohermes californicus* (Walker) (Megaloptera: Corydalidae). *Pan-Pacific Entomologist,* vol. 46, pp. 142-150.

## ■ SNAKEFLIES (Family Raphidiidae)

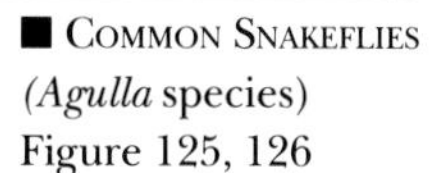

■ COMMON SNAKEFLIES
*(Agulla* species)
Figure 125, 126

The elongate prothorax and projecting head, suggesting the raised fore quarter of a snake, are responsible for the common name of these insects. Several species are found in our area. Adults are sometimes common in the springtime on shrubbery and trees, mainly in hilly areas.

Adults feed on small soft-bodied insects, including young scale insects, aphids, and mites, and thus are of benefit to the farmer and gardener. They will eat larger insects if the latter are injured and unable to escape or defend themselves. The adults themselves are about 3/4 inch (20 mm) long.

125. Common snakefly *(Agulla* species). Photograph by C. Hogue.

126. Common snakefly larva *(Agulla* species). Photograph by J. Hogue.

The larvae have similar tastes and will devour caterpillars and other insect larvae as well. Young larvae sometimes live under empty shells of scale insects. Mature larvae hibernate in accumulated dry leaves, in vegetation, on the ground, under bark, and in spaces in porous wood.

REFERENCES. Acker, T. S. 1966. Courtship and mating behavior in *Agulla* species (Neuroptera: Raphidiidae). *Annals of the Entomological Society of America,* vol. 59, pp. 1-6.

Woglum, R. S., and E. A. McGregor. 1959. Observations on the life history and morphology of *Agulla astuta* (Banks) (Neuroptera: Raphidiodea: Raphidiidae). *Annals of the Entomological Society of America,* vol. 52, pp. 489-502.

## ■ GREEN LACEWINGS (Family Chrysopidae)

■ COMMON GREEN LACEWING *(Chrysoperla plorabunda)* Figure 127

The behavior of this species has been extensively studied. During the warmer times of the year adults rest near porch lights to which they have been attracted during their nocturnal wanderings. The light may cause the lacewing's eyes to reflect like brilliant golden beads, and the name "Golden Eyes" is sometimes given to these fragile green-tinted insects. Courting pairs exhibit abdominal jerking, which sets up auditory signals important to mating.

The larvae are called "aphid lions" because they have powerful appetites for aphids and similar injurious plant pests, which they catch with their sickle-shaped jaws. The young larvae are so voracious that the eggs, which are laid in a cluster on the surface of a leaf, would not survive if it were not for the fact that each is placed on a long slender silken stalk, out of the reach of precocious newly hatched individuals. The adults, which average 1/2 inch (13 mm) in length, have been seen feeding on aphids and on honeydew.

Other species of green lacewings occur throughout our area, but they are less common than *Chrysoperla plorabunda;* they are all very similar in appearance and are indistinguishable to the uninstructed eye.

OTHER SCIENTIFIC NAMES. *Chrysopa carnea* (actually the name of a European species).

REFERENCES. Bowden, J. 1979. Photoperiod, dormancy and the end of flight activity in *Chrysopa carnea* Stephens (Neuroptera: Chrysopidae). *Bulletin of Entomological Research,* vol. 69, pp. 317-330.

Henry, C. S. 1979. Acoustical communication during courtship and mating in the Green Lacewing *Chrysopa carnea* (Neuroptera: Chrysopidae). *Annals of the Entomological Society of America,* vol. 72, pp. 68-79.

New, T. F. 1975. The biology of Chrysopidae and Hemerobiidae (Neuroptera), with reference to their usage as biocontrol agents: A review. *Transactions of the Royal Entomological Society of London,* vol. 127, pp. 115-140.

Smith, R. 1922. The biology of the Chrysopidae. *Memoirs of the Cornell University Agricultural Experiment Station,* no. 58, pp. 1291-1372.

Toschi, C. A. 1965. The taxonomy, life histories, and mating behavior of the green lacewings of Strawberry Canyon (Neuroptera: Chrysopidae). *Hilgardia,* vol. 36, pp. 391-433.

## ■ BROWN LACEWINGS (Family Hemerobiidae)

THESE NEUROPTERANS are very similar in structure and habits to green lacewings, but they may be distinguished mainly by their slightly smaller size (they are 5/16 in., or 8 mm, long) and brownish color (Figure 128). In addition, the wings of brown lacewings are covered by microscopic bristles that collectively give them a velvety or dull appearance.

The larvae are predaceous, like those of green lacewings; they may pile bits of debris on their backs to give them a measure of camouflage.

Several kinds live in the basin. A common species is *Hemerobius pacificus.*

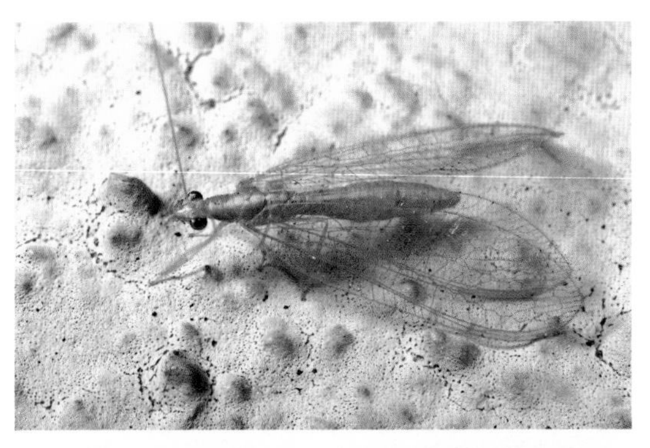

127. Green lacewing (*Chrysoperla* species). Photograph by C. Hogue.

128. Brown lacewing. Photograph by M. Badgley.

REFERENCES. Neuenschwander, P. 1975. Influence of temperature and humidity on the immature stages of *Hemerobius pacificus*. *Environmental Entomology*, vol. 4, pp. 215-220.

Oswald, J. D. 1988. A revision of the genus *Sympherobius* Banks (Neuroptera: Hemerobiidae) of America north of Mexico with a synonymical list of the world species. *Journal of the New York Entomological Society*, vol. 96, pp. 390-451.

### ■ ANTLIONS (Family Myrmeleontidae)

THESE INSECTS are best known to most of us for the funnel-shaped pits (Figure 129) dug by the larvae of a few genera (primarily *Myrmeleon*). In these species, the larva lies buried at the bottom of the pit with only its sickle-shaped jaws protruding from the soil, ready to catch any unsuspecting prey—often an ant—that happens to tumble into the trap. The traps are typically dug in fine sandy soil in protected places. The slope of the funnel rests at the critical angle of repose for sand, so that the sides readily give way under the feet of the would-be escapee, sliding it always back down into the terrible jaws at the bottom of the pit. The larva makes matters worse for the prey by throwing sand onto it.

Antlions are also called "doodle bugs." Country children of earlier generations knew that one can entice the larva to leave its pit by pronouncing the magic phrase, "Doodle bug, doodle bug, come out of your hole." Actually, the expirations that accompany these words, which must be spoken very close to the pit, dislodge some sand, and the doodle bug—believing it has made a capture—becomes excited and crawls into view.

Adult antlions (Figure 130) are long (body length is 1 3/8 to 2 in., or 35 to 50 mm) and are gracefully slender creatures with four similarly sized speckled wings; their antennae are short and clubbed. Although they are feeble fliers, they often come to lights.

Several species belonging to the genera *Brachynemurus* and *Myrmeleon* occur in the basin.

REFERENCES. Herbert, H. 1936. Insects of prey. *Nature*, vol. 27, no. 2, pp. 80-83.

Stange, L. A. 1970. Revision of the ant-lion tribe Brachynemurini of North America. *University of California Publications in Entomology*, vol. 55, pp. 1-192.

## CADDISFLIES (Order Trichoptera)

THESE INSECTS look a good deal like moths, but they lack the long coiled sucking mouthparts that most moths have, and they have hairs on the wings instead

129. Pits of antlion larvae (probably *Brachynemurus* species). LACM photograph.

130. Antlion adult. Photograph by P. Bryant.

of scales. They occur near water and are frequently attracted to lights at night.

The larvae, which are aquatic bottom dwellers, are well known to stream fishermen as "caseworms." Those of most species live in some sort of protective case or tube made of silk, with bits of leaves, twigs, sand grains, pebbles, or other objects incorporated into the material to give the larvae additional physical protection and camouflage. The shape and method of construction of the case is characteristic for a species or group of species, and the variety in these "mobile homes" is extensive: they may be purse-shaped, tubular, curved, snail-shaped, or rectangular, and there are even types with sticks set in an ascending square framework that mimics a little log cabin.

Just a handful of caddisfly species live in the basin itself; they are to be found in the foothills of the San Gabriels, Santa Monicas, and Santa Anas, but they sometimes intrude into the lower elevations along the few rivers and streams emerging from the hills and mountains. None exist far from running water (the larvae of our species do not inhabit ponds or other bodies of standing water). The following species are fairly conspicuous, although none have common names.

131. Caddisfly (*Hesperophylax incisus*). Photograph by C. Hogue.

132. Larva of caddisfly (*Hesperophylax* species) in its case. Drawing by T. Ross.

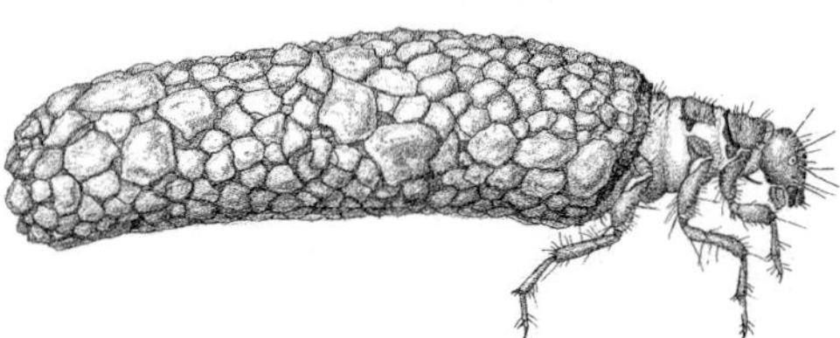

■ *Gumaga* (formerly *Sericostoma*) *griseola* (family Sericostomatidae). This is a small gray species (length about 3/8 in., or 9 mm), which may occur in abundance in the Westwood Hills and elsewhere. The larvae build long slender cases of tightly bound sand grains.

■ *Helicopysche borealis* (family Helicopsychidae) and *Hydropsyche philo* (family Hydropsychidae). These widespread species appear in the basin on occasion. Both develop in streams. The larva of *Helicopysche borealis* uses large sand grains to build a case shaped like a snail shell; the adult is 1/4 inch (6 mm) long and has antennae that are not as long as the body. In contrast, *Hydropsyche philo* larvae are free-living and build silk nets on the undersides of rocks to ensnare tiny organisms passing through on the current; the adult is larger (1/2 in., or 13 mm), and its antennae are longer than its body. Adults of both species are mottled gray.

■ *Hesperophylax incisus* (family Limnephilidae; Figures 131 and 132). This fairly large species (length from head to wing tips is 3/4 in., or 20 mm) is handsomely colored: the fore wings of the adult are a rich tan with a longitudinal silvery stripe; the head and the tips of the hind wings are covered with silky hair. It is primarily a mountain species. The larva lives in swift streams where it builds a compact cylindrical case of small pebbles.

REFERENCE. Wiggins, G. B. 1977. *Larvae of the North American caddisfly genera (Trichoptera).* Toronto: University of Toronto.

# 11 MOTHS AND BUTTERFLIES (Order Lepidoptera)

THE LEPIDOPTERA is a large and generally homogeneous order of holometabolous insects; its members are characterized by the presence of microscopic overlapping scales on the surfaces of the wings. These scales are easily dislodged and will adhere to the fingers like fine powder if the wings are rubbed. The pattern on the wings is actually a mosaic of colored scales; without them the wing membrane is semitransparent and colorless.

With the exception of a few primitive moths that have mandibles, all moths and butterflies have a uniquely fashioned set of mouthparts as adults. These consist of a long pair of slender grooved filaments that interlock so that a canal is formed between them. The insect uses this tongue or "proboscis" like a soda straw (Figure 264), dipping the tip into the liquid food and siphoning it up. When not in use, the tongue is kept coiled up like a watch spring under the head (Figure 219). In many species these mouth structures are reduced, and the adult does not feed.

The larvae of moths and butterflies are the familiar caterpillars. These are soft, somewhat wormlike creatures of diverse form and color. In addition to three pairs of short jointed legs on the thorax that correspond to the normal legs of other insects and the adult lepidopterans, caterpillars have a varying number of short stubby unsegmented legs, called prolegs (see Figure 134, for example), on certain of their abdominal segments, including the last. The prolegs are tipped with rows of tiny hooks, allowing the caterpillars to cling tightly to twigs and leaves. Caterpillars walk by sequentially releasing and regrasping the substrate with their prolegs; those with prolegs on only the segments at the rear of their bodies walk with a looping movement (see Figure 192).

Despite the fact that butterflies and moths are classified together in the Lepidoptera, the two groups are by no means equivalent. Moths, which constitute the vast majority of Lepidoptera, are more primitive and more generalized in structure than butterflies and are primarily nocturnal. Butterflies make up a relatively small side branch of the main lepidopteran family tree and are specialized mainly for a diurnal existence. Nevertheless, the division between butterflies and moths is popularly considered equal, and many people ask "What is the difference between moths and butterflies?" There are several answers to this question. None of the distinctions always holds true, but, if a specimen satisfies most of one or the other set of conditions in Table 5, a correct identification can usually be made.

## ■ RAISING CATERPILLARS

THERE IS NO BETTER WAY to witness the miracle of insect metamorphosis than to raise moths and butterflies from their caterpillars. Furthermore, most species of Lepidoptera are known only from the adult stage, and to identify the species of a caterpillar it is usually necessary to rear it through to the adult.

Most caterpillars do not require a great deal of care. Success in rearing them can be attained by following the rules outlined in the appendix on keeping insects in captivity (Appendix B). It is best to start with caterpillars that are nearly or already full grown rather than larvae just hatched from eggs. In any case, one must be especially careful to provide leaves of the correct food plant species.

A simple cage for rearing specimens can be made by rolling nylon window screen material into a cylinder and covering the open ends with the lids from cylindrical cardboard ice-cream cartons or with the halves of a plastic soft drink bottle cut apart. Fresh sprigs or twigs of the food plant should be inserted frequently. Not all species spin cocoons or form hanging chrysalids; in fact, most moth larvae burrow into soil to pupate. Therefore, when the caterpillars approach maturity (when they stop feeding and begin to wander), the bottom portion of the cage should be

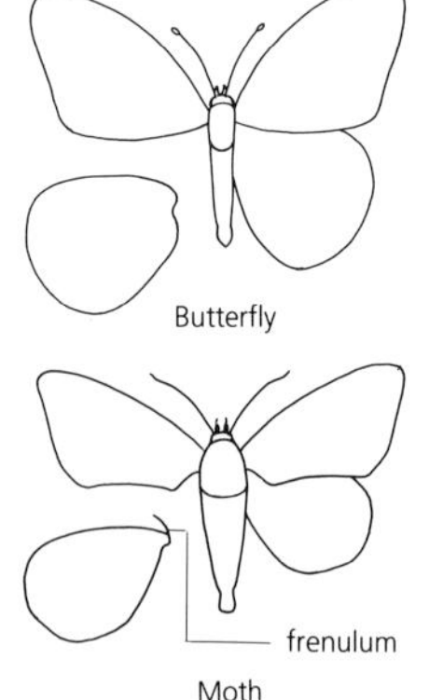


Table 5. General differences between butterflies and moths.

| | BUTTERFLY | MOTH |
|---|---|---|
| WINGS | Held together vertically over the back when at rest (Figure 215) | Held roof-like over the body when at rest (Figure 198) |
| PUPA | Almost always a naked chrysalid (Figure 214) rather than enclosed in a cocoon | Often enclosed in a cocoon (Figures 139, 193) or in the soil |
| ANTENNAE | Slender, knobbed or swollen at tip to form a club | Varied in form (feathery to hair-like) but rarely knobbed |
| BODY | Usually broad in proportion to wings (except in Skippers) | Relatively slender in proportion to wings |
| HABITS | Diurnal | Mostly nocturnal |
| COLOR | Brightly colored | Dull or somber colored |
| WING COUPLING | Hind wing with enlarged lobe at base in contact with fore wing | Hind wing usually coupled to fore wing with a frenulum (small spine) |

filled to a depth of about 2 inches (5 cm) with loamy soil topped with dry leaf litter.

A few days after the larva enters the ground, the pupa will have formed and may be dug up and laid on the surface; the adult will emerge normally, and the pupa is less prone to develop harmful molds than those left underground. It will be necessary, however, to sprinkle the exposed pupa lightly with water once in a while to prevent desiccation.

The time required for development and emergence after pupation varies greatly with the species of butterfly or moth. The adults of some species will come out after a few days; others will pass a single summer or winter in the pupal state; still others may remain dormant for several years. If you do not know when to expect emergence, you must be always ready. Provide some rough surface for the newly emerged adult to climb upon (a cylinder of toweling paper is good) so that it may properly fill out its wings.

Many lepidopterous larvae are too small to be kept in a cage the size of an ice-cream carton. These are best reared in a large simple stoppered plastic pill vial. Food will keep well since the humidity remains high, but the chamber will have to be cleaned frequently to avoid undesirable mold growth.

REFERENCES. Crotch, W. J. B. 1956. The know-how. Part 1 in *A silkmoth rearer's handbook.* London: Amateur Entomological Society.

McFarland, N. 1964. Notes on collecting, rearing and preserving larvae of macrolepidoptera. *Journal of the Lepidopterists' Society,* vol. 18, pp. 201-210.

______. 1964. Additional notes on rearing and preserving larvae of macrolepidoptera. *Journal of the Lepidopterists' Society,* vol. 19, pp. 233-236.

## ■ BUTTERFLY GARDENING

AN ESPECIALLY REWARDING WAY to promote beauty in gardens and parks is to plant host plant species for the butterfly caterpillars and flowering varieties that serve as nectar sources for the adults. Hosts that will nourish caterpillars and that are easy to establish in our area are the Passion Vine *(Passiflora coerulea)* for Gulf Fritillaries, Sweet Fennel *(Foeniculum vulgare)* for the Anise Swallowtail, and *Cassia* for the Senna Sulfur. A spare corner of the garden can be used to foster certain weeds that are hosts to butterfly larvae, such as the Cheeseweed *(Malva parviflora)* for the Painted Lady, Pigweed *(Chenopodium* species) for Pygmy Blues, and Wild Mustard *(Brassica rapa sylvestris)* for Cabbage Whites.

Adult butterflies (and other insects) are attracted to most flowers but especially to light fragrant nectar-rich kinds such as Butterfly Bush *(Buddleia* species), zinnias, and lantana. (For the sound of humming bees on warm days plant eucalyptus, Brazilian Pepper *(Schinus terebrinthefolius),* bottlebrush *(Callistemon),* and ice plants.)

REFERENCES. Donahue, J. P. 1977. Take a butterfly to lunch: A guide to butterfly gardening in Los Angeles. *Terra,* vol. 14, no. 3, pp. 3-12.

Tekulsky, M. 1985. *The butterfly garden.* Boston: Harvard Common.

Xerces Society and Smithsonian Institution, creators. 1990. *Butterfly gardening: Creating summer magic in your garden.* San Francisco: Sierra Club.

## MOTHS

THE VERY WORD "MOTH" strikes terror in the heart of the householder. In defense of woolen sweaters and blankets, we annihilate every hapless specimen that chances to enter the home. The real culprits are two tiny inconspicuous species, which usually go unnoticed while innocent vegetarians are needlessly killed.

The moths described here (including the true Clothes Moth) are only a few of the hundreds of species found in the Los Angeles Basin.

REFERENCES. Holland, W. J. 1968. *The moth book.* New York: Dover. [A reprint of the original 1903 edition.]

McFarland, N. 1965. The moths (Macroheterocera) of a chaparral plant association in the Santa Monica Mountains of southern California. *Journal of Research on the Lepidoptera,* vol. 4, pp. 43-74.

Various authors are involved in a continuing series entitled *Moths of America* and published by E. W. Classey and the Wedge Entomological Research Foundation, London.

### ■ WILD SILK MOTHS (Family Saturniidae)

■ CEANOTHUS SILK MOTH
*(Hyalophora euryalus)*
Figures 133, 134

This is a spectacular moth, with its pleasing, warm, reddish-brown color and large size (wing expanse up to 5 in., or 13 cm). It may appear in the suburbs but usually occurs only in the foothill areas and canyons, where males are sometimes seen flying to light in early spring.

During the summer the larvae feed primarily on species of mountain wild lilac *(Ceanothus)* in the chaparral plant association (it is in the chaparral that the adults are most common). In captivity the caterpillars will also live on willow, manzanita *(Arctostaphylos),*

133. Ceanothus Silk Moth. Photograph by C. Hogue.

134. Ceanothus Silk Moth larva. Photograph by C. Hogue.

Laural Sumac *(Rhus laurina),* and other shrubs and trees. When mature, they attain giant size (a length of up to 4 in., or 10 cm) and resemble fat powdery green sausages set with yellow and bluish tubercles. The gray, flask-shaped cocoons are spun on the host plant in the fall.

To attract members of the opposite sex, an unmated female emits a chemical scent from glands that she protrudes from the tip of her abdomen. This scent travels great distances on air currents and is a powerful sexual stimulant to the males; they lose no time flying to "calling" females and quickly mate. During copulation the female withdraws the glands, and she then loses any further attractiveness to other males.

The adults have atrophied mouthparts and do not feed. They die shortly after completing their sole function, which is to mate and reproduce.

OTHER SCIENTIFIC NAMES. *Samia rubra; Platysamia euryalus.*

135. Polyphemus Moths. Photograph by J. Hogue.

136. Polyphemus Moth larva. Photograph by A. Johnson.

■ **Polyphemus Moth**
*(Antheraea polyphemus)*
Figure 135, 136

The distinctive feature of this moth is the large eye spot on each hind wing. The species occurs only sporadically in the basin, usually in the vicinity of the larvae's favored food plants—Coast Live Oak, birch, and elm.

Mature larvae have strongly convex segments and are a pale translucent green with diagonal silver lines on the sides and pearl-colored tubercles on the back. They feed through the summer and early fall.

The cocoon is spun in the fall and drops from the tree onto the ground. Bits of leaf or small twigs are usually incorporated into the cocoon walls, and these help to camouflage it among the litter at the base of the host plant.

Adults fly from May to October. Like the Ceanothus Silk Moth, Polyphemus Moths are very large (with wing expanses of more than 5 in., or 13 cm), have reduced nonfunctional mouthparts, and are attracted to lights at night. The females also emit a powerful male attractant.

OTHER SCIENTIFIC NAMES. *Telea polyphemus.*

137. Domestic Silk Moth. Photograph by C. Hogue.

138. Domestic Silk Moth larva. Photograph by C. Hogue.

139. Domestic Silk Moth cocoon. Photograph by C. Hogue.

## ■ SILK MOTHS (Family Bombycidae)

■ DOMESTIC SILK MOTH
*(Bombyx mori)*
Figures 137-139

Neither this moth nor its caterpillar (known as the "silkworm") will be seen out of doors here in the basin or anywhere else, for this is a totally domesticated insect that cannot survive without man's constant care. The species has been selectively bred for centuries to improve the quality of its silk. But in the process it has lost its self sufficiency: although its wings remain, they are stunted and weak and no longer serve their original purpose of flight.

Man's long interest in this insect comes from the larva's ability to spin a very long continuous fiber of unrivaled strength, smoothness, and luster. When the fiber is wound into thread and woven into fabric, we have silk, the "queen of textiles."

Sericulture (the commercial rearing of silkworms) is almost as old as history itself. The discovery of silk dates back at least four thousand years and is attributed to the Chinese empress Si-ling, who accidentally let a cocoon that she was fondling fall into a cup of hot tea. According to legend, as she retrieved it, she discovered the strands of the cocoon when they became loose between her fingers. She experimented with the weaving of this fiber, and from the results evolved the silk industry.

The modern processing of silk begins with "filature," the unwinding of the cocoons. They are first floated in pans or troughs of hot water. The strands are then wound onto spools and later cleaned, dried, and subjected to various throwing and looping maneuvers to form threads of sufficient size and strength to be woven into cloth.

Today, sericulture and trade in raw silk are still important industries, primarily in Oriental countries, in spite of the availability of many popular synthetic substitutes. A number of efforts to establish the industry in California and other parts of North America have all failed for one reason or another.

The life cycle of the Domestic Silk Moth is familiar to generations of school children as the classic example of insect development and metamorphosis. The eggs (sometimes called "seed") are still readily available to teachers and others wishing to see the transformations of this famous insect (for sources, see "General Bibliography and Resources" in the appendixes).

The smooth spherical eggs are at first yellowish-white but turn dark as the embryo develops. The caterpillars are powdery white, naked, and have a short horn on the rear. They will remain healthy and grow steadily only if they are fed the leaves of the mulberry tree *(Morus)*. They mature in about forty-five days, when they have reached a length of 2 inches (50 mm).

The oval white or yellow cocoons are spun in a cage corner or on a clump of twigs. After twelve to sixteen days, the adult emerges through one end of the cocoon, which has been softened by a caustic secretion from the pupa. The adult has a wing expanse of 1½ inches (40 mm).

REFERENCES. Cooper, E. 1961. *Silkworm and science, the story of silk.* New York: Harcourt, Brace, World.

Essig, E. O. 1945. Silk culture in California. California Agricultural Experiment Station, Circular no. 363, pp. 1-15.

Yokoyama, T. 1963. Sericulture. *Annual Review of Entomology,* vol. 8, pp. 287-306.

## ■ SPHINX MOTHS (Family Sphingidae)

THESE MOTHS feed while in flight, hovering before flowers as they dip their tongues in and out of the nectaries like hummingbirds. For this reason, mem-

140. Tobacco Hornworm moth. Photograph by P. Bryant.

141. Tobacco Hornworm larva. Photograph by C. Hogue.

142. Tobacco Hornworm pupa. Photograph by P. Bryant.

bers of the family are sometimes called "hummingbird moths" ("hawk moths" is yet another name). The more common name, Sphinx moths, stems from the fright posture of the larva: with its fore quarter reared back and head tucked under, it suggests the famous Sphinx edifice in Egypt.

The larvae of nearly all species are sometimes called "hornworms" because of the erect spine or horn at the posterior end. Though formidable in appearance, the spine is harmless and is apparently only "for show," to discourage predators.

---

■ TOBACCO HORNWORM
*(Manduca sexta)*
Figures 140-142

Many people do not know that the large green "tomato worms" (Figure 142) that reduce their backyard tomato vines to a state of rubble in the summer later turn into large sphinx moths. When mature (4 in., or 10 cm, long), these caterpillars crawl to the ground and burrow 2 to 6 inches (5 to 15 cm) below the surface to pupate. The pupa is notable for its large size (2 in., or 5 cm, or more) and curious "jug handle" appendage, which is actually a case for the developing proboscis of the adult. When extended, this proboscis may be 3 1/2 to 4 inches (9 or 10 cm) long and provides the moth with an effective means for stealing nectar from deep-throated flowers.

The mottled gray fore wings of the adult span 4 to 4 1/2 inches (10 to 11.5 cm). The hind wings are white with black bands. Another identifying feature is a row of six large orange spots on each side of the abdomen.

There are two generations of this species per year in the basin. Fall pupae remain in the ground until the following summer; summer pupae produce adults after only a week or two. In addition to tomato, the caterpillar feeds on many plants of the nightshade family, including potato, tobacco, Indian Tobacco *(Nicotiana glauca)*, and Jimpson Weed *(Datura)*.

OTHER SCIENTIFIC NAMES. *Protoparce sexta, Phlegonthius sexta.*

REFERENCES. Heinrich, B. 1971. The effect of leaf geometry on the feeding behavior of the caterpillar of *Manduca sexta* (Sphingidae). *Animal Behavior,* vol. 19, pp. 119-124.

Yamamoto, R. T., and G. S. Fraenkel. 1960. The specificity of the tobacco hornworm, *Protoparce sexta,* to solanaceous plants. *Annals of the Entomological Society of America,* vol. 53, pp. 503-507.

---

■ WHITE-LINED SPHINX
*(Hyles lineata)*
Figures 143, 144

Also called the Striped Morning Sphinx, this is one of our commonest moths. Its names refer to the broad oblique stripe running from the base to tip of the fore

143. White-lined Sphinx. Photograph by J. Hogue.

144. White-lined Sphinx larva. Photograph by C. Hogue.

wing; the stripe is interrupted by numerous transverse white streaks. The hind wing is pink with black at the base and margins. During the summer, adults often may be caught at rest near the lights of storefronts, even in the metropolitan area. The moth is fair-sized, measuring 2½ to 3 inches (65 to 75 mm) from wing tip to wing tip.

The large larva (length to 3½ in., or 90 mm) comes in black as well as green color phases: the black version has a row of pale spots bordered above and below with black along its back; the green larva has three yellow lines on its back and broken black lines on its sides.

The caterpillars may literally swarm on desert vegetation in the spring. Like the larva of the Tobacco Hornworm, the caterpillar goes into the ground to pupate; the pupa lacks the "jug handle" of its relative, however, even though the adult has a fairly long tongue. The list of larval food plants is almost infinite; a common host in the city is fuchsia.

OTHER SCIENTIFIC NAMES. *Celerio lineata.*

## ■ TUSSOCK MOTHS (Family Lymantriidae)

■ WESTERN TUSSOCK MOTH
*(Orgyia cana)*
Figures 145-147

The furry caterpillar of this species feeds on a variety of plants in the basin, most commonly the rosaceous types (pome fruits, pyracantha) but also Coast Live Oak *(Quercus agrifolia)*. Full-grown larvae are a little over 1 inch (25 mm) long, with numerous red, blue, and yellow spots and various tufts of hairs, the most characteristic being the dense erect group of four on the back of the front abdominal segments. These hairs easily dislodge and occasionally cause a skin reaction or allergy in sensitive persons contacting them. The caterpillars incorporate these hairs into their cocoons.

The male moth (Figure 145) is ordinary looking, with dark fore and hind wings (its wing span is 1 to 1 3/8 in., or 25 to 35 mm). But the female (Figure 146) resembles a furry white bag; it is stubby-winged and flightless and never crawls far from its cocoon. It lays its eggs in gray felt-covered masses. The adults are active from May to July and again, from a second brood, in September and October. The species passes the winter in the egg stage. Larvae are most common in March, April, and May.

The Western Tussock Moth is distributed sporadically through the basin, although it is most common in the foothill canyons and on the surrounding slopes; it also occurs along the coast.

OTHER SCIENTIFIC NAMES. *Hemerocampa vetusta.*

REFERENCES. Atkins, E. 1958. The western tussock moth, *Hemerocampa vetusta* (Bdr.), on citrus in southern California. *Journal of Economic Entomology,* vol. 51, pp. 762-765.

Riotte, J. C. E. 1973. Uber *Orgyra (O.) gulosa* und *Orgyra (O.) cana* (Lep.: Lymantriidae). *Entomologische Zeitschrift,* vol. 83, pp. 129-140.

## ■ OWLET MOTHS (Family Noctuidae)

THIS IS THE LARGEST FAMILY of Lepidoptera; there are scores of species in the basin. They are the common, dull-colored moths or "millers" seen around lights throughout most of the year. Some species even fly during the coldest nights in January.

Another group of Lepidoptera, the measuring worm moths, are also common locally around lights. But measuring worm moths have relatively slender bodies and rest with their wings upright (in the position characteristic of butterflies) or at right angles to the body and appressed to the substrate. In contrast, owlet moths have heavier bodies and rest with the

145. Western Tussock Moth male. Photograph by P. Bryant.

146. Western Tussock Moth female. LACM photograph.

147. Western Tussock Moth larva. LACM photograph.

wings held roof-like over the body.

A further characteristic of owlet moths is the tympanum or "ear" located below the hind wing in the wall of the thorax. This organ is composed of a thin taut membrane from which auditory nerve fibers run to the central nervous system. Recent investigations have shown that the owlet moth "ear" does function as a hearing organ and is tuned in particular to the ultrahigh-frequency sounds emitted by bats. When recorded bat noises were trained in the direction of study individuals, the moths performed evasive flight maneuvers.

The caterpillars of many species of the Noctuidae are plant pests. The armyworms and many of the fruit worms, stalk borers, and loopers, which are agriculturally and horticulturally important types, belong to this family. One group in particular, the cutworms, are notable for the surreptitious manner in which they cause injury: these caterpillars curl up and hide in the soil during the daytime and feed on roots or neatly nip off plants at ground level under the protection of darkness.

Most species pupate underground in an earthen cell. Upon emerging, the moth has to push through several inches of soil to reach the surface. Some of the more important Noctuidae occurring in our area are described in the following sections.

REFERENCES. Crumb, S. 1956. The larvae of the Phalaenidae. U.S. Department of Agriculture, Technical Bulletin, no. 1135.

Godfrey, G. 1972. A review and reclassification of larvae of the subfamily Hadeninae (Lepidoptera, Noctuidae) of America north of Mexico. U.S. Department of Agriculture, Technical Bulletin, no. 1450.

Roeder, K. D., and A. E. Treat. 1961. The detection and evasion of bats by moths. *American Scientist,* vol. 49, no. 2, pp. 135-148.

---

■ CORN EARWORM
*(Helicoverpa zea)*
Figures 148, 149

Cooks and chefs discover this species when they open a fresh ear of corn and find one or two brownish-striped caterpillars that are 1 1/4 inches (32 mm) long unashamedly stealing the terminal kernels of the cob and dropping copious quantities of unsightly brown fecal matter in the process. Corn Earworms destroy or devalue thousands of dollars worth of sweet corn in this manner each year.

The moth has a wing span of about 1 1/4 inches (32 mm) and is generally light yellowish-brown with dark spots on the fore wings and a brown marginal band on the hind wings. Females deposit their eggs on the silk of young corn ears, and the larvae gain access to the kernels by eating through the developing tassel. The striped caterpillars also cause severe damage to tomatoes, beans, bell peppers, cotton, geraniums, and other plants.

OTHER SCIENTIFIC NAMES. *Heliothis zea.*

REFERENCE. Hardwick, D. F. 1965. The corn earworm complex. *Memoirs of the Entomological Society of Canada,* no. 40, pp. 1-246.

148. Corn Earworm adult. Photograph by M. Badgley.

149. Corn Earworm larva. Photograph by P. Bryant.

■ CUTWORMS
Figures 150–165

Our common cutworms are listed below. All are general feeders and will eat most kinds of low-growing herbaceous vegetation, including grasses. The adults are all drab brown or gray moths with pale translucent hind wings. The fore wings have varied markings, but in many species there are noticeable round and kidney-shaped spots near the leading edge.

■ Black Cutworm *(Agrotis ipsilon;* Figures 150, 151). The larva, which is 1 1/4 to 1 3/4 inches (32 to 44 mm) long, is greasy gray to brown on top, with faint lighter-colored stripes, and is covered with rounded granules of different sizes. The adult (wing span is 1 1/4 to 2 in., or 32 to 50 mm) is called the Ipsilon Dart because of a small black wedge-shaped mark (like the Greek letter upsilon) in the middle of the fore wing just beyond the kidney-shaped spot.

■ Variegated Cutworm *(Peridroma saucia;* Figures 152, 153). A distinct pale yellow or orange dot on the top of several of the front segments distinguishes this larva (length 1 1/2 in., or 39 mm); frequently, a dark "W" mark or triangle is also found on the back of the eighth abdominal segment. In the adult (wing span 1 1/2 to 2 1/8 in., or 39 to 54 mm), the rounded wing spot is indefinite, but the kidney-shaped spot is sharp, and there is a dark blotch extending inwards from the leading edge of the wing.

■ Army Cutworm *(Euxoa auxiliaris;* Figures 154, 155). This cutworm (length 1 in., or 25 mm) is pale yellowish gray-brown with pale stripes and fine splotches of white and brown along its back; there are no prominent marks. The adult (wing span $1\frac{1}{2}$ in., or 39 mm) has very distinct fore wing markings, especially a dark bar running from the base to the center of the wing.

■ Dark-sided Cutworm *(Euxoa messoria;* Figures 156, 157). This cutworm is dull gray and darker on the sides than on the top (length $1\frac{1}{4}$ in., or 32 mm). A distinct white line runs down each side of the caterpillar just below the spiracles (the airholes). The Reaper Dart, as the adult is called (wing span $1\frac{3}{8}$ to $1\frac{1}{2}$ in., or 35 to 39 mm), has clear zig-zag transverse lines on the fore wings and well-defined rounded and kidney-shaped spots.

■ Granulate Cutworm *(Agrotis subterranea;* Figures 158, 159). This is a dingy brown cutworm (length $1\frac{1}{4}$ to $1\frac{1}{2}$ in., or 32 to 39 mm) that has a broad buff-gray stripe on its back that is subdivided into triangular areas on each segment and underlined by a narrow dark stripe on each side of the body; the skin is covered with bluntly conical granules of various sizes. The adult (wing span $1\frac{1}{4}$ to $1\frac{1}{2}$ in., or 32 to 39 mm) is known as the Subterranean Dart. It is identified by

150. Black Cutworm adult. Photograph by C. Hogue.

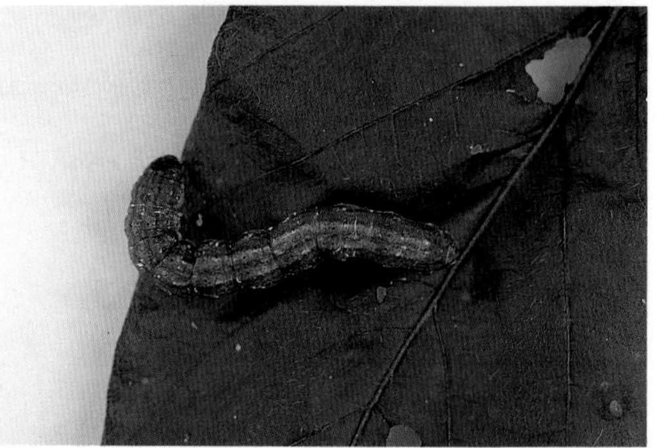

151. Black Cutworm larva. Photograph by G. Godfrey.

152. Variegated Cutworm adult. Photograph of LACM specimen by C. Hogue.

153. Variegated Cutworm larva. Photograph by C. Hogue.

154. Army Cutworm adult. Photograph of LACM specimen by L. Reynolds.

155. Army Cutworm larva. Photograph by J. Lafontein.

the black bar connecting the round and kidney-shaped spots on the fore wing. The species is common in dichondra lawns.

■ Armyworm *(Pseudaletia unipuncta;* Figures 160, 161). This is a greenish brown or gray larva (length 1 in., or 25 mm) with three yellowish stripes along the back and a broad, darker-yellow stripe on each side. The fore wing of the adult (wing span 1 3/8 to 1 7/8 in., or 35 to 47 mm) is tan flecked with black and often tinged with orange. A short dark line extends obliquely inward from the wing tip.

■ Beet Armyworm *(Spodoptera exigua;* Figures 162, 163). This larva (length 1 in., or 25 mm) is pale or olive green on top with a dark stripe along its back and a yellow stripe on either side of this; the entire underside is olive, pale yellow, or cream. The adult is much smaller than those of other cutworms (its wing span is 1 to 1 1/8 in., or 25 to 28 mm) and has a white round fore wing spot with an orange dot in the center.

■ Yellow-striped Armyworm *(Spodoptera ornithogalli;* Figures 164, 165). There is a pair of elongate black triangles on the back of most segments of this larva; the triangles are heavier in color on the segments toward the rear. The larva is usually marked on each side with a bright orange or yellow stripe just outside of these triangles. Young (and some mature) larvae

156. Dark-sided Cutworm adult. Photograph of LACM specimen by L. Reynolds.

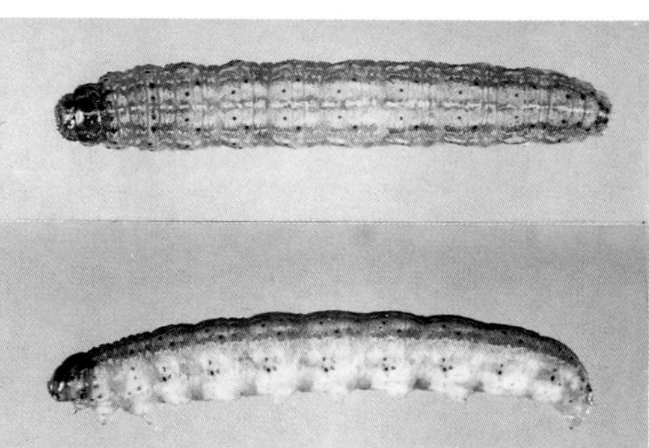

157. Dark-sided Cutworm larva, top and side views. Photograph by J. Lafontein

158. Granulate Cutworm adult. Photograph by C. Hogue.

159. Granulate Cutworm larvae, top and side views. Photograph by L. Brown.

160. Armyworm adult. Photograph of LACM specimen by L. Reynolds.

161. Armyworm larvae, side and top views. Photograph by L. Brown.

162. Beet Armyworm adult. Photograph by C. Hogue.

163. Beet Armyworm larva. Photograph by M. Badgley.

164. Yellow-striped Armyworm adult. Photograph by C. Hogue.

165. Yellow-striped Armyworm larva. Photograph by C. Hogue.

are largely black; the mature form is 1 1/2 inches (38 mm) in length. The fore wing of the moth (wing span 1 1/2 in., or 38 mm) has a complex harlequin pattern. There is a conspicuous oblique yellowish area extending in from the front margin of the wing and gradually disappearing beyond the wing's center.

REFERENCES. Todd, E. L., and R. W. Poole. 1980. Keys and illustrations for the armyworm moths of the noctuid genus *Spodoptera* Guenée from the Western Hemisphere. *Annals of the Entomological Society of America,* vol. 73, pp. 722-738.

Walton, W. R., and C. M. Packard. 1951. The armyworm and its control. United States Department of Agriculture, Farmer's Bulletin, no. 1850, pp. 1-10.

---

## ■ Loopers

Figures 166–169

These insects are called "loopers" because the larvae lack the two front pairs of abdominal prolegs and are forced to walk by arching the body in inchworm style (loopers have three pairs of abdominal prolegs, and sometimes vestiges of two additional pairs, instead of only two as in measuring worms). Adults usually have erect tufts of scales on the back of the thorax and abdomen.

Three species of loopers are common in the basin.

■ Alfalfa Looper *(Autographa californica;* Figure 166). The moth (wing span 1 1/4 to 1 1/2 in., or 32 to 38 mm) is generally mottled gray with a silver mark resembling the Greek letter "gamma" near the middle of the fore wing. It normally flies to flowers at dusk, but it sometimes pursues its activities in broad daylight. The larva (length 1 in., or 25 mm), which is a general feeder, is dark olive-green with a pale head and dark lines along the back and sides of its body.

■ Chocolate Looper *(Autographa biloba;* Figure 167). This moth (wing span 1 3/4 in., or 45 mm) has rich dark-brown fore wings marked with a distinct bilobed silver spot, which resembles a man's house slipper in silhouette. The larva (length 1 in., or 25 mm) is a general feeder on low-growing plants. It is variegated medium-green and white and has conspicuous body hairs.

■ Cabbage Looper *(Trichoplusia ni;* Figures 168, 169). This species is generally like other loopers, but the moths (wing span 1 5/8 in., or 41 mm) are more brownish. The fore wing is mottled light and dark brown; the central spots are separate. The larva (length 1 1/4 in., or 32 mm) is generally green and similar to that of the Alfalfa Looper but somewhat more clearly marked; the darker upper half of the body is separated from the lighter lower half by a fairly wide stripe. It prefers plants in the mustard family, especially cabbage, but will feed on the leaves of a variety of other plants. In southern California, the species is

166. Alfalfa Looper adult. Photograph by C. Hogue.

167. Chocolate Looper adult. Photograph by C. Hogue.

168. Cabbage Looper adult. Photograph by C. Hogue.

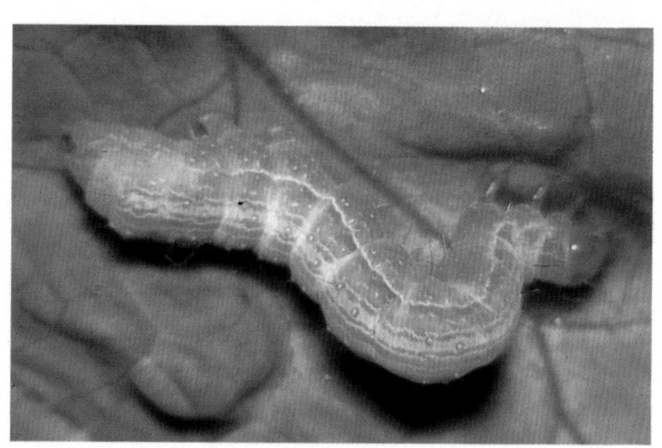

169. Cabbage Looper larva. Photograph courtesy of Los Angeles County Agricultural Commissioner's Office.

regarded as one of the most damaging of the insects that attack agricultural crops.

OTHER SCIENTIFIC NAMES. The Cabbage Looper has been referred to as *Autographa brassicae.*

REFERENCES. Eichlin, T. D. 1975. Guide to the adult and larval Plusiinae of California (Lepidoptera: Noctuidae). Occasional papers in Entomology of the California Department of Food and Agriculture (Division of Plant Industry-Laboratory Services), no. 21, pp. 1-73.

Eichlin, T. D., and H. B. Cunningham. 1978. The Plusiinae (Lepidoptera: Noctuidae) of America north of Mexico, emphasizing genitalic and larval morphology. Agricultural Research Service, U.S. Department of Agriculture, Technical Bulletin, no. 1567, pp. 1-122.

Shorey, H. H., and R. L. Hale, Jr. 1965. Cabbage looper, a principal pest of agricultural crops in California. *California Agriculture,* vol. 19, pp. 10-11.

Shorey, H. H., L. A. Andres, and R. L. Hale, Jr. 1962. The biology of *Trichoplusia ni* (Lepidoptera: Noctuidae), part 1: Life history and behavior. *Annals of the Entomological Society of America,* vol. 55, pp. 591-597.

---

■ BLACK WITCH
*(Ascalapha odorata)*
Figure 170

Because of its great size (a wing span of 6 to 7 in., or 15 to 18 cm), this moth might be mistaken for a member of the giant silk moth family, or even a bat. However, it is merely a giant cutworm or owlet moth.

Its caterpillar will not be found in the basin, although the moth occasionally appears around lights in the late summer or fall (August to October). During the 1984 Olympic Games, I saw large numbers of Black Witch moths in Los Angeles. Dozens of individuals, attracted to the stadium lights, flew into the Coliseum at night and took rest there under the bleachers; during the day they could be seen resting on the tunnel walls. Fortunately, they did not interfere with the sporting events. But they did provide an additional spectacle, especially when they were picked out of the air at dusk by the local sparrow hawks.

Both sexes have dark brown wings marked with transverse dark and light squiggly lines. There is a small round clear area midway along the leading edge of the fore wing and a partial eye-like spot at the apex of the hind wing. Females are much lighter in color than males and give off a violet reflection when rotated under a light.

The species is probably not a permanent resident of the continental United States, although it may persist for several successive generations in the Los Angeles area on ornamental acacias. The Black Witch is actually a tropical insect and is very common in Mexico, Central America, and northern South America. It habitually migrates northward, however, when warm humid weather prevails. Specimens (mostly males) have been taken, usually in a worn condition, in the extreme northern states and Canada. At home in the tropics, the caterpillars feed on acacias and cassias. They are nocturnally active, resting during the day on the bark or branches of the host tree.

The mature larva is large (almost 2 1/2 in., or 63 mm, long), plump, and cylindrical although slightly wider at the fourth segment. There are broad black broken stripes on the sides of its back and a very incomplete dark band running along its sides through the large black spiracles.

OTHER SCIENTIFIC NAMES. *Erebus odora, Otosema odora.*

REFERENCES. Comstock, J. A. 1936. Notes on the early stages of *Erebus odora* L. (Lepidopt.). *Bulletin of the Southern California Academy of Sciences,* vol. 35, pp. 95-98.

Grant, M. 1984. The 'bugging' of the '84 Olympics. *The San Diego Union* (newspaper), Friday, 3 August 1984.

Sala, F. P. 1959. Possible migration tendencies of *Erebus odora* and other similar species. *Journal of the Lepidopterists' Society,* vol. 13, pp. 65-66.

---

■ IRENE UNDERWING
*(Catocala irene)*
Figures 171, 172

Underwing moths, so named because of the brilliant color of the hind wings, are rare in our area compared to the eastern United States. This moderately large species (its wing span is 2 1/2 to 3 in., or 63 to 76 mm) has bright red or orange hind wings, which it hides when at rest under fore wings marked precisely like rough tree bark. It usually is seen only when it comes to light or when accidentally alarmed during the daytime—the flashing color of the hind wings attracts the eye.

The larva is large (body length 1 5/8 in., or 41 mm) and mottled brown. It has five pairs of prolegs and a thick marginal fringe, which is held in contact with the substrate when the larva is at rest. This posture has the effect of erasing any semblance of a body shadow and enhancing the camouflage provided by the coloration of the larva's back. The larval food plant is willow *(Salix).*

---

■ WALNUT UNDERWING
*(Catocala piatrix)*
Figures 173, 174

This large underwing moth (its wing span is 2 7/8 to 3 in., or 73 to 76 mm) flashes bright orange hind wings in flight. The fore wings are bark-patterned like those of other species of the genus.

170. Black Witch. Photograph by C. Hogue.

171. Irene Underwing. Photograph by C. Hogue.

172. Irene Underwing larva. Photograph by D. Hawks.

173. Walnut Underwing. Photograph of LACM specimen by C. Hogue.

On the hottest days of summer (in July and August) adults may been seen flying during the day in the neighborhood of the larval food plant, which is the California Walnut *(Juglans californica)*. The larva is similar to that of the Irene Underwing.

■ MOON UMBER
*(Zale lunata)*
Figures 175, 176

The coloration of the adult of this species is variable; a dark form is shown in Figure 175. The more common light form has wide splotches of silver intruding on the wings from their margins. Both varieties are medium-sized moths, with wings spans of 1 3/8 to 2 inches (35 to 50 mm). The caterpillar feeds primarily on willow but also on oak, elm, pyracantha, and other woody plants. It is pinkish brown when full grown (body length nearly 2 in., or 50 mm), with a pair of conspicuous humps on the next to the last abdominal segment. When mature it has a full set of short abdominal legs (that is, five pairs), but it walks with a looping motion. This behavior plus the lack of a marginal fringe distinguishes it from the larvae of underwing moths.

■ PALM MOTH
*(Litoprosopus coachella)*
Figure 177

This moth is a resident of Los Angeles because its food plants, the fan palms *(Washingtonia filifera* and *W. robusta)*, are such widely grown parkway trees. The larva feeds on the reticulated fibers at the base of the stems and on the flowers and fruits. It sometimes makes itself a pest by entering homes in search of a site in which to spin its loose cocoon. It may cut fabrics (curtains or clothing in storage) with its mandibles during its domestic wanderings.

The wings of the adult are entirely cream with a satin sheen; they span 3/4 to 1 inch (20 to 26 mm). Their only conspicuous marks are twin black 0's on the margin of the hind wings near the inner angle.

REFERENCE. Comstock, J. A. 1956. Is this a new, and giant clothes moth? *Bulletin of the Southern California Academy of Sciences,* vol. 55, pp. 51-53.

## ■ TIGER MOTHS (Family Arctiidae)

TIGER MOTHS resemble owlet moths structurally and even possess a thoracic ear. But tiger moths usually sport gaudy color patterns. The caterpillars are typically covered with a dense growth of stiff hairs and are known popularly as "wooly-bears." Usually the caterpillar incorporates some of these hairs into its cocoon, which it spins above ground in some protected spot. The hairs are not poisonous and seldom irritate human skin.

174. Walnut Underwing larva. Photograph by D. Hawks.

175. Moon Umber, dark form. Photograph by C. Hogue.

176. Moon Umber larva. Photograph by C. Hogue.

177. Palm Moth larva (left; top view) with fan palm flower and adult. Photographs by R. Pence.

There are several common species of tiger moths living in the basin; the caterpillars of all but one of them are solid black or dark brown and are indistinguishable from each other except in detail.

■ ACREA MOTH
*(Estigmene acrea)*
Figures 178, 179

This moth is medium sized (its wing span is 2 to 2 1/2 in., or 50 to 64 mm) and is probably the commonest tiger moth in the United States and in the basin, where it appears during the summer. The sexes differ in the color of the hind wings, which are orange-yellow in the male and white in the female. The abdomens of both are orange-yellow with a median row of transverse black spots. The caterpillars, which will eat almost any plant, may at times be so numerous that they are considered pests.

The larva is a typical wooly-bear (slightly over 2 in., or 50 mm, long), but its hairs are rather sparse. Its integument is also fairly light and has colored spots; the centers of the spiracles are white. In the fall, the caterpillars may be seen crossing highways in agricultural districts in large numbers. When mature, they overwinter in plant debris. Pupation occurs in the spring in a loose gray cocoon constructed of silk and larval hairs. The larva is called the Salt Marsh Caterpillar because it infests and destroys grasses in marshy areas on the Atlantic Coast.

Males extend abdominal glands, which produce chemicals that act as a mating call. Both females and males respond to this stimulation, and aggregations of Acrea Moths may form as a result.

REFERENCE. Willis, M. A., and M. C. Birch. 1982. Male lek formation and female calling in a population of the arctiid moth *Estigmene acrea. Science,* vol. 218, pp. 168-170.

■ VESTAL TIGER MOTH
*(Maenas vestalis)*
Figure 180

The Vestal Tiger Moth is immediately recognizable by the conspicuous red fore legs of the adult, which is otherwise pure white except for a few scattered black flecks on the wings. It is slightly smaller than the Acrea Moth (wing span about 2 in., or 50 mm). The adult flies during the spring and is common at lights. The black wooly-bear larva feeds on a large variety of soft-leaved garden and field plants.

■ PAINTED ARACHNIS
*(Arachnis picta)*
Figures 181, 182

The wings of this species are distinctively marked with broad wavy gray lines over a white background on the fore wings and a red background on the hind wings. The adults, which are moderate sized (wing span 2 in., or 50 mm), may be seen at night in late September to

178. Acrea Moth female. LACM photograph.

179. Acrea Moth larva. Photograph by J. Hogue.

180. Vestal Tiger Moth. Photograph by C. Hogue.

181. Painted Arachnis. Photograph by C. Hogue.

October resting by porch lights or on store fronts.

The mature larva is about 1 $1/2$ inches (40 mm) long and is densely covered with stiff black hairs; the head is black. The caterpillar feeds nocturnally on a great variety of weedy plants, including wild radish, Wandering Jew, and *Acanthus.* It hides during the day, sometimes retreating into the soil, and it rolls into a ball when disturbed. It develops during the winter and then is somewhat dormant (although active, it does little feeding) until late the following summer, when it pupates; on a warm fall evening, the adult emerges. Individual caterpillars occasionally pupate immediately after maturing and pass the summer in the pupal stage.

---

■ EDWARD'S GLASSY-WING
*(Hemihyalea edwardsii)*
Figure 183

This is a large tiger moth (wing span 2 $1/2$ in., or 64 mm) that is most common in mountain canyons, the habitat of its principal food plant, the Coast Live Oak *(Quercus agrifolia).* The adult is present in the fall. The wooly-bear larva, which is up to 2 $3/4$ inches (70 mm) long and has an oversized glossy brown head, hides in bark crevices during the day and crawls up the trunk to feed on the leaves after dark. Pupation occurs in the typical tiger moth cocoon of silk and larval hairs, which is attached to the twigs and leaves of the host plant.

---

■ SPOTTED HALISIDOTA
*(Halisidota maculata)*
Figures 184, 185

This is one tiger moth that has a distinctive caterpillar. Though still classed as a wooly-bear, the larva has a wide belt of soft yellow hairs around the middle of the body and, to the front and rear, black zones with tufts of extra long white hairs. It is about 1$1/2$ inches (39 mm) long when mature. Larvae may be very numerous on willows in the fall in some areas. The adults are generally yellowish brown in color with a marbled pattern on the fore wings that is a slightly darker brown; the wing span is about 1$1/2$ to 1$3/4$ inches (38 to 45 mm).

---

■ MEXICAN TIGER MOTH
*(Notarctia proxima)*
Figure 186

The fore wings of this moth are crisscrossed with white lines so that only discrete triangles and squares of black color remain. They span about 1$1/4$ inches (32 mm); the hind wings of the males are white, and those of the females are red. Both sexes are often attracted to light during the summer. The wooly-bear larva is a general feeder on low-growing herbaceous plants.

OTHER SCIENTIFIC NAMES. This species and the following (Hewlett's Tiger Moth) were formerly placed in the genus *Apantesis.*

182. Painted Arachnis larva. Photograph by C. Hogue.

183. Edward's Glassy-wing. Photograph by C. Hogue.

184. Spotted Halisidota. Photograph of mounted specimen by D. Tiemann.

185. Spotted Halisidota larva. Photograph by C. Hogue.

186. Mexican Tiger Moth male. Photograph by C. Hogue.

■ HEWLETT'S TIGER Moth *(Notarctia hewletti)* Figure 187

This moth is very similar to the Mexican Tiger Moth in size and pattern. However, this species is variable in color, and the hind wings of both sexes are yellow (sometimes tinged with red) with a greater amount of black than those of the Mexican Tiger Moth; also, the white bars of the fore wings are much narrower, and the pattern of black polygons is more fragmented because of the presence of additional fine streaks of white. The uncommon adult flies during late spring; males come to light in the very early morning, and females are diurnal.

The larva is a general feeder on low-growing herbaceous plants and is yet another of the wooly-bear type.

OTHER SCIENTIFIC NAMES. The species is closely related to the more northern Ornate Tiger Moth *(N. ornata),* which our local type was cited to be in the first edition of this book.

## ■ DIOPTID MOTHS (Family Dioptidae)

■ CALIFORNIA OAK MOTH *(Phryganidia californica)* Figures 188-190

Much of the decline of the Coast Live Oak in the Los Angeles area is attributable to the ravages of this species. The caterpillars, which have a conspicuous smooth brown or reddish head, are especially destructive in dry years, when they may defoliate trees over large areas. They also attack Cork Oak.

The pupae are naked and shining whitish or yellowish with black streaks. They hang head downwards like butterfly chrysalids, often in large numbers, from the under surfaces of leaves, limbs, trunks of trees, and nearby objects.

The wings of the adult, which expand up to 1¼ inches (32 mm), are a uniform pale gray brown with darker veins; the males have a faint yellowish patch near the middle of the fore wing. Although development is more or less continuous throughout

187. Hewlett's Tiger Moth. Photograph of LACM specimen by C. Hogue.

188. California Oak Moths copulating; male at right. Photograph by C. Hogue.

189. California Oak Moth larva. Photograph by J. Donahue.

190. California Oak Moth pupae. Photograph by J. Donahue.

the year, the adults of a conspicuous spring brood are usually seen fluttering weakly around the host trees in June or July.

REFERENCES. Brown, L. R., and C. O. Eads. 1965. California oak moth. Pages 44-47 in A technical study of insects affecting the oak tree in southern California, edited by L. Brown and C. Eads. California Agriculture Experiment Station, Bulletin no. 810.

Harville, J. P. 1955. Ecology and population dynamics of the California Oak Moth *Phryganidia californica* Packard (Lepidoptera: Dioptidae). *Microentomology*, vol. 20, pp. 83-166.

Horn, D. T. 1974. Observations on primary and secondary parasitoids of California Oakworm, *Phryganidia californica*, pupae (Lepidoptera: Dioptidae). *Pan-Pacific Entomologist*, vol. 50, pp. 53-59.

Milstead, J. E. 1959. Observations on the host spectrum of the California oakworm, *Phryganidia californica* Packard (Lepidoptera: Dioptidae). *Pan-Pacific Entomologist*, vol. 65, pp. 50-57.

## ■ MEASURING WORM MOTHS
(Family Geometridae)

THE COMMON NAME of this family refers to the mode of locomotion employed by the caterpillars. Because it does not have legs on the intermediate abdominal segments, the larva can hold on to the substrate only with the thoracic legs at the front end of the body and with two pair of abdominal legs at the back. To advance, the rear set of legs is released and brought up behind the fore set, the body arching up strongly (or "looping") at the same time. Then the front set is released and the front portion of the body is stretched forward. With this gait, the caterpillar appears to be "inching" along, measuring the extent of its progress. Unlike loopers, measuring worms have only two pair of abdominal legs and lack even any vestiges of the front three pair that are found in loopers and other noctuid larvae.

The adults of most species somewhat resemble owlet moths in that they are drably colored and possess a pair of hearing organs (pockets with auditory membranes; in measuring worm moths, these organs are on either side of the first abdominal segment rather than on the thorax). Most measuring worm species have slender bodies as adults and rest with the wings against the substrate. Owlet moths, in contrast, have thicker bodies and hold their wings in a roof-like fashion over the body.

■ OMNIVOROUS LOOPER
*(Sabulodes aegrotata)*
Figures 191-193

This is a prolific species that breeds locally on a wide variety of plants throughout most of the year. The caterpillars, which are light yellowish-green and marked with irregular broken longitudinal lines, are common on English Ivy and are a pest on avocado. The species is much more abundant in city gardens than in areas of native vegetation. During the day, they usually hide in a loose webby shelter that they spin in a leaf fold or between two leaves; pupation takes place in a similarly flimsy white cocoon. Feeding is primarily a nighttime activity.

The moth is medium-sized (wing span 1¾ in.,

191. Omnivorous Looper adult. Photograph by C. Hogue.

192. Omnivorous Looper larva. Photograph by M. Badgley.

193. Omnivorous Looper pupa in cocoon. LACM photograph.

or 44 mm) and generally a pale creamy brown with dark speckles throughout the upper wing surfaces. There are also two faint widespread transverse lines across the wings. When the moth is at rest, the wings are held out to the sides, with the edges pressed tightly against the substrate.

OTHER SCIENTIFIC NAMES. *Sabulodes caberata.*

REFERENCE. Bailey, J. B., M. P. Hoffman, and K. N. Olsen. 1988. Blacklight monitoring of two avocado insect pests. *California Agriculture,* vol. 42, no. 2, pp. 26-27.

---

■ McDunnough's Pero
*(Pero macdunnoughi)*
Figure 194

This moth, which has a wing span of 1½ to 1⅝ inches (39 to 42 mm), sometimes rests with its wings together over the back in butterfly fashion, although it also sits with the wings to the sides and the tips crumpled like dead leaves. It is nocturnal and has hair-like antennae, so there is no confusion as to its proper place among the moths. The larva is another general feeder on trees and shrubs; but it is usually found on Red Berry Buckthorn *(Rhamnus crocea),* Flat Top *(Eriogonum fasciculatum),* and California Sage *(Artemisia californica)* among the native plants, and on apricot and privet among the cultivated.

REFERENCE. Poole, R. W. 1987. A taxonomic revision of the New World moth genus *Pero* (Lepidoptera: Geometridae). U.S. Department of Agriculture, Technical Bulletin, no. 1698, 257 pp.

## ■ TINEID MOTHS (Family Tineidae)

---

■ Webbing Clothes Moth
*(Tineola bisselliella)*
Figure 195

Of the several species called "clothes moths" because they do damage to fabrics, this is by far the most common. In nature, and doubtlessly before the advent of man, this insect scavenged on dry animal debris, including fur, skin, feathers, and horn. When the moth is in our homes, it attacks clothes, carpets, rugs, upholstered furniture, and other objects of animal origin (silk, paper, or other substances of plant origin are usually not eaten, and neither are synthetic fabrics, such as nylon, unless they are woven in combination with wool).

It is the larva and not the adult clothes moth that eats the fabric. The mouthparts of the adult would never permit it to chew such material, whereas the jaws of the caterpillar are adapted for the purpose. The colorless larva, which is tiny (less than ¼ in., or 6 mm, long), is especially attracted by stains on wool; once it finds a suitable spot, it settles down and feeds until a hole forms.

The adult Webbing Clothes Moth is also very small (its length at rest is ⅜ in., or 9 mm; its wing span is about ½ in., or 13 mm). It has shiny unmarked satin

194. McDunnough's Pero. Photograph by C. Hogue.

195. Webbing Clothes Moth adult and larvae. Photograph by R. Pence.

196. Case-bearing Clothes Moth larvae in cases. Photograph courtesy Los Angeles County Agricultural Commissioner's Office.

wings; its body is covered with brilliant scales, and there is a fluff of reddish golden hair on the head and long fringe on the hind wings.

---

■ CASE-BEARING CLOTHES MOTH
*(Tinea pellionella)*
Figure 196

The larva of this clothes moth is distinguished from that of its more common relative, the Webbing Clothes Moth, by the case it carries. The structure is an elongate flattened sac that is made of silk and is slightly splayed at the open end. The adult is small (wing span 1/4 in., or 6 mm) and resembles the Webbing Clothes Moth in its shiny grayish-brown to gray color; it has an obscure, slightly darker spot in the outer third of its fore wing.

■ MEXICAN JUMPING BEAN MOTH
*(Cydia saltitans)*
Figure 197

The larva of this moth supplies the "jump" in the Mexican jumping bean. The bean is actually the seed of the Arrowplant *(Sebestiana pringlei)*, which is native to northwestern Mexico; seeds that are infested by the small grub-like *C. saltitans* larvae will lurch as the caterpillars turn and jerk violently in their confined quarters. The action is enhanced by heating the beans, and they may actually jump off the surface.

The larva gets inside the bean by eating its way into the young soft pod soon after the pod forms at the base of the flower. The bean ceases to jump when the larva dies or pupates.

The moth that emerges from the bean is small (1/2 in., or 13 mm, wing span) and mottled gray in color. The species is a very close relative of the injurious Codling Moth *(Cydia pomonella)*, whose larva lives in apples and walnuts.

Jumping Beans, which are also called "devil's beans" or *brincadores* ("jumpers" in Spanish), are imported in considerable numbers and sold as curiosities in a few places in the United States. When in season, in June and July, they usually may be purchased on Olvera Street in Los Angeles.

REFERENCES. Hess, L. 1945. The jump in the jumping bean. *Nature,* vol. 38, pp. 473-475.

Truxal, F. S. 1964. Los Brincadores (The Jumpers). *Los Angeles County Museum Quarterly,* vol. 3, pp. 14-15.

■ CODLING MOTH
*(Cydia pomonella)*
Figure 198

The Codling Moth, which was introduced into California in 1872 from other parts of the United States, is originally from Europe. It is a small moth (wing span 5/8 to 3/4 in., or 9 to 19 mm) and is generally bluish-gray in color. The apex of the fore wing has a conspicuous coppery spot containing two irregular golden lines; the rest of the wing is marked by numerous dark wavy lines.

The dark-headed, pale or pinkish caterpillars may obtain a length of 1 inch (25 mm) or more. They burrow into a variety of fruits, primarily apples, pears, and English walnuts, causing major economic damage.

Locally, there may be as many as three broods per year, beginning in the spring with the emergence of adults that hatch from the cocoons of overwintering larvae. The females lay their eggs on the new tree growth, which provides food for the larvae of the first brood. Injury to the fruits is caused by feeding of the larvae of subsequent broods.

197. Mexican Jumping Bean Moth adult (LACM specimen) and cutaway view of larva in bean (a reconstruction). LACM illustration.

198. Codling Moth. Photograph by M. Badgley.

199. Greater Wax Moth adult (LACM specimen) and larvae. Photographs by L. Reynolds and L. Brown (larvae).

■ GREATER WAX MOTH
*(Galleria mellonella)*
Figure 199

The larva of this wax moth lives in Honey Bee hives, boring through the comb, feeding on pollen and waste materials in the cells, and destroying young bees in the process. Because of this habit, the insect is on the beekeeper's black list, although it thrives only in abandoned or weakened bee colonies (bees in strong colonies normally destroy the larvae). Newly hatched larvae feed on stored honey and pollen. When more mature, the caterpillars tunnel into other cells, eating wax and organic debris. They also make silk tunnels throughout the hive.

When full grown, larvae creep into a corner of the hive or into another protected niche nearby and spin tough white cocoons of silk mingled with their own dry fecal pellets. They later pupate in these cocoons.

The moth is medium-sized (its wing span is about 1 in., or 50 mm) and generally gray in color; the outer margin of the wing is notched just below the tip. It is usually never seen except by the apiculturist. However, it occasionally infests the honeycombs of wandering swarms that have settled in chimneys, attics, or other nooks and crannies around houses. From this source, it sometimes strays indoors, perhaps attracted by houselights, and makes itself a nuisance.

The species is not native. No doubt it was introduced from the Old World along with its Honey Bee hosts.

REFERENCES. Nielsen, R. A., and D. Brister. 1979. The greater wax moth: Adult behavior. *Annals of the Entomological Society of America,* vol. 70, pp. 101-103

_______. 1979. Greater wax moth: Behavior of larvae. *Annals of the Entomological Society of America,* vol. 72, pp. 811-815.

Wyndham, R. 1959. The curious ways of the wax moth. *Nature,* vol. 52, pp. 189-191.

■ NAVEL ORANGE WORM
*(Amyelois transitella)*
Figure 200

Despite its common name, this moth does more damage to walnuts and almonds than to oranges. It is small (wing span 7/8 in., or 23 mm), with pale elongate grayish wings marbled with brown and black. There is an oblique interrupted line across the base of the wings; toward the front of the wing, the line is broadened and dark, forming a spot. The larva is 3/4 inch (20 mm) long and pink.

The females place their eggs on the green husks of the nut. The newly hatched larvae then tunnel through the shell and feed on the meat.

200. Navel Orange Worm moth. Photograph of LACM specimen by C. Hogue.

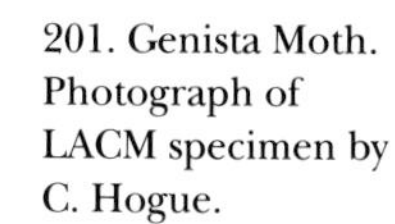

201. Genista Moth. Photograph of LACM specimen by C. Hogue.

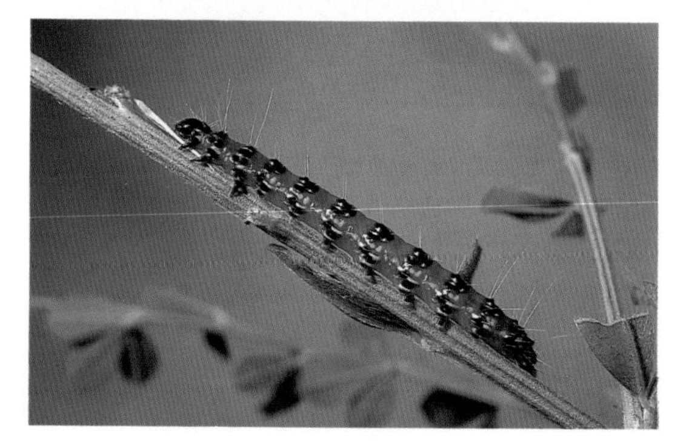

202. Genista Moth larva. Photograph by C. Hogue.

■ GENISTA MOTH
*(Uresiphila reversalis)*
Figures 201, 202

This is a large pyralid (its wing span is 1⅛ in., or 28 mm), with reddish-brown to dull brown fore wings; the tips are darkened as is a curve of dots crossing the outer third of the wing and a large spot near the leading edge. The hind wings are yellow with a dark border that is broad at the wing tip and becomes narrower and disappears near the wing's center. Adults are attracted to lights.

When mature the larva is medium-sized (length 1⅛ in., or 28 mm) and colorfully marked. It is generally yellow with three rows of black papillae that are white at the upper and lower edges; each papilla bears a long white hair. A series of lemon-yellow dots runs along the side of the body just below the spiracles, and beneath each spiracle is a white dash. Larvae feed on the leaves, shoots, and bark of various leguminous shrubs, especially Genista brooms such as Scotch broom, an invasive weed. They are gregarious and live in webs. They pupate off the host, usually on the ground, in loosely woven cocoons.

REFERENCE. Comstock, J. A., and C. M. Dammers. 1933. Notes on the life histories of two California lepidopterous insects. *Bulletin of the Southern California Academy of Sciences,* vol. 32, pp. 27-37.

---

■ PANTRY MOTHS
Figures 203–205

Several small species of pyralid moths commonly occur in the home. They usually originate in the kitchen, where the larvae infest stored cereals, flour, corn, oatmeal, pet food, bird seed, herbs and spices, and other dry food products of plant origin. Their presence is often indicated by copious masses of webbing on the inside of food containers and on the food's surface. The larvae can chew their way through cellophane and thin plastic bags.

The adults of the various species are fairly easily identified by their size and wing color pattern, but the larval stages can only be distinguished by the arrangement of minute hairs on the body (see Corbet and Tams, 1943).

■ Meal Moth *(Pyralis farinalis;* Figure 203). The adult's wing span varies from 5/8 to 7/8 inch (15 to 23 mm). The fore wings bear a central light brown zone bounded on both sides by darker brown; thin white lines separate the three color fields. The larvae are dirty white except for the dark brown head and first thoracic segment. They live in cases or tunnels of silk mixed with food debris; they prefer stale damp material, but all stored vegetable products are susceptible to their ravages.

■ Indian Meal Moth *(Plodia interpunctella;* Figure 204). The wing span of this species is only about 5/8 inch (15 mm). The fore wings are divided into two color fields: a small silvery white or gray region near the base and a larger outer bronzy region.

Only the head of the larva is pale yellowish brown; the rest of the body is white. The caterpillars can be found upon an endless assortment of foodstuffs, always in a mass of silken webbing.

■ Mediterranean Flour Moth *(Anagasta kuehniella;* Figure 205). The habits and food preference of the larva of this pantry moth are similar to those of the Indian Meal Moth, as is its coloration: its body is pinkish or whitish, and its head is dark. But the larva

203. Meal Moth. Photograph by C. Hogue.

204. Indian Meal Moth. Photograph by C. Hogue.

205. Mediterranean Flour Moth. Photograph by M. Badgley.

of this species lives in silken tubes rather than in masses of webbing.

The wing span of the adult is slightly less than 1 inch (24 mm). The fore wings are variegated gray with obscure wavy lines.

REFERENCES. Corbet, A. S., and W. H. T. Tams. 1943. Keys for the identification of the Lepidoptera infecting stored food products. *Proceedings of the Zoological Society of London* (series B), vol. 113, pp. 55-148.

Donohoe, H. 1937. Indian meal moth in California. *Journal of Economic Entomology,* vol. 30, pp. 680-681.

Hinton, H. W. 1943. The larvae of the Lepidoptera associated with stored products. *Bulletin of Entomological Research,* vol. 34, pp. 163-212.

Tzanakakis, M. E. 1959. An ecological study of the Indian-meal moth *Plodia interpunctella* (Hübner) with emphasis on diapause. *Hilgardia,* vol. 29, pp. 205-245.

## ■ INCURVARIID MOTHS (Family Incurvariidae)

■ YUCCA MOTH
*(Tegeticula maculata)*
Figure 206

This moth is well known to students of ecology for the part it plays in the pollination of yucca plants. The only yucca growing naturally in our area is the Common Yucca or Quixote Plant *(Yucca whipplei);* the species, which is a conspicuous member of the coastal sage plant association, has a large basal cluster of sharp-tipped, narrow leaves from the center of which grows a tall flower stalk.

In the early spring the Yucca Moth is attracted to the flowers of this plant. With a piercing ovipositor, females insert their eggs in the ovary of the flower, where the seeds will later develop. Upon emerging from these eggs, the caterpillars feed on some of the seeds of each pod. The remaining seeds mature normally, because the flower was pollinated by the female at the time of her egg-laying visit. The female moths even have mouthpart structures especially modified to facilitate the gathering and transport of pollen. This moth-yucca relationship constitutes a case of *symbiosis,* a partnership between two different kinds of organisms wherein both species benefit by way of their complementary needs. In this case, the relationship is obligatory, as the plant is unable, or rarely able, to pollinate itself by any other means.

The moths are small (their wing span is $^3/_4$ to $^7/_8$ in., or 20 to 23 mm), and their fore wings are elongate and pure white with black markings at the tips. On spring nights, the moths may be attracted to lights that are in the vicinity of flowering yuccas. The larvae burrow into the soil at the base of the plant to pupate.

The larvae of the Bogus Yucca Moths *(Prodoxus* species) live in the bases of yucca seed capsules and in the main flowering stalk, not in the seeds. To pupate, the larvae burrow into the pithy yucca stalk rather than into the ground. The Bogus Yucca Moths play no part in the plant's pollination, but because of their similarity and proximity to true Yucca Moths they are often confused with them.

REFERENCES. Kelley, T. E. 1986. Inter- and intralocular distribution of yucca moth larvae in *Yucca whipplei*

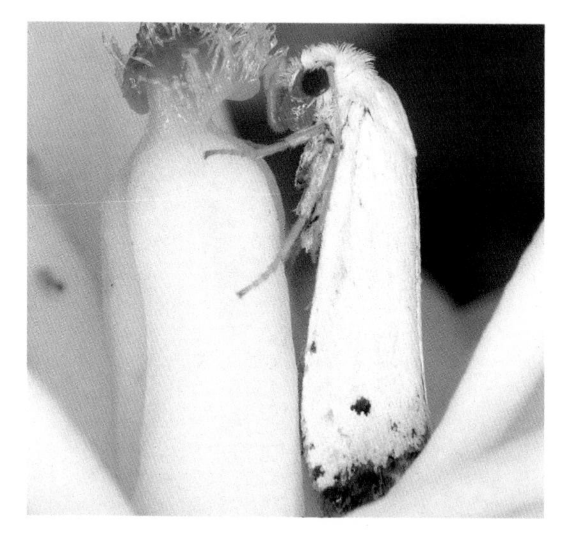

206. Yucca Moth female placing pollen on style of yucca flower. Photograph by D. Frack.

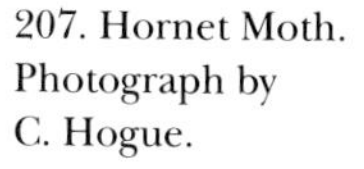

207. Hornet Moth. Photograph by C. Hogue.

(Agavaceae). *Bulletin of the Southern California Academy of Sciences,* vol. 85, pp. 173-176.

Powell, J. A., and R. A. Mackie. 1966. Biological inter-relationships of moths and *Yucca whipplei. University of California Publications in Entomology,* vol. 42, pp. 1-59.

Riley, V. C. 1892. The yucca moth and yucca pollination. 3rd annual report of the Missouri Botanical Garden, pp. 99-158.

## ■ CLEAR-WINGED MOTHS (Family Sesiidae)

■ HORNET MOTH
*(Synanthodon robiniae)*
Figure 207

This moth is a near-perfect mimic of a wasp. There is little question that the species derives strong protection from its resemblance in form, color, and behavior to wasps of the genus *Polistes.* The larva burrows in the live wood of poplar and willow. When mature, it chews an opening to the outside and then withdraws back into the tree to pupate.

The mature pupa squirms out of the exit to

emerge. Dry empty cases are often seen protruding from branches and are a sign of the moth having completed a generation in the host.

OTHER SCIENTIFIC NAMES. *Paranthrene robiniae.*

## ■ CARPENTER MOTHS (Family Cossidae)

■ **CARPENTER WORM** *(Prionoxystus robiniae)* Figure 208

This wood-boring caterpillar may require two to three years to complete its development. It is large (its body length is 2 to 3 in., or 51 to 76 mm, long) and lacks the lobed legs of other lepidopterous larvae. It is creamy white with a dark brown head and several short stout hairs on each body segment. It feeds at first in the sap wood of its host and then migrates to the heartwood. The larva expels large masses of sawdust and fecal droppings from its exit holes, and this material accumulates beneath the tree and indicates that damage is occurring.

In the basin, the Carpenter Worm is commonly found in English elm, maple, willow, and cottonwood. Over a period of years, the tunneling of several larvae may greatly reduce the vitality of a large hardwood tree or even kill it. A tree may be reinfested year after year, and the burrows may expose the tree to further damage from wood-rot fungi.

The moth, which emerges in April, is large (wing span 3 in., or 75 mm) with mottled gray-black fore wings; the body is very heavy, and the hind wings of males are orange.

REFERENCE. Solomon, J. D., and C. J. Hay. 1974. Annotated bibliography of the carpenter worm, *Prionoxystus robiniae.* United States Department of Agriculture, Forest Service, General Technical Report SO-4, pp. 1-13.

## ■ PLUME MOTHS (Family Pterophoridae)

THE ADULTS OF MEMBERS OF THIS FAMILY frequently arouse the curiosity of local citizens, who do not recognize them as moths. At rest they hold their slender rolled-up wings at right angles to the body and form the figure of a cross (Figure 209).

Plume moths get their name from the shapes of the wings, which are divided in a fan-like fashion; each fore wing is notched at the end, and each hind wing is deeply incised into three paddle-shaped lobes. The adults are further characterized by their long slender legs, each hind pair with two pair of large spurs on the undersides.

Several species and genera occur locally. They are all small, the largest only a little more than 3/4 inch

208. Carpenter Worm moth. Photograph of LACM specimen by C. Hogue.

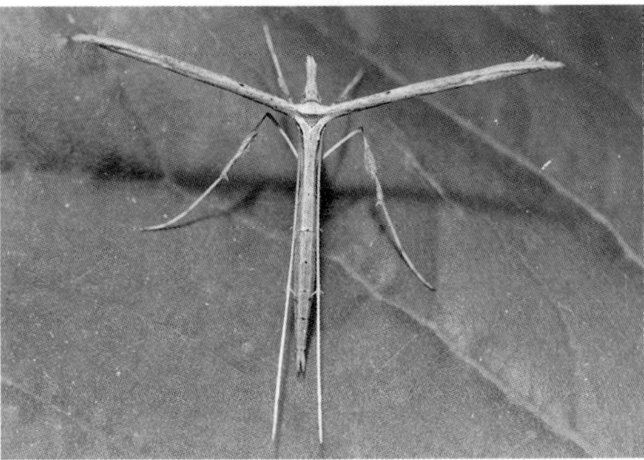

209. Plume moth *(Oidaematophorus longifrons)*. LACM photograph.

(20 mm) in body length and 1/2 to 3/4 inch (13 to 20 mm) in wing span.

A local representative is the Ragweed Plume Moth *(Adaina ambrosiae)*.

REFERENCE. Goeden, R. D., and D. W. Ricker. 1976. Life history of the Ragweed Plume Moth, *Adaina ambrosiae* (Murtfeldt), in southern California. *Pan-Pacific Entomologist,* vol. 52, pp. 251-255.

## BUTTERFLIES

BUTTERFLIES—ALONG WITH BIRDS AND FLOWERS—have always been especially appreciated by lovers of natural beauty. Even in ancient times, butterflies and their metamorphoses were known and admired, so it is not surprising that the butterfly was adopted as the symbol of the soul, or Psyche, in Greek religion and poetry. In classic art, Psyche—the wife of Cupid—is often represented as a girl with butterfly wings.

A surprisingly large number of butterfly species live in the basin, although almost all are much less common today than in former times. It is speculated that this reduction in numbers has come about through loss of habitat from urbanization and possibly also through heavy and indiscriminate use of insecticides during the two decades following the end of World War II.

REFERENCES. Audubon Society (R. M. Pyle, consulting lepidopterist). 1981. Field guide to North American butterflies. New York: A. A. Knopf.

Comstock, J. A. 1927. Butterflies of California. Published by the author, Los Angeles.

Emmel, T. C., and J. F. Emmel. 1973. *The butterflies of southern California.* Science Series, no. 26. Los Angeles: Natural History Museum of Los Angeles County.

Garth, J. S., and J. W. Tilden. 1986. California butterflies. California Natural History Guide Series. Berkeley, Calif.: University of California.

Mattoni, R. 1990. Butterflies of greater Los Angeles [color poster]. Beverly Hills, Calif.: Lepidoptera Research Foundation [c/o Rudolph Mattoni].

Scott, J. A. 1986. *The butterflies of North America.* Stanford, Calif.: Stanford University.

Tilden, J. W., and A. C. Smith. 1986. *A field guide to western butterflies.* Peterson Field Guide Series. Boston: Houghton Mifflin.

## ■ SWALLOWTAIL BUTTERFLIES
(Family Papilionidae)

■ WESTERN TIGER SWALLOWTAIL *(Papilio rutulus)* Figures 210, 211

This large species (its wings span up to 4 in., or 10 cm), with its brilliant yellow wings crossed by black stripes, is one of our most conspicuous butterflies.* Although more prevalent in riparian situations in rural and wild areas, it is not uncommon in urban and suburban residential districts.

The mature caterpillars are all green except for two "eye" spots on the back of the swollen thorax and a transverse black-margined yellow band a short distance behind the eye spots on the tapering abdomen; there are small lavender spots on each segment. The principal local food plants for the larvae are poplar, willow, sycamore, and alder *(Alnus* species). The chrysalis of the Western Tiger and other swallowtail butterflies rests in an upright position against a tree trunk or other vertical structure. It is attached at the bottom or tail end by a button of silk and is supported by a thread-like girdle around the middle, an arrangement that suggests a window washer standing on a building ledge and leaning back on his safety belt.

*A similar relative, the Pale Swallowtail *(Papilio eurymedon),* has white rather than yellow wings.

210. Western Tiger Swallowtail. Photograph by C. Hogue.

211. Western Tiger Swallowtail larva. Photograph by C. Hogue.

■ ANISE SWALLOWTAIL
*(Papilio zelicaon)*
Figures 212-214

This yellow and black species is not as common in our area nowadays as it was in the past, probably because of fragmentation of populations of its principal food plant in the basin, which is Sweet Fennel *(Foeniculum vulgare)*, a tall introduced licorice-flavored weed that grows on hillsides and in vacant lots. (The plant is often erroneously called anise because of its similarity to that commercial species.) Garden-grown parsley and carrot tops also serve as hosts, and larvae are occasionally found feeding on *Citrus* species. Before the introduction of these foreign plants, the larvae fed on various native umbellifers, such as *Lomatium, Heracleum,* and *Tauschia* species.

The mature caterpillar is predominantly green with black bands that are interrupted by orange spots. When disturbed it rears up, and two orange-colored fleshy horns appear behind the head. These structures produce a musty odor, which presumably is distasteful or noxious to predators.

With a wing expanse of up to 3½ inches (9 cm), the species is a little smaller than the Western Tiger Swallowtail. It may complete several generations in a year, but it passes the winter in the pupal stage.

There are brown, gray, and green forms of the

chrysalis. The color usually matches that of the substrate (brown and gray on wood and tree trunks, green on living plants).

REFERENCES. Coolidge, K. R. 1910. A California orange dog. *Pomona College Journal of Entomology,* vol. 2, pp. 333-334.

Shapiro, A. M., and K. K. Masuda. 1980. The opportunistic origin of a new citrus pest. *California Agriculture,* vol. 34, no. 6, pp. 4-5.

Sims, S. R. 1980. Dispause dynamics and host plant suitability of *Papilio zelicaon* (Lepidoptera: Papilionidae). *American Midland Naturalist,* vol. 103, pp. 375-384.

## ■ BRUSH-FOOTED BUTTERFLIES
(Family Nymphalidae)

■ MONARCH
*(Danaus plexippus)*
Figures 3, 215-218

The Monarch butterfly is probably our most widely known insect. It is conspicuous and easily recognized because of its large size (wing span 3 to 4 in., or 75 to 100 mm) and handsome orange wings with wide black veins and black margins marked with small white spots.

A conspicuous black oval spot on the second vein from the inner margin of the hind wing distinguishes the male Monarch from the female. The spot is actually a pouch-like structure containing cells that produce biologically active chemical scents (pheromones), whose function is not completely understood. Before mating, the male inserts a quantity of female-stimulating fluid, which is secreted by special glands located at the tip of the abdomen, into these pouches; this fluid then diffuses during the mating process.

Female Monarchs lay their eggs only on the various species of milkweed (locally those of the genus *Asclepias),* and the leaves of this plant form the exclusive diet of the caterpillars. The latter sport a zebra style pattern of narrow alternating bands of black, white, and pale yellow. A pair of long fleshy filaments projects from the second body segment, behind the head; two shorter filaments project from near the rear of the body. Larval development requires ten to twenty days.

The chrysalis (Figure 217) is a beautiful object, a waxy jade green structure with brilliant gold spots and tiny black dots; it has been dubbed "the little green house with the golden nails," by one poetic entomologist.

The butterfly is known to be distasteful and even poisonous to birds and other potential predators. It acquires this unpleasant character from toxins

212. Anise Swallowtail. Photograph by J. Hogue.

213. Anise Swallowtail larva. Photograph by M. Badgley.

214. Anise Swallowtail pupa. Photograph by J. Hogue.

naturally present in the milkweed, the food plant of the caterpillar. For many years it was believed that the Viceroy butterfly *(Basilarchia archippus)* was itself a palatable insect that mimicked the Monarch's coloration to avoid being eaten by predators; the Viceroy/ Monarch relationship was frequently cited as a classic example of "Batesian mimicry." But recent studies have shown that the Viceroy is a noxious species in its own right, and that the Viceroy and Monarch represent a case of "Mullerian mimicry" rather than Batesian mimicry. (The two kinds of mimicry, which are

215. Monarch female. Photograph by J. Johnson.

216. Monarch larva. Photograph by J. Johnson.

217. Monarch pupa. LACM photograph.

218. Monarchs roosting in overwintering site. Photograph by W. Sakai.

named after their discoverers, are widespread among insects species throughout the world. Although the Viceroy is not found in the Los Angeles area, it deserves mention here because its story has so often been told in our textbooks and classrooms.)

The Monarch is famous for its yearly migrations. Each fall large numbers move out of their northern breeding grounds to overwinter in certain areas in the southern United States and Mexico. The distance covered by individuals is remarkable: one tagged specimen released in New England was recovered 2,900 miles (4,650 km) to the south in the main overwintering sites in Michoacan, Mexico.

Monarchs that pass the winter along the California and Baja California coast have come from throughout the region west of the Rocky Mountains. The butterflies form roosting aggregations (Figure 218) in large trees, primarily eucalyptus. Groves of these so-called butterfly trees are found all along the coast from northern Mendocino County, California, to Ensenada, Baja California; among them is "Butterfly Park" (officially George Washington Park) in Pacific Grove, California, where a city ordinance provides for the protection of the Monarchs during their annual stay. The value of roosting sites is also recognized by California state law (Assembly Bill 1671, 1987).

In the Los Angeles Basin there are only a few Monarch colonies; these are located in the Santa Monica Mountains, at Ballona Creek, in the city of Santa Monica, and on the Palos Verdes Peninsula. The butterflies usually begin arriving in September, and they stay until February or March, when the return migration begins. Individuals fly northward or eastward, breeding and stopping to lay eggs on milkweed along the way. The offspring of these transients continue the flight to the main breeding ground in the far north as soon as they mature.

The advantage of this migration to the Monarch Butterfly, an essentially tropical insect, is obvious: it escapes the harsh northern winter. But the factors that start and guide the Monarch on its journeys are not so apparent and are still largely unknown.

Many of the overwintering sites of this butterfly are threatened by urban development. Because of this, the Monarch is listed on the Bonn Convention, an international treaty protecting migratory animals.

REFERENCES. Brower, L. P. 1985. The yearly flight of the Monarch butterfly. *Pacific Discovery,* vol. 38, pp. 4-12.

Lane, J. 1985. California's Monarch butterfly trees. *Pacific Discovery,* vol. 38, pp. 13-15.

Malcolm, S. B., M. P. Zalucki, editors. 1993. *Biology and conservation of the Monarch butterfly.* Science Series, no. 38. Los Angeles: Natural History Museum of Los Angeles County.

Nagano, C. D., and C. Freese. 1987. A world safe for Monarchs. *New Scientist,* vol. 114, no. 1554, pp. 43-47.

Ritland, D. B., and L. P. Brower. 1991. The Viceroy is not a batesian mimic. *Nature,* vol. 350, pp. 497-498.

Tuskes, P. M., and L. P. Brower. 1978. Overwintering ecology of the Monarch butterfly, *Danaus plexippus* L., in California. *Ecological Entomology,* vol. 3, pp. 141-153.

Urquhart, F. A. 1987. *The Monarch butterfly: International traveller.* Toronto: University of Toronto.

Urquhart, F. A., and N. R. Urquhart. 1977. Overwintering areas and migratory routes of the Monarch butterfly *(Danaus p. plexippus;* Danaidae; Lepidoptera) in North America with special reference to the western population. *Canadian Entomologist,* vol. 109, pp. 1583-1589.

---

■ **Gulf Fritillary**
*(Agraulis vanillae)*
Figures 219-221

The metallic silver spots on the underside of the wings (Figure 219) immediately identify this species; its wing span is 2 to 3 inches (50 to 70 mm). The butterfly's presence is a sure sign that a passion flower

219. Gulf Fritillary that has just emerged from its chrysalis. Photograph by C. Hogue.

220. Gulf Fritillary, upper surface. Photograph by C. Hogue.

221. Gulf Fritillary larva. Photograph by C. Hogue.

vine *(Passiflora* species) grows nearby. The caterpillars feed exclusively on this vine, which is native to Latin America and a popular garden ornamental there. The larvae feed only on the type with soft five-lobed leaves *(P. coerulea)*, however, and not on the thicker three-lobed garden hybrid *(P. alato-coerulea)*.

The Gulf Fritillary undoubtedly followed the passion flower vine when it was brought to California by the early Mexican immigrants. The adult butterflies seldom wander far from the plant, and a number of them may play about it on warm summer days.

During courtship the male performs a wing-clapping display in which he alights next to the female and repeatedly brings his wings together, often catching her antennae between them. These movements probably cause transmission of chemical signals that are responsible for mating success.

The mature caterpillar is about 1 5/8 inches (41 mm) long, spined, and colored slate gray or purplish on the back with burnt orange stripes along the sides. The mottled gray chrysalis hangs from almost any horizontal object near the food plant, commonly the wires and braces of a chain-link fence through which the vine is growing. In its color and form, the chrysalis closely resembles a dry, curled leaf.

REFERENCES. Coolidge, K. R. 1924. *Agraulis vanillae* Linn. on the Pacific Coast (Lepid.; Nymphalidae). *Entomological News,* vol. 35, pp. 22-23.

Dimock, T. E. 1986. Hidden variation in *Agraulis vanillae incarnata* (Nymphalidae). *Journal of Research on the Lepidoptera,* vol. 25, pp. 1-14.

Rotowski, R. L., and J. Schaefer. 1984. Courtship behavior of the Gulf Fritillary, *Agraulis vanillae* (Nymphalidae). *Journal of the Lepidopterists' Society,* vol. 38, pp. 23-31.

---

■ PAINTED LADY
*(Vanessa cardui)*
Figures 222, 223

This species has the widest distribution of any butterfly in the world and is migratory like the Monarch; its migration routes and frequencies, however, are not as well defined nor as well studied. Mass flights of this species through the Los Angeles area have occurred fairly often. A very heavy flight took place in the winter and spring of 1958, when tens of thousands of Painted Ladies passed through the basin in a northwest direction. The place of origin of these migrants is not known, but they probably develop during wet years in northern Mexico. Evidence for a return flight in the fall is scanty.

In the Los Angeles Basin, the spiny caterpillars feed on hollyhocks and on Cheeseweed *(Malva parviflora),* a common weed of vacant lots. They are usually found nestled under a loose silk web in the cupped leaves of these plants. Until totally consumed, the leaf receptacle forms a place of abode for the larva. Thistles and nettles are hosts for the larvae in other areas. The adults have an average wing expanse of around 2 inches (50 mm).

OTHER SCIENTIFIC NAMES. The Painted Lady, and the West Coast and Virginia Ladies, are sometimes placed in the genus *Cynthia.*

REFERENCES. Abbott, C. H. 1962. A migration problem—*Vanessa cardui* (Nymphalidae), the painted lady butterfly. *Journal of the Lepidopterists' Society,* vol. 16, pp. 229-233.

Field, W. D. 1971. *Butterflies of the genus* Vanessa *and of resurrected genera* Bassaris *and* Cynthia *(Lepidoptera: Nymphalidae).* Smithsonian Contributions in Zoology, vol. 84, 105 pp. Washington, D.C.: Smithsonian Institution.

Shapiro, A. M. 1980. Evidence for a return migration of *Vanessa cardui* in northern California (Lepidoptera: Nymphalidae). *Pan-Pacific Entomologist,* vol. 56, pp. 319-322.

Tilden, J. W. 1962. General characteristics of the movement of *Vanessa cardui* (L.). *Journal of Research on the Lepidoptera,* vol. 1, pp. 43-49.

---

■ WEST COAST LADY
*(Vanessa annabella)*
Figure 224

This species is more a resident of our area than the Painted Lady and is much more common except in years of mass migrations of the Painted Lady. Vacant lots are a favorite haunt of the frisky adults, but they may be encountered almost anywhere. The species sometimes migrates. The variably colored caterpillar closely resembles that of the Painted Lady and feeds on similar plants: it eats a variety of plants in the nettle (Urticaceae) and malva (Malvaceae) families. Young larvae fold spun silk webbing across the leaf upon

222. Painted Lady. Photograph by P. Bryant.

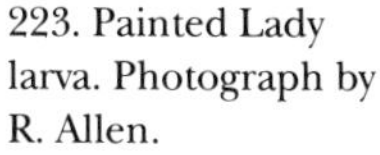

223. Painted Lady larva. Photograph by R. Allen.

which they are feeding, forming a kind of nest.

The West Coast Lady is a little smaller (average wing span $1^{3}/_{4}$ in., or 45 mm) than the Painted Lady. It is a species that is easily attracted to butterfly gardens by planting hollyhocks or encouraging Cheeseweed *(Malva parviflora)*.

OTHER SCIENTIFIC NAMES. *Vanessa carye.*

REFERENCE. Dimock, T. E. 1978. Notes on the life cycle and natural history of *Vanessa annabella* (Nymphalidae). *Journal of the Lepidopterists' Society*, vol. 32, pp. 88-96.

---

■ VIRGINIA LADY
*(Vanessa virginiensis)*
Figures 225, 226

Although about the same size as the Painted Lady and West Coast Lady, this third member of the "painted lady group" is perhaps the most distinctive with its two oversized eyespots on the underside of the hind wing. Its distribution is the most limited, and it is scarce in the basin in comparison to either of the other two species. The structure and food preferences of the immature stages are similar to those of its two close relatives.

---

■ RED ADMIRAL
*(Vanessa atalanta)*
Figures 227-229

The wings of this butterfly, which span approximately 2 inches (50 mm), are mostly black; a vivid red band runs diagonally across the fore wing, and the hind wing has a similarly colored border. Its caterpillars feed on the leaves of Nettle *(Urtica* species) and Pellitory *(Parietaria* species) in the foothills. They hide in the leaves, which they roll about themselves and seal with silk; the pupa may also be concealed in a leafy bower of this sort. In the city, they have been found feeding on the common ground cover, Baby's Tears *(Soleirolia* species).

Comstock (1927) used the alternate name Alderman Butterfly for this species because to him its bold colors were "suggestive of the ancient aldermanic costume that was in vogue in the British Isles [in Anglo Saxon times] when men first began to note and record the various animals and insects."

REFERENCE. Comstock, J. A. 1927. Butterflies of California. Published by the author, Los Angeles.

---

■ MOURNING CLOAK
*(Nymphalis antiopa)*
Figures 230-232

The Mourning Cloak is fairly common in the basin, even in metropolitan areas, because the usual food plants of its larvae are widely grown ornamental trees—elm, willow, and poplar. I have also found the larvae feeding on the Floss Silk Tree *(Chorisia speciosa)*.

The caterpillars are black with long spines; a

224. West Coast Lady. Photograph by C. Hogue.

225. Virginia Lady. Photograph by P. Bryant.

226. Virginia Lady larva. Photograph by R. Mattoni.

row of dull brick-red spots runs down the back. The larvae live in communal webs until nearly mature, when they leave the nest to feed alone and finally to search out a pupation site. At this time, they are large (length 1½ in., or 40 mm) and attract attention as they cross sidewalks and driveways. They often settle on eaves, window sills, and porch ceilings as well as on their host plants. I have seen the chrysalids hanging from certain houses in such profusion that they suggest Christmas decorations. The pupa is gray or pur-

227. Red Admiral. Photograph by P. Bryant.

228. Red Admiral larva. Photograph by P. Bryant.

229. Red Admiral pupa. Photograph by P. Bryant.

230. Mourning Cloak. Photograph by C. Hogue.

231. Mourning Cloak larva. Photograph by C. Hogue.

232. Mourning Cloak pupa. Photograph by C. Hogue.

plish-beige with dark markings; the back of the abdomen is heavily spined.

The adult is large (its wing span is 2 1/4 to 2 1/2 in., or 70 to 83 mm) and lovely. The tops of the wings are a deep purplish-black with a broad yellow border; inside of the jagged margins of this border runs a series of blue spots.

Adult Mourning Cloaks are inquisitive creatures. They are attracted to unlikely resting places, such as lawn furniture, and will even perch on a person who stands perfectly still. I have amazed onlookers on more than one occasion by holding my hand out in the direction of a circling specimen and having it promptly land on my outstretched finger. This trick takes advantage of the butterfly's natural habit of guarding its territory from a prominent position. In so doing it often displays considerable aggressiveness, chasing other butterflies and even birds out of its vicinity. In 1988, I observed a Mourning Cloak in the patio of the Natural History Museum diving at pigeons and diverting their movements.

This is one of the few butterflies that overwinters as an adult. It may be found on cold winter days nestled in the crevices of tree bark or in other protected places. Its torpor is not deep, however, and individuals will venture forth to enjoy a flight and take some nectar should a warm day come along.

---

■ BUCKEYE
*(Precis coenia)*
Figure 233

The large eye spots on the upper surfaces of the wings immediately identify this moderate-sized species (its wing span is 2 to 2 1/2 in., or 50 to 65 mm). It belongs to a primarily tropical group, called the Peacock Butterflies because their eye spots resemble the pattern on the tips of the feathers of peacocks.

Buckeyes display a fighting bent when other butterflies come too close. The intruder is submitted to a series of forceful rapid wing blows, which are often successful in driving it off.

Two small forward-pointing horns on the head distinguish the caterpillar from similar spined forms. It feeds on the leaves of a variety of plants; locally the commonest are snapdragons and plantain *(Plantago* species), a common lawn weed.

OTHER SCIENTIFIC NAMES. *Junonia coenia.*

---

■ CHALCEDON CHECKER-SPOT
*(Euphydryas chalcedona)*
Figures 234-236

Though rarely seen in the basin's flatlands, this species may be quite abundant in the surrounding foothills, visiting flowers in the spring and early summer. Adults have a wing span of from 1 1/4 to 1 3/4 inches (32 to 45 mm); the wings are covered in a gay, extremely variable checkerboard pattern of black, red, and pale yellow squares and spots. The spined larvae pass the winter protected under leaves and debris on the ground or in other secluded niches. They feed on a variety of plants of the Figwort family (Scrophulariaceae), but locally they are particularly fond of Sticky Monkey Flower *(Diplacus longiflorus)*, a common native shrub of the coastal sage plant community. The chrysalis is pale blue with black flecks and orange beads on the abdomen.

REFERENCES. Dammers, C. M. 1940. *Euphydras chalcedona* Dbldy. and Hew. *Bulletin of the Southern California Academy of Sciences*, vol. 39, pp. 123-125.

Lincoln, D. E., T. S. Newton, P. R. Ehrlich, and K. S. Williams. 1982. Coevolution of the checkerspot butterfly *Euphydryas chalcedona* and its larval food plant *Diplacus aurantiacus:* Larval response to protein and leaf resin. *Oecologia*, vol. 52, pp. 216-223.

Mooney, H. A., K. S. Williams, D. E. Lincoln, and P. R. Ehrlich. 1981. Temporal and spatial variability in the interaction between the checkerspot butterfly, *Euphydryas chalcedona* and its principal food source, the California shrub, *Diplacus aurantiacus. Oecologia*, vol. 50, pp. 195-198.

---

■ **California Sister**
*(Limenitis bredowii)*
Figures 237-239

This butterfly is moderately large (wing span 2¼ in., or 57 mm) and beautifully marked with a white median band and large orange apical spot on the fore wing. Comstock (1927) describes the California Sister as "a regal butterfly, of dignified demeanor and exclusive habits. It spends much of its time in solitary flight about the upper branches of the live-oaks, or perched on high vantage points where it can survey the woodland life below."

The butterfly is most common during the summer in the Santa Monica Mountains, in the canyons and foothills of the San Gabriels, and wherever the Canyon Oak *(Quercus chrysolepis)* grows. The caterpillar feeds on the leaves of this and related live oaks. It is dark green above and lighter green to brown or pink below, with six pairs of fleshy tubercles. With its brown color, white splotches, and irregular shape, the chrysalid resembles a dead leaf.

REFERENCES. Comstock, J. A. 1927. Butterflies of California. Published by the author, Los Angeles.

Comstock, J. A., and C. M. Dammers. 1931. The metamorphosis of *Heterochroa bredowii californica* Butl. (Lepid.). *Bulletin of the Southern California Academy of Sciences*, vol. 30, pp. 83-87.

233. Buckeye.
Photograph by P. Bryant.

234. Chalcedon Checker-spot.
Photograph by J. Hogue.

235. Chalcedon Checker-spot larva.
Photograph by C. Hogue.

236. Chalcedon Checker-spot pupa.
Photograph by C. Hogue.

237. California Sister. Photograph by J. Levy.

238. California Sister larva. Photograph by K. Wolfe.

239. California Sister pupa. Photograph by K. Wolfe.

**■ Lorquin's Admiral**
*(Limenitis lorquini)*
Figures 240, 241

This butterfly is easily mistaken for the California Sister, because it is generally similar in basic color pattern, size (its wing span is 2 1/4 in., or 57 mm), and habitat. But the Lorquin's Admiral has orange bars along the apex of the fore wing rather than a large round orange spot.

Comstock (1927) aptly describes this Admiral's habits: "It frequents water courses and moist wooded areas, being especially abundant in river bottomlands that are overgrown with willow (the larval food plant). The flight of this boldly marked butterfly is characterized by a series of short twitching motions, with the wings held nearly flat, interspersed with leisurely volplaning."

The species once occurred in the Baldwin Hills but is probably extirpated from there now because of the destruction of its habitat by land development. The canyons of the San Gabriel and Santa Monica mountains are now its principal refuge in our area.

The larval food plant is willow *(Salix* species). The mature caterpillar is olive-brown, with yellowish mottling in the thoracic area and a whitish patch above on the middle abdominal segments. A pair of fringed horns projects forward from the second segment.

REFERENCE. Comstock, J. A. 1927. Butterflies of California. Published by the author, Los Angeles.

## ■ SATYRS (Family Satyridae)

**■ California Ringlet**
*(Coenonympha california)*
Figure 242

This delicate creamy-white butterfly is a precocious member of the oak woodland fauna, appearing—sometimes in abundance—in the early days of spring. Its feeble halting flight and small size (its wing span is 1 1/8 in., or 28 mm) easily separate it from the "whites" described below.

The larva is long and slender, without spines or hairs, and it matches in shades of brown and green the grasses upon which it feeds.

## ■ WHITES AND SULFURS (Family Pieridae)

**■ European Cabbage Butterfly**
*(Pieris rapae)*
Figures 243-245

Except for black wing tips and small black spots, this butterfly is all white; its wing span is 2 inches (50 mm). Although it is our commonest butterfly, it is not a native species. It was accidentally introduced into Quebec from Europe around 1860 and quickly spread throughout the United States, reaching the Los Angeles area probably sometime late in the same century.

240. Lorquin's Admiral. Photograph by P. Bryant.

241. Lorquin's Admiral larva. Photograph by C. Hogue.

242. California Ringlet. Photograph by P. Bryant.

The caterpillars, which are pale green and somewhat worm-like in form, feed on various plants of the mustard family, especially cabbage (their appetite for this vegetable is so great that they consume thousands of dollars worth yearly). Locally, however, the principal food plants of the larvae are weeds of the family Brassicacaea, such as Wild Mustard *(Brassica rapa* subspecies *sylvestris)* and Wild Radish *(Raphanus sativus)*. The caterpillars are also to be found in the flower garden on nasturtiums. They produce green to yellow chrysalids that match their background in color.

REFERENCE. Hovanitz, W. 1962. The distribution of the species of the genus *Pieris* in North America. *Journal of Research on the Lepidoptera,* vol. 1, pp. 73-83.

---

■ COMMON WHITE
*(Pieris protodice)*
Figure 246

When it is in flight, this butterfly could be confused with the European Cabbage Butterfly; at rest, however, its checkered pattern of black spots immediately distinguishes it from its close relative. There is some difference in the coloration of the sexes: the males lack dark markings on the hind wings and are only moderately speckled; the females are extensively marked on the upper surfaces of both pairs of wings.

The caterpillars vary in color from deep to pale bluish to green, but all have four yellow lengthwise stripes and are thickly covered with black dots. They feed on a wide variety of plants in the Brassicacaea family, especially wild mustard in our area.

Although much less common than the European Cabbage Butterfly, this species is also an intruder from the Old World. It was first reported in Quebec about 1859, and it rapidly spread over most of North America.

---

■ NICIPPE YELLOW
*(Eurema nicippe)*
Figure 247

Deep orange wings with broad black borders immediately set this species apart. There is also a small transverse spot, which—with a bit of the same imagination needed to see the "face" in the California Dogface—looks like a half-closed eye and has given it the name "Sleepy Yellow." Its wing span is 1½ inches (40 mm).

The Nicippe Yellow is a tropical butterfly that has spread northward into our area. The green wormlike caterpillars normally feed on the exotic sennas *(Cassia* species); other legumes, such as clover, act as secondary host plants. Although it was once abundant, it is now rare in the basin.

---

■ CALIFORNIA DOGFACE
*(Colias eurydice)*
Figure 248

Within its black border, the wing of the male California Dogface gives off a singular purplish iridescence, which changes in intensity depending on the angle from which it is viewed. Some people see a dog's face in profile in the pattern of the fore wing. The outline of the wings and contrasting gaudy colors of the fore and hind wings simulate the pansy flower in the minds of other observers, who have nicknamed the butterfly "Flying Pansy." The females are pure sulfur yellow except for a round dark spot near the center of the

243. European Cabbage Butterfly. Photograph by J. Levy.

244. European Cabbage Butterfly larva. Photograph by C. Hogue.

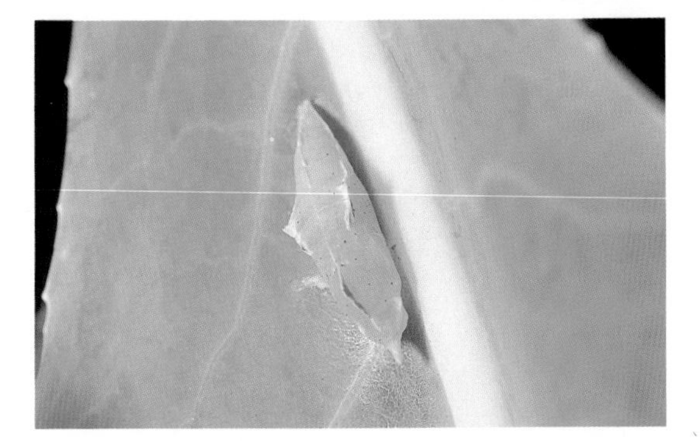

245. European Cabbage Butterfly pupa. Photograph by C. Hogue.

246. Common White female. Photograph by P. Bryant.

leading edge of the fore wing. The butterfly's wings span 2 to 2 1/4 inches (50 to 55 mm).

Though normally a mountain species, the California Dogface occasionally strays into the hilly northern areas of the basin. The caterpillar, which is 1 inch (25 mm) long, has a white line on its side that is edged with orange below and black above; the general color is dull green above the line and light green beneath. It feeds on the leaves of the False Indigo Plant *(Amorpha californica)*.

In 1972, the California Legislature designated the California Dogface the official state insect.

REFERENCE. California State Government Code. 1972. Section 424.5: State insect.

---

■ ORANGE SULFUR
*(Colias eurytheme)*
Figure 249

The caterpillars of this butterfly have a voracious appetite for alfalfa, and this has earned them the enmity of farmers throughout most of the western United States. Clover also serves as a food plant, and the larvae will feed on many other members of the pea family.

The yellow-orange adults, which have a wing span of 2 inches (50 mm), are a familiar sight in the alfalfa-growing areas of the Imperial and Coachella valleys to the southeast of the basin. During the warm days of the growing season, hundreds of individuals can be seen flying about the fields and milling back and forth across the highway. Many are caught in the radiators of passing automobiles and may even pile up heavily enough to impair the cars' function.

Many female specimens have white rather than orange wings, a genetic variant.

The species is also known as the Alfalfa Butterfly. The very similar Clouded or Common Sulfur *(Colias philodice)*, which is bright lemon yellow (without the orange tint of *C. eurytheme)*, is not found in the basin.

REFERENCES. Michelbacher, A., and R. Smith. 1943. Some natural factors limiting the abundance of the Alfalfa Butterfly. *Hilgardia,* vol. 15, pp. 369-397.

Silberglied, R. E., and O. R. Taylor, Jr. 1978. Ultraviolet reflection and its behavioral role in the courtship of the sulfur butterflies *Colias eurytheme* and *C. philodice* (Lepidoptera, Pieridae). *Behavioral Ecology and Sociobiology,* vol. 3, pp. 203-243.

---

■ SENNA SULFUR
*(Phoebis sennae)*
Figures 250-252

This is a tropical butterfly that inhabits our territory only sporadically. It is able to pass a succession of generations locally on cultivated "big-leafed" cassias,

247. Nicippe Yellow. Photograph by of LACM specimen by C. Hogue.

248. California Dog-face. Photograph by P. Bryant.

249. Orange Sulfur. Photograph by P. Bryant.

but it appears to be unable to survive the winter and is therefore not a permanent resident.

The adults are fairly large sulfurs (wing span 2 in., or 50 mm) that are brilliant lemon yellow or yellow-green and have a characteristic rapid dodging flight. The caterpillar is green and similar to that of the Cabbage White but larger (length 1½ in., or 40

250. Senna Sulfur. Photograph of LACM specimen by C. Hogue.

251. Senna Sulfur larva. Photograph by P. Bryant and J. Levy.

252. Senna Sulfur pupa; uncast larval skin is visible at top of chrysalis, near attachment point. Photograph by P. Bryant and J. Levy.

mm). There are numerous minute black nodules scattered over the surface of the body and a yellow stripe on the side that is bordered above with blue points. The chrysalis ranges in color from pink to green and has a yellow and green stripe on its abdomen. It has a swollen appearance because of the greatly enlarged wing cases.

## ■ BLUES, COPPERS, AND HAIRSTREAKS
(Family Lycaenidae)

■ **Pygmy Blue**
*(Brephidium exilis)*
Figure 253

This is one of the world's smallest butterflies; the wing expanse of even the largest individuals barely exceeds 1/2 inch (13 mm). The upper surface is medium brown, and the hind wing is margined with a row of dark spots, which on the under side are defined by circles of metallic scales. The tiny grub-like caterpillars feed on Salt Bush *(Atriplex* species) and Pigweed *(Chenopodium* species).

The Pygmy Blue is seen mainly in wild areas, especially where the salt-loving food plants grow—in alkaline valleys and flats, in coastal salt marshes, and along beach bluffs.

■ **Marine Blue**
*(Leptotes marina)*
Figures 254, 255

This is another butterfly that is common in local parks and gardens because its larva feeds on the buds and blossoms of ornamental shrubs and vines *(Plumbago* species, Wisteria Vine, sweet pea, and other members of the Pea Family). Though it is a fairly close relative of the tiny Pygmy Blue, it is larger: its wing span is slightly less than 1 inch (24 mm). The upper wing surface is pale lilac blue and slightly transparent, so that the ripple-like pattern of dark bands on the under side shows through faintly.

253. Pygmy Blue. Photograph by P. Bryant.

■ EL SEGUNDO BLUE *(Euphilotes battoides allyni)* Figure 256

Although this butterfly had been collected by entomologists for some time, it was not recognized as a distinct entity and named until 1975. It is a subspecies of a wide-ranging butterfly—the Square-spotted Blue *(Euphilotes battoides)*—and it currently survives in an extremely limited habitat in the Los Angeles area. A small population lives a precarious existence in two small remnants of the formerly expansive El Segundo Dunes near the beach to the west of Los Angeles International Airport. Because its survival is uncertain, the subspecies was in 1976 officially designated as endangered under the federal Endangered Species Act. This status prohibits collection of the butterfly or degradation of its habitat and will, it is hoped, prevent the El Segundo Blue's extinction.

The butterfly has one generation per year. Adults fly from mid July to early September. The emergence of the adults coincides with the peak flowering of the buckwheats *Eriogonum parvifolium* and *cinereum,* the food plants of the larvae.

The small caterpillars (body length 1/2 in., or 13 mm) are slug-like and feed on buckwheat blossoms, whose color and pattern they mimic. Larvae are present from early August through late September. Pupation occurs in the debris and sand at the base of the plant. The pupa is small, about the size of a rice grain.

The adult is a small butterfly (wing span 3/4 to 1 in., or 20 to 25 mm). Males are brilliant metallic blue above, and females are dull brown. The undersides of the wings of both sexes are medium gray with numerous white-bordered squarish black spots.

This butterfly has become a symbol of the fragility of nature in the face of urban development. Its legal status as an endangered species has delayed construction of a golf course on most of its home territory, and a small portion of the land to be developed has been set aside as a sanctuary.

A related but less famous butterfly is the Palos Verdes Blue *(Glaucopsyche lygdamus palosverdesensis;* Figure 257), which was unique to the Palos Verdes Peninsula. Named in 1977, it became extinct only six years later as a result of destruction of its habitat and resultant loss of its foodplant, the locoweed, through urban development. The butterfly had been recognized as in danger of extinction by the U.S. Fish and Wildlife Service in 1980. A court action in April 1987 found the City of Palos Verdes accountable for the loss, and a fine of $28,000 was imposed on the city.

254. Marine Blue. Photograph by P. Bryant.

255. Marine Blue larva. Photograph by P. Bryant.

256. El Segundo Blue. Photograph by R. Mattoni.

257. Palos Verde Blue, an extinct Los Angeles Basin butterfly. Photograph of LACM specimen by C. Hogue.

OTHER SCIENTIFIC NAMES. *Shijimiaeoides battoides allyni.*

REFERENCES. Arnold, R. A. 1983. Ecological studies of six endangered butterflies (Lepidoptera, Lycaenidae): Island biogeography, patch dynamics, and the design of habitat preserves. *University of California Publications in Entomology,* vol. 99, pp. 1-161 [see pp. 80-98].

_______. 1987. Decline of the endangered Palos Verdes Blue Butterfly in California. *Biological Conservation,* vol. 40, pp. 203-217.

Moran, J. 1987. Peninsula city charged with killing butterfly. *Los Angeles Times,* 3 April 1987, part 2, pp. 1-3.

Nagano, C. D. 1981. California coastal insects: Another vanishing community. *Terra,* vol. 19, no. 4, pp. 27-30.

Powell, J. A. 1981. Endangered habitats for insects: California coastal sand dunes. *Atala,* vol. 6, pp. 41-55.

■ COMMON HAIRSTREAK
*(Strymon melinus)*
Figure 258, 259

This species belongs to a group of small butterflies that typically display a hair-like appendage projecting from the rear margin of the hind wing. This feature, and the pale slate-gray upper wing surfaces marked only by a crimson spot on the hind wing at the base of the hairs or "tails," distinguish the Common Hairstreak from its much less common local relatives. The undersides of the wings also show orange spots at the base of the tail and are crossed by a conspicuous black line with a white margin.

The adult, which has a wing span of 3/4 to 1 1/8 inches (20 to 28 mm), flies short distances in energetic bursts. Then, after landing, it walks nervously about, with its wings held tightly together above the back, stopping frequently to rub the hind pair together with a "grinding" motion; the significance of this peculiar behavior is not known.

The small slug-like larvae bore into the fruits and seeds of a great variety of plants. In some parts of the species' range, the food plants include cultivated beans and sweet peas, causing it to be considered a pest. In our area, the principal hosts are probably common mallow and hibiscus, around which the adults are frequently seen. In my garden I have observed females ovipositing on the buds of hibiscus in September; the larvae feed on the petals of the developing flower. I have also seen larvae developing on corn plants.

## ■ SKIPPERS (Family Hesperiidae)

■ FIERY SKIPPER
*(Hylephila phyleus)*
Figures 260, 261

Also called the Lawn Skipper, this species is small (wing span 1 to 1 3/8 in., or 25 to 33 mm), yellowish-brown, and moth-like and is exceedingly abundant in residential districts. Its scurrying flight over lawns and

258. Common Hairstreak. Photograph by P. Bryant.

259. Common Hairstreak larvae. Photograph by R. Mattoni.

260. Fiery Skipper. Photograph by P. Bryant.

261. Fiery Skipper larva. Photograph by C. Hogue.

its habit of sunning with wings partially spread open, the fore and hind pair separated, are characteristics that readily identify it.

The larva (length $1\,^7/_8$ in., or 47 mm) feeds on lawn grasses, preferring Bent Grass and Bermuda Grass *(Cynodon dactylon)*. It is distinguished from the common lawn cutworms by skin that is densely set with short hairs and by its large dark head, which is set off by a constricted prothoracic segment. Although abundant, it is seldom seen because it conceals itself in a silken tube woven through the underlayer of the grass. The pupae are also found in these tubes.

Adults are territorial; I have even observed them chasing hummingbirds.

REFERENCE. Shapiro, I. D. 1977. Interaction of population biology and mating behavior of the Fiery Skippers, *Hylephila phylaeus* (Hesperiidae). *American Midland Naturalist,* vol. 98, pp. 85-94.

---

■ FUNERAL DUSKYWING
*(Erynnis funeralis)*
Figures 262, 263

This skipper is large (wing span $1\,^3/_8$ to $1\,^5/_8$ in., or 35 to 42 mm) and black and is characterized by the white fringe bordering its hind wings. It is common in hilly areas through the summer, buzzing from perch to perch in open areas of sparse herbaceous vegetation.

The larva is pale green and, like the larva of the Fiery Skipper, has an outsized brown head with paired lumps on top; its thorax is narrow and constricted into a neck-like structure.

Funeral Duskywing larvae feed on various leguminous plants. I have found them in the northern foothills on Deerweed *(Lotus scoparius),* and this is probably the usual food plant in our area. They may be found by searching for the loose chamber of leaves and silk in which they hide themselves.

REFERENCE. Burns, J. M. 1964. *Erynnis funeralis.* Pages 173-182 in Evolution in skipper butterflies of the genus *Erynnis,* edited by J. Burns. *University of California Publications in Entomology,* vol. 37, pp. 1-216.

---

■ WANDERING SKIPPER
*(Panoquina errans)*
Figure 264

This species exists only in coastal salt marshes and is another butterfly, like the El Segundo Blue, with narrow requirements for life that make it vulnerable to extinction. The sparse populations that are known to persist at Ballona Creek and near the Venice Canals are dependent on Salt Grass *(Distichlis spicata)* growing in soil moistened by spring tides. The caterpillars apparently consume salts in their diet of this grass but are metabolically able to tolerate it.

262. Funeral Duskywing. Photograph by P. Bryant.

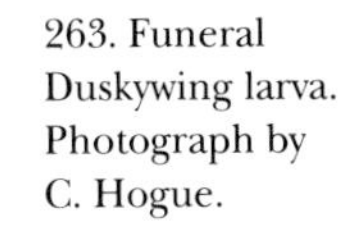

263. Funeral Duskywing larva. Photograph by C. Hogue.

264. Wandering Skipper with proboscis inserted in flower. Photograph by P. Nordin.

The adults are typical medium-sized skippers (wing span 3/4 in., or 20 mm) but have rather pointed fore wings. There are pale spots in the center of both surfaces of the otherwise medium-brown wings.

The mature larva is deep green, with a dark stripe on the middle of its back and a band of yellowish-white along its side. The naked pupa is attached vertically to the food plant with a girdle and a terminal silk button; it can be recognized by its long pointed head projection.

REFERENCE. Wells, S. M., R. M. Pyle, and N. M. Collins. 1983. Wandering skipper. Pages 417-419 in *The IUCN invertebrate red data book*. Gland, Switzerland: IUCN.

# 12 GNATS, MIDGES, AND FLIES (Order Diptera)

THE SCIENTIFIC NAME of this order of insects means "two-winged" in Greek and describes the chief feature of its members' anatomy that distinguishes them from other insects: only the fore wings—the *mesothoracic* pair—are developed; evolution has modified the hind or *metathoracic* wings into small knobbed sense organs or "balancers" *(halters)*. These organs are thought to function like the gyroscopic attitude instrument of an airplane, providing the fly with information on its flight position.

Flies generally are excellent aeronauts, and the fastest flying insects are included among them. Horse flies have been clocked at around 45 miles (75 km) per hour; certain warble or bot flies are probably capable of maintaining speeds of 60 miles (100 km) per hour or more in short bursts.

Flies also move their wings up and down very rapidly. House Flies have been recorded to have as many as 200 wing beats per second; mosquitoes, 600 beats per second; and biting midges of the family Ceratopogonidae, 1,000 beats per second. The beat is often so rapid that it actually becomes a vibration, like that of a tuning fork, and gives rise to a hum or buzz; this accounts for the irritating whine of the mosquito.

The metamorphosis of flies is of the complete type, and they are in the Subclass Holometabola. The larvae—at least of the majority of the more highly evolved flies—are worm-like creatures without heads or appendages and are called "maggots." Prior to pupation, maggots form neither protective cocoon nor earthen cell. Rather, a unique hard capsule develops around the pupa and keeps it safe from desiccation and injury in the way that the hull of a seed surrounds and protects the kernel. This structure, called a "puparium," is derived from the last skin of the larva, which is detached during the pupal molt but not discarded. In some species, the adult fly escapes from the puparium by popping off one end with a balloon-like organ on its head.

The word "fly" usually brings to mind the House Fly *(Musca domestica)* or similar species. Actually, the order Diptera includes some 80,000 diverse forms throughout the world that can be divided into two very broad groups. The first and more primitive group contains the gnats and midges. These are generally small, frail-bodied, and long-legged flies, typified by the mosquito. The second group, the "flies proper," contains relatively robust flies with short legs—the common House Fly is the best example.

REFERENCES. Cole, F. R. 1969. *The flies of western North America.* Berkeley, Calif.: University of California.

McAlpine, J. F., editor. 1981, 1987, 1989. Manual of the Nearctic Diptera, 3 volumes. Research Branch, Agriculture Canada, Monographs, nos. 27, 28, and 32.

Oldroyd, H. 1964. *The natural history of flies.* New York: Norton.

## CRANE FLIES (Family Tipulidae)

THESE FLIES, which are frequently seen indoors resting on walls, ceilings, and windows, are often mistaken for giant mosquitoes. Some people believe that they catch and eat mosquitoes and call them "mosquito hawks." But crane flies are not closely related to mosquitoes in spite of the similarities in appearance, nor do they prey upon them (the short soft mouthparts of crane flies make them incapable of biting). An added characteristic of the family are long, ungainly legs that break off easily.

The two species described below are among the many species of crane flies living in the basin.

REFERENCE. Alexander, C. P. 1967. The crane flies of California. *Bulletin of the California Insect Survey,* vol. 8, pp. 1-269.

■ COMMON CRANE FLY
*(Tipula planicornis)*
Figures 265, 266

This is a medium-sized light brown species (body length 1/2 to 3/4 in., or 13 to 20 mm), which is seen outdoors resting with legs outstretched on walls and window panes or hanging in vegetation. It is attracted to lights and often finds its way indoors.

The stout worm-like larvae (called "leather jackets" because of their thick dark skin) live in damp loose soil or leaf mold and feed on the roots of herbaceous plants. In the spring, when such food supplies and moisture abound, large larval populations may develop and produce swarms of adults.

■ GIANT CRANE FLY
*(Holorusia hespera)*
Figure 267

This is one of the world's largest flies: the body length of the average female often reaches 1 3/8 inches (35 mm), and the wings may span nearly 3 inches (75 mm). The body is reddish brown, and there are white areas on the sides of the thorax.

The fly is most often seen during the springtime in grassy areas near streams. The large cylindrical worm-like larvae (body length 2 in., or 50 mm) may be semiaquatic or aquatic, living in shore vegetation and trash or on the bottom of small stream pools.

OTHER SCIENTIFIC NAMES. *Holorusia grandis; H. rubiginosa.*

REFERENCES. Arnaud, P. H., Jr., and G. W. Byers. 1990. *Holorusia hespera,* a new name for *Holorusia grandis* (Bergroth) (=*Holorusia rubiginosa* Loew) (Diptera: Tipulidae). *Myia,* vol. 5, pp. 1-9.

Kellogg, V. L. 1901. The anatomy of the larva of the giant crane fly *(Holorusia rubiginosa). Psyche,* vol. 9, pp. 207-213.

265. Common Crane Fly. Photograph by C. Hogue.

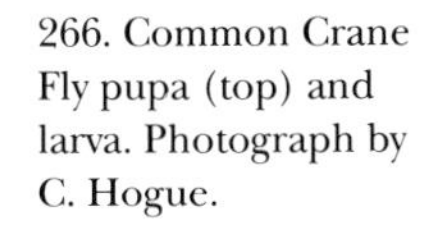

266. Common Crane Fly pupa (top) and larva. Photograph by C. Hogue.

267. Head and thorax of Giant Crane Fly. Photograph by C. Hogue.

## MOSQUITOES (Family Culicidae)

NO CATEGORY OF INSECT has received more attention from scientists, physicians, politicians, generals, sportsmen, and citizens at large than mosquitoes. The reason is apparent—in addition to annoying us with their bite and buzz, mosquitoes transmit a long list of our worst diseases. At the top of this list is malaria. For centuries this affliction has outranked warfare as a source of human suffering, and it continues today to kill and depress the energy of thousands.

Thanks to an intensive control program extending over the past seventy years, malaria has practically vanished in the United States. As recently as 1916, the Central Valley of California had a death rate due to malaria of 14.2 per 100,000 population. Today, a malaria case there is a novelty and is always traceable to some special circumstance, usually involving a single carrier who has infected a few mosquitoes in a restricted location. The main share of credit for the victory over malaria in our state is due the California Legislature of 1915, which passed the Mosquito Abatement District Act under which mosquito control could be carried on by organized districts throughout the state. There are now numerous agencies in the state organized for mosquito control purposes.

In the Los Angeles Basin, the health danger from malaria is virtually nonexistent. The public health menace of mosquitoes persists here, however, primarily because of their ability to transmit the viruses responsible for the Saint Louis and Western Equine varieties of encephalitis, which are diseases characterized by inflammation of the brain. These and other diseases caused by mosquito-borne viral encephalitides are commonly called "sleeping sicknesses" but should not be confused with tropical African Sleeping Sickness, a very different disease caused by a species of trypanosome protozoan and transmitted by tsetse flies.

The Western Encephalitis Mosquito *(Culex tarsalis)* and the Southern House Mosquito *(Culex quinquefasciatus)* are suspected to be vectors of the encephalitis viruses from wild birds to man. The birds carry the pathogens in their blood but do not manifest the disease. Mosquitoes feeding on infected birds pick up the viruses and may later inject them with their saliva into the blood stream of a healthy person. The most serious epidemic of encephalitis in California's recent history occurred during the summer of 1952, when 792 human cases were reported. Between 1983

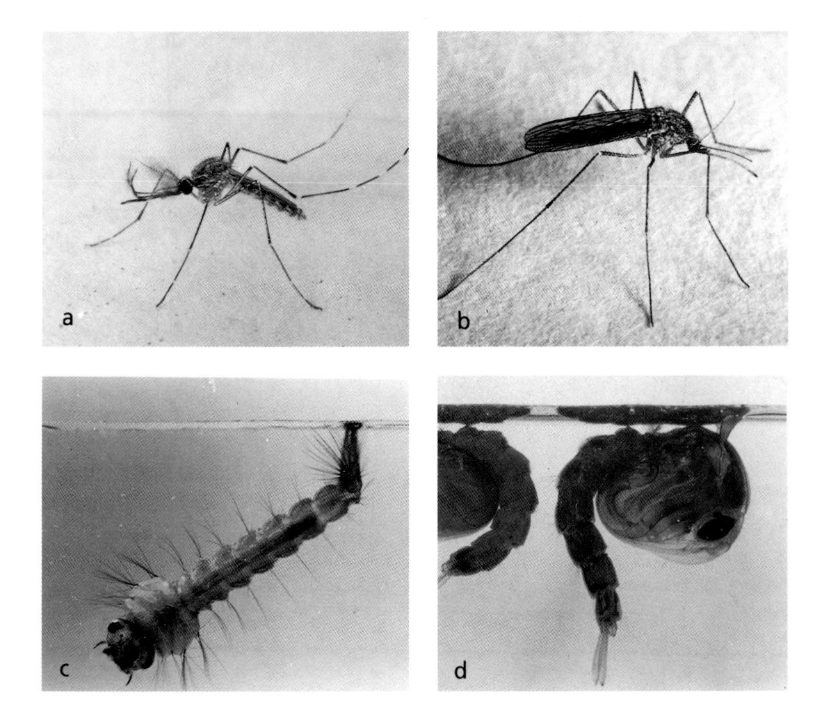


268. Mosquito male (a), female (b), larva (c), and pupa (d). Photographs by J. Honey.

and 1987, there were only 25 human cases of the encephalitis in the greater Los Angeles area.

Not all mosquitoes are important disease carriers, even though most suck the blood of man and other vertebrate animals. Only the female bites; the male, which is recognizable by its bushy antennae (Figure 268a), feeds on nectar and fruit juices or not at all.

These dipterans are good fliers. Although most species stay in the vicinity of their birthplace, some stray many miles on dispersal flights. The characteristic buzzing sound that they emit during flight is caused by high-frequency vibration of the wings.

All mosquitoes require standing or slowly moving water in which to develop, and this abounds in the basin—in fish ponds, untended swimming pools, rain puddles, and abandoned containers such as jars, cans, and old tires. Residual water in the basin's storm drains is a major breeding site for mosquitoes.

The fertilized female lays her eggs directly on the water film or upon surfaces that will later become covered with water. The larvae (Figure 268c) are active creatures commonly called "wrigglers." They often rest head down just below the water surface, taking in air with the elongated breathing tube at the tail end of the body as they feed on microscopic organisms suspended in the water. Feeding is essentially a straining operation, which is carried out by rapidly vibrating brush-like mouthparts. After four molts the larvae transform into pupae, called "tumblers," which to the unaided eye look like small seeds, each with an elongate curved tail with paddles on the end (Figure 268). Pupae do not feed and usually lie quietly just below the water surface. However, they are unusual among insect pupae in that they can be very active: if they are disturbed, their tails move back and forth rapidly and propel them through the water away from danger.

Several species of mosquitoes are found in the basin; all are small gnats, and their identification requires examination of minute structures with a microscope and the aid of technical literature (see the references at the end of this section). The most common species are listed below in order of their public health importance and abundance.

■ Southern House Mosquito *(Culex quinquefasciatus)*. This species breeds in foul water such as is found in septic tanks, drains, and gutters. The adults are domestic and are frequently found indoors, where they readily bite people. This species is a significant vector in the encephalitis virus transmission cycle.

■ Western Encephalitis Mosquito *(Culex tarsalis)*. This mosquito breeds in almost any accumulation of water. The adults are strong fliers and readily bite human beings and birds. This species is also a significant vector in the transmission of encephalitis.

■ Foul Water Mosquito *(Culex peus,* formerly *C. stigmatosoma)*. This mosquito, like the previous two species, breeds in polluted water, although it is also found in temporary accumulations of fresh water. Adults of this common species hardly ever feed on human blood, but they have been shown to be effective vectors of viruses in the laboratory.

■ Tule Mosquito *(Culex erythrothorax)*. This species breeds in all kinds of large shallow ponds that have a heavy growth of emergent vegetation, especially tules. The females readily bite people.

■ Winter Mosquito *(Culiseta inornata)*. This is a species that breeds in most kinds of ground pools and in artificial containers like tin cans, wading pools, and old tires. The adults will bite humans but prefer other hosts.

■ Cool Weather Mosquito *(Culiseta incidens;* Figure 269). Breeding sites for this species include almost any accumulation of water. Adults bite people freely but prefer the blood of fowl and domestic animals. The species is readily recognized by its relatively large size for a mosquito (body length about 1/4 in., or 6

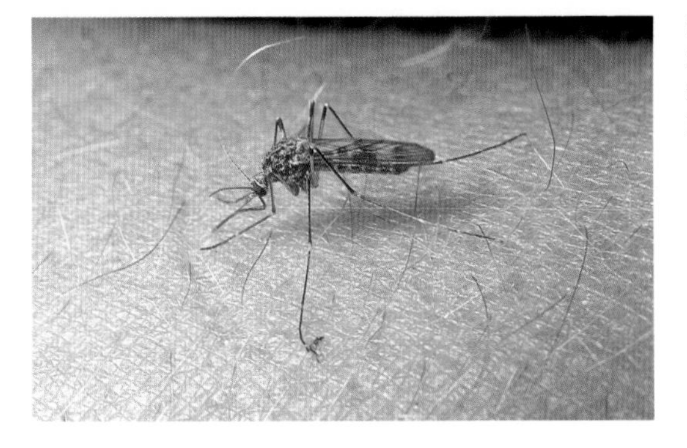

269. Cool Weather Mosquito sucking blood from a human hand. Photograph by C. Hogue.

mm) and speckled wings. It is most abundant during the cooler months (December through February).

■ Western Tree Hole Mosquito *(Aedes sierrensis)*. The larvae of this mosquito are found primarily in rotholes in tree trunks (especially oaks and cottonwoods). Although the adults are major pests elsewhere, locally they do not seem to be readily attracted to humans.

■ Field Mosquito *(Anopheles franciscanus)*. This mosquito breeds in clear ground pools with abundant algal growth. The adults prefer the blood of domestic stock but will occasionally attack people.

■ Southern California Malaria Mosquito *(Anopheles hermsi)*. This species has the same general habits as the Field Mosquito. It is the vector of local malaria cases. Curiously, it was not named until 1989.

REFERENCES. Barr, A. R., and P. Guptavanij. 1989. *Anopheles hermsi* n. sp., an unrecognized American species of the *Anopheles maculipennis* group (Diptera: Culicidae). *Mosquito Systematics,* vol. 20, pp. 352-356.

Bohart, R. M., and R. K. Washino. 1978. *Mosquitoes of California,* 3rd edition. Berkeley, Calif.: University of California.

Carpenter, S. J., and W. J. LaCasse. 1955. *Mosquitoes of North America (north of Mexico).* Berkeley, Calif.: University of California.

Darsie, R. F., Jr., and R. A. Ward. 1981. Identification and geographical distribution of the mosquitoes of North America, north of Mexico. *Mosquito Systematics,* supplements, no. 1, 313 pp.

Pierce, W. D., W. E. Duclus, and M. Y. Longacre. 1945. Mosquitoes of Los Angeles and vicinity. *Sanitarian,* vol. 7, pp. 718-726.

Reisen, W. K., R. P. Meyer, C. H. Tempelis, and T. J. Spoehel. 1990. Mosquito abundance and bionomics in residential communities in Orange and Los Angeles Counties, California. *Journal of Medical Entomology,* vol. 27, pp. 356-367.

Ternes, A., editor. Mosquitoes unlimited (special issue). *Natural History,* July 1991.

## BITING GNATS

In addition to the mosquitoes, there are members of three additional families (Simuliidae, Ceratopogonidae, and Rhagionidae) of gnat-like flies with blood-sucking habits in our area. Fortunately, these gnats are only minor pests here because they occur primarily in mountainous areas where there are few people. As more residences are built in the Santa Monica Mountains and other locales near the biting gnats' breeding places, complaints against these insects may increase. As with mosquitoes, only the females of the biting gnats bite.

### ■ BLACK FLIES (Family Simuliidae)

Also called "buffalo gnats" because of their humpbacked form (Figure 270), these small flies commonly attack people and domestic animals or birds. Locally they are most common in the Santa Monica Mountains and foothills. The larvae are found in tremendous numbers in streams, usually in masses on smooth stones, vegetation, or other objects in areas where the current is swiftest (Figure 271). When disturbed, the larva, which is slimy to the touch, curls its body into a U-shape and may detach from the substrate, releasing a long silk drag line.

This is one of the few members of the Diptera whose larva forms a cocoon to protect the pupa. The cocoon is attached to an underwater object, and at the time of emergence the adult rises to the surface in an air bubble.

Adults usually find their victims during the daytime, in a shady place where the air is still. Some people exhibit severe skin reactions to the bite, but no diseases are transmitted by these flies in our area.

Our black fly fauna consists of only a few species in the genus *Simulium,* principally *S. virgatum* (Figure 270), *argus, vittatum,* and *piperi.*

REFERENCE. Hall, F. 1972. Observations on black flies of the genus *Simulium* in Los Angeles County, California. *California Vector Views,* vol. 19, pp. 53-58.

### ■ PUNKIES (Family Ceratopogonidae)

These minute gnats deserve their pidgin-English name, "no-see-ums," for they are indeed almost too small to be seen, even when their presence is painfully felt: the effects of the bite are far out of proportion to the size of the biter. Specimens of most species barely reach

270. Black fly *(Simulium virgatum)*. Drawing by T. Ross.

271. Larvae of a black fly *(Simulium* species) on stone in shallow water. Photograph by C. Hogue.

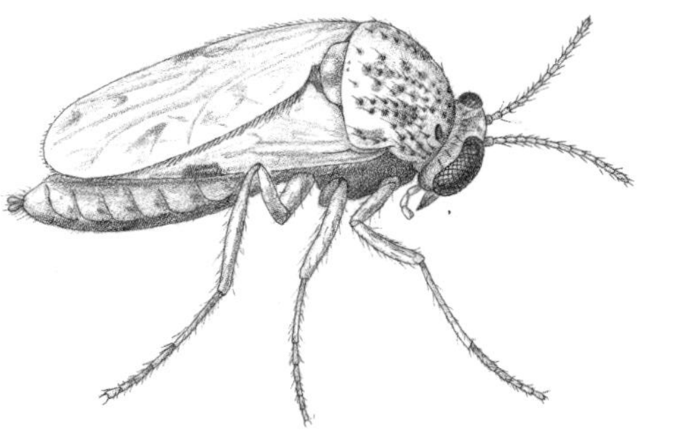

272. Punkie *(Culicoides variipennis occidentalis)*. Drawing by T. Ross.

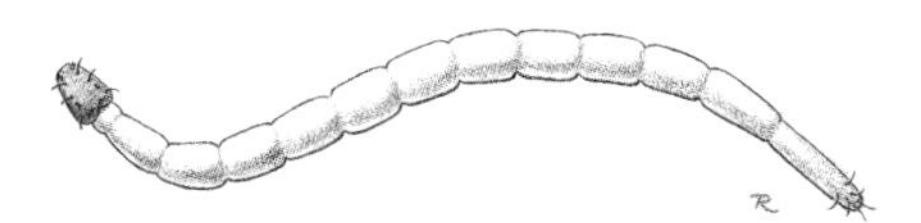

273. Larva of punkie *(Culicoides* species). Drawing by T. Ross.

$^1/_{32}$ inch (1 mm) in length and can easily pass through ordinary window screen.

The effects of a punkie's bite are a short-lived inflammatory swelling followed by intense itching, which may continue for a week or longer. A water-filled blister may form at the site of the bite.

There are numerous species of several genera in our part of the state. However, only members of the genera *Leptoconops* and *Culicoides* are blood-sucking. The former genus contains the Valley Black Gnat *(L. carteri,* formerly *L. torrens)* and the Bodega Gnat *(L. kerteszi* complex), biters that are notorious pests in other parts of California. Among the major biters to be found in the Los Angeles area are *Leptoconops foulki,* which is known from the Santa Ana River, and *Culicoides variipennis occidentalis* (Figure 272), which lives near the coast.

The immature stages of the local punkies are terrestrial but require moisture. The immatures of *Leptoconops foulki* have been found in damp surface sand in river bed depressions. Those of *Culicoides variipennis occidentalis* live in saline or alkaline soil bordering ponds or marshes.

Elongate and eel-like in form, the larvae (Figure 273) live as predators or scavengers in the interstices between soil grains. The pupae are similar in appearance to those of mosquitoes.

REFERENCES. Clastrier, T., and W. W. Wirth. 1978. The *Leptoconops kerteszi* complex in North America (Diptera: Ceratopogonidae). U.S. Department of Agriculture, Technical Bulletin, no. 1573, pp. 1-58.

Sjogren, R. D., and T. D. Foulk. 1967. Colonization studies of *Leptoconops kerteszi [Leptoconops foulki]* biting gnats. *Mosquito News,* vol. 27, pp. 394-397.

Smith, L. M., and H. Lowe. 1948. Black gnats of California. *Hilgardia,* vol. 18, pp. 157-183.

Whitsel, R. H., and R. F. Schoeppner. 1966. Summary of a study of the biology and control of the valley black gnat, *Leptoconops torrens* Townsend (Diptera: Ceratopogonidae). *California Vector Views,* vol. 13, pp. 17-25.

Wirth, W. W. 1952. The Heleidae of California. *University of California Publications in Entomology,* vol. 9, pp. 95-266.

Wirth, W. W., and A. P. Assis de Moraes. 1979. New records and new species of biting midges from salt marshes in California and Mexico (Diptera: Ceratopogonidae). *Pan-Pacific Entomologist,* vol. 55, pp. 287-298.

Wirth, W. W., and R. H. Jones. 1957. The North American subspecies of *Culicoides variipennis* (Diptera, Heleidae). U.S. Department of Agriculture, Technical Bulletin, no. 1170, pp. 1-35.

## ■ SNIPE FLIES (Family Rhagionidae; *Symphoromyia* species)

SNIPE FLIES are about the size of House Flies (body length under 1/4 in., or 6 mm) but are brownish gray or blackish and have unmarked thoraxes (House Flies are browner in color and have stripes on the back of the thorax). They are hairy to varying degrees (some are quite hirsute), and in all species the main antennal segment is kidney-shaped (Figure 274).

Large numbers may be encountered in the uplands, especially near water courses in the springtime, when their biting habits are pronounced. The bites, which raise itching red welts, are extremely irritating, and hypersensitive humans may experience shock reactions after even a few bites. Wild hosts include large mammals such as deer.

Immature forms have been described for several species, but there is no information about the habits or habitats of the larvae of most species. The larvae are maggot-like and feed on organic detritus or small organisms in the soil. Males aggregate and form loose swarms, which the females enter to pair. Development occurs in the spring, and the larvae of our local snipe flies probably develop in the damp soil of stream banks.

The commonest local biting species is unnamed.

REFERENCES. Hoy, J. B., and J. R. Anderson. 1966. Snipe flies *(Symphoromyia)* attacking man and other animals in California. Proceedings and Papers of the 13th Conference of the California Mosquito Control Association, pp. 61-64.

_______. 1978. Behavior and reproductive physiology of blood-sucking snipe flies (Diptera: Rhagionidae: *Symphoromyia)* attacking deer in northern California. *Hilgardia,* vol. 46, pp. 113-168.

Turner, W. J. 1974. A revision of the genus *Symphoromyia* Frauenfeld (Diptera: Rhagionidae), part 1: Introduction.

274. Snipe fly *(Symphoromyia* species). Drawing by T. Ross.

Subgenera and species-groups. Review of biology. *Canadian Entomologist,* vol. 106, pp. 851-868.

______. 1979. A case of severe human allergic reaction to bites of *Symphoromyia* (Diptera: Rhagionidae). *Journal of Medical Entomology,* vol. 15, pp. 138-139.

## NONBITING GNATS AND MIDGES

PERIODIC OUTBREAKS OF TINY BLACK FLIES occur frequently in the basin and attract attention from homeowners, who despair of picking gnats from their breakfast cereal. Although these little insects do not suck blood or carry disease, they can be extremely annoying. Only a few species in the three families (Sciaridae, Bibionidae, and Chironomidae) are involved.

### ■ ROOT GNATS (Family Sciaridae)

■ BLACK GNAT
*(Bradysia impatiens)*
Figure 275

This is the tiny (1/16 in., or 1 mm, long) black gnat that flits in your face while you are watching television and that always seems to get stuck in fresh paint. The larva lives in decaying plant material, such as compost, peat, and sphagnum; it also commonly infests the roots and stems of various herbaceous plants. The insect may develop in the media used for potted plants, which explains its mysterious appearance indoors.

### ■ MARCH FLIES (Family Bibionidae)

■ LITTLE BLACK MARCH FLY
*(Dilophus orbatus)*
Figure 276

The adults of this small black fly (body length 1/4 in., or 6 mm) are poor fliers, and they are often described as "those little black bumbling gnats." They are active throughout the year and common over lawns. Females have blackish wings and are distinguishable in flight from males, which have transparent wings. Males fly in a dancing pattern up and down; females fly slowly and undirectionally.

The larvae are primarily scavengers and live in decaying organic materials in the soil and among the roots of grasses and other plants. They may occur in large numbers under turf, where they do little or no direct damage. They are recognizable by a leathery gray warty skin, a distinct brown head, and an enlarged posterior pair of spiracles.

REFERENCES. Hardy, D. E. 1961. The Bibionidae of California. *Bulletin of the California Insect Survey,* vol. 6, pp. 177-195.

Rotramel, G. 1969. Orientation and coupling in *Dilophus orbatus* (Diptera: Bibionidae). *Pan-Pacific Entomologist,* vol. 45, p. 74.

275. Black Gnat female. Photograph by M. Badgley.

276. Little Black March Fly. Drawing by T. Ross.

## ■ WATER MIDGES (Family Chironomidae)

WATER MIDGE LARVAE develop in shallow areas of lakes, ponds, and streams where there is a heavy growth of aquatic plants. Adults may emerge in such numbers as to become a distinct nuisance around wastewater stabilization lagoons and homes and resorts that are near lakes and streams. Their overabundance can often be traced to pollution or human alteration of the habitat.

Occasionally, larvae living in storage and distribution systems for potable water may pass through pipes and taps, causing concern to householders. They are also known to interfere with the activated sludge process of water treatment in sewage plants.

Adults (Figure 277) are very similar to mosquitoes in size and general appearance and are often mistaken for them. But water midges lack the biting mouthparts of mosquitoes and have hairy or naked rather than scaled bodies and wings. The larvae are worm-like and are usually white; they have short filamentous gills under the rear portion of the body and a single projection behind the head. Some larvae, called "blood worms" (Figure 278), are bright red from hemoglobin in their blood, a rare condition in insects.

Small clouds of males are frequently seen hovering in the air over or near water. At times they form larger clouds that look like smoke over trees or tall structures; these aggregations are attractive to females and are the chief mating strategy of many species. Tremendous numbers may also gather around lights on warm summer evenings.

Common local species are *Chironomus attenuatus* and *Micropsectra nigripila.*

REFERENCES. Ali, A., M. S. Mulla, B. A. Federici, and F. W. Pelsue. 1977. Seasonal changes in chironomid fauna and rainfall reducing chironomids in urban flood control systems. *Environmental Entomology,* vol. 6, pp. 618-622.

Ali, A., M. S. Mulla, A. R. Pfuntner, and L. L. Luna. 1978. Pestiferous midges and their control in a shallow residential-recreational lake in southern California. *Mosquito News,* vol. 38, pp. 528-535.

Grodhaus, G. 1967. Identification of chironomid midges commonly associated with waste stabilization lagoons in California. *California Vector Views,* vol. 14, no. 1, pp. 1-12.

_______. 1975. Bibliography of chironomid midge nuisance and control. *California Vector Views,* vol. 22, pp. 71-81.

## MOTH FLIES (Family Psychodidae)

■ BATHROOM FLY
*(Clogmia albipunctata)*
Figures 279, 280

This gnat is small (wing span about 1/8 in., or 3 mm) and furry looking, with speckled wings. The antennae are white, and there is a white spot at the tip of the pointed wings. The fly rests head up with the wings flat and angled slightly out from the body.

The Bathroom Fly is often noticed indoors in damp areas—on the walls of bathrooms, showers, lavatories, and washrooms.

The brown worm-like larva develops in the sludgy organic muck that accumulates outdoors in shallow pools and tree holes and, under artificial conditions, in sink traps, drains, and dead-flow areas in the household plumbing.

*Psychoda alternata,* a smaller (wing span 1/16 in., or 1 mm) gray-colored relative of the Bathroon Fly, frequents the same damp habitats. It may rest with the head in any direction and with the wings folded rooflike over the back.

OTHER SCIENTIFIC NAMES. The species has also been called the "Drain Fly" and has gone under the scientific name *Telmatoscopus albipunctatus.*

277. Water midge female. Photograph by C. Hogue

278. Larva ("blood worm") of water midge *(Chironomus* species). Photograph by P. Bryant.

279. Bathroom Fly. Photograph by C. Hogue.

280. Bathroom Fly larva. Photograph by C. Hogue.

■ Big Black Horse Fly
*(Tabanus punctifer)*
Figures 281, 282

As a result of the disappearance, chemical treatment, or pollution of their aquatic breeding places, horse flies are now rare in the basin. The larvae of the Big Black Horse Fly still occur occasionally in marshy ponds, along lake or stream margins, and in sloughs, irrigation ditches, and similar watery habitats, usually in wet mud or decaying vegetation just at the water line.

The adults are large robust flies nearly 3/4 inch (20 mm) in body length. The male possesses very large eyes, which meet on the midline of the head, making it appear to be nearly all eye; the back of the thorax is black except for a fringe of white hairs along the side and rear borders. The female differs in that the eyes are separated and the back of the thorax is all white or pale cream.

Females of this species glue their elongate eggs in dense masses to bullrush stems, coarse grasses, trunks of small trees, and other vegetation overhanging water surfaces, at heights varying from 1 to 3 feet (30 to 90 cm). Upon hatching, the larvae drop into the water, where they feed on other aquatic invertebrates (snails and relatively sedentary bottom and shore-dwelling insects, including other horse fly larvae), which they kill with a venomous bite. When full grown, the larvae are worm-like, about 1 7/8 inches (47 mm) long, with rounded lobes circling the front third of each abdominal segment. Pupation occurs out of the water in damp shore trash or on loose soil.

Because they possess a voracious appetite for the blood of horses and cattle, the female flies may be extremely bothersome, especially when numerous. They have been observed biting rhinoceroses, tapirs, and hippopotamuses at the Los Angeles Zoo. They occasionally bite humans, with painful results. Natural saccharine fluids, such as fruit juices and nectar from flowers, nourish the nonbiting males and also serve as a diet supplement for the females.

REFERENCES. Middlekauff, W. W., and R. S. Lane. 1980. Adult and immature Tabanidae (Diptera) of California. *Bulletin of the California Insect Survey,* vol. 22, pp. 1-99.

Webb, J. L., and R. W. Wells. 1924. Horseflies: Biologies and relation to western agriculture. U.S. Department of Agriculture, Department Bulletin, no. 1218, p. 1-36. [See especially pp. 10-20.]

281. Big Black Horse Fly male. Photograph by P. Bryant.

282. Horse fly larva in water. Photograph by C. Hogue.

## FLOWER FLIES (Family Syrphidae)

NUMEROUS SPECIES OF THESE COMMON FLIES occur in the basin. Typically, the adults (Figure 283) hover about flowers and, upon landing, reveal a wasp-like black and yellow abdomen. The maggot-like larvae of most forms are beneficial because they prey on other, often injurious, insects (Figure 284). They may live on plants, as do our most prevalent species (which are in the genera *Syrphus, Metasyrphus, Scaeva,* and *Allograpta),* and do considerable good destroying aphids in gardens.

Other syrphids, such as the Drone Fly and Cactus Fly, have very different habits in the immature stages.

REFERENCE. Campbell, R. E., and W. M. Davidson. 1924. Notes on aphidophagous Syrphidae of southern California. *Bulletin of the Southern California Academy of Sciences,* vol. 23, pp. 3-9.

---

■ DRONE FLY
*(Eristalis tenax)*
Figures 285, 286

This fly receives its common name from its resemblance to the drone Honey Bee (the male of *Apis mellifera).* Moderately large (length 5/8 in., or 15 mm) and similar to a bee in color, the Drone Fly also behaves like the bee, buzzing loudly when restrained. It is, of course, still a fly and has no sting.

283. Flower fly. Photograph by C. Hogue.

284. Flower fly larva feeding on aphids. Photograph by C. Hogue.

Adults may be abundant in flower fields and may occasionally enter homes. The larvae live in water, usually in sluggish streams or small stagnant ponds that are foul with organic matter; they may also breed in fresh liquid cow manure. Because of their extremely long, extendable posterior breathing tube, the larvae are called "rat-tailed maggots."

■ CACTUS FLY *(Copestylum mexicana)* Figure 287

This is a giant member of the flower fly family (its body length is 5/8 to 3/4 in., or 15 to 20 mm), with a shiny smooth purplish-black body; it is occasionally seen in the spring and summer in areas where cacti grow. The larvae are large (body length 3/4 in. or 20 mm) pale cylindrical maggots that feed in the rotting, soupy interior of dead and decaying tissues of cacti.

OTHER SCIENTIFIC NAMES. *Volucella mexicana.*

## ROBBER FLIES (Family Asilidae)

■ BUMBLE BEE ROBBER FLY *(Mallophora fautrix fautricoides)* Figure 288

This fly's large size (length 5/8 in., or 15 mm) and robust shape, its powerful flight, and a dense covering of yellow and black hairs create a resemblance to the bumble bee. Both males and females are predaceous, feeding on the blood of other insects, which they

285. Drone Fly larva. Drawing by T. Ross after various sources.

286. Drone Fly. Photograph by J. Hogue.

287. Cactus Fly. Photograph by C. Hogue.

288. Bumble Bee Robber Fly. Photograph by C. Hogue

often capture on the wing, drag to a perch, and drain of fluid at their leisure. A taste for Honey Bees, which seem to be preferred prey of southern California's Bumble Bee Robber Flies, has earned them the name "bee killers."

Very little is known of the life cycle of this species. The larvae probably develop in the soil; it is not certain whether they feed on animal or vegetable material.

REFERENCE. Alcock, L. 1974. Observations on the behavior of *Mallophora fautrix* Osten Sacken (Diptera: Asilidae). *Pan-Pacific Entomologist,* vol. 50, pp. 68-72.

## SOLDIER FLIES (Family Stratiomyidae)

■ WINDOW FLY
*(Hermetia illucens)*
Figures 289, 290

With its buzzing flight, this moderately large fly (about 5/8 to 3/4 in., or 15 to 20 mm, long) resembles a black wasp; it is shiny black in color, with dusky wings and a pair of translucent areas ("windows") at the base of the abdomen. A thin longitudinal black line between the windows creates the illusion of a narrow waist. The adults are sluggish until induced to fly. They may be attracted into buildings by lights, but they do not molest human food.

The larva is robust, tapered in outline, and somewhat flattened, with a tough brown leathery skin covered with numerous short bristles. It is terrestrial and breeds in various organic substances, commonly decaying fruits and vegetables. In urban areas, the larva occurs in garden soil, in piles of compost and ground mulch, and near garbage pails that have been placed on soil. It is also found in defunct Honey Bee combs.

There is no certainty that putrifying materials are actually the larva's food. However, there is evidence that other fly larvae that are present in such decaying media, or Honey Bee larvae in hives, may be preyed upon by the Window Fly larvae.

In spite of its waspish look and aggressiveness, the fly neither bites nor stings people.

REFERENCE. Furman, D. P., R. D. Young, and E. P. Catts. 1959. *Hermetia illucens* (Linnaeus) as a factor in the natural control of *Musca domestica* Linnaeus. *Journal of Economic Entomology,* vol. 52, no. 5, pp. 917-921.

289. Window Fly. Photograph by C. Hogue.

290. Window Fly larva. Photograph by C. Hogue.

291. Large Bee Fly. Photograph by C. Hogue.

292. Black-winged bee fly *(Hemipenthes sinuosa jaennickiana)*. Photograph by C. Hogue.

## BEE FLIES (Family Bombyliidae)

MOST BEE FLIES ARE RECOGNIZED by their furry bee-like bodies and nectar-feeding habits (hence their name) and also by the bold patterns of dark splotches and spots on their wings. They are frequently seen on warm days hovering in front of flowers or resting on the ground with wings held at an angle and flat to the sides. They feed on nectar with a long slender rigid proboscis.

A very common local species is the Large Bee Fly *(Bombylius major;* Figure 291), a medium-sized fly with a body length of 1/4 to 3/8 inch (6 to 10 mm) and a wing span of 5/8 to 7/8 inch (15 to 22 mm). It can be recognized by the dark brown shading on the front third of its wings and the dark rounded spot in the clear portion adjacent to this shading. The abdomen is black with a fringe of white hairs. The larvae probably live in the nests of solitary bees upon whose young they prey.

There are also a few kinds of bee flies in our area with largely black wings. These are *Hemipenthes sinuosa jaennickiana* (Figure 292), *Anthrax analis,* and *Villa atrata.* Virtually nothing is known about the biology of any of these species, but the larvae of *H. sinuosa jaennickiana* are probably hyperparasites (parasites on other parasitic species) of Hymenoptera (ants, bees, and wasps) and other dipterans.

REFERENCE. Hull, F. M. 1973. *Bee flies of the world.* Washington, D.C.: Smithsonian Institution.

## TACHINID FLIES (Family Tachinidae)

MOST SPECIES OF THIS GROUP resemble overgrown House Flies but are covered with extra-heavy stiff bristles. The thorax is usually a solid color and is not banded lengthwise as it is in other domestic flies. Tachinid flies are often seen flying closely over or crawling through low vegetation in search of insect larvae (cutworms and the like) upon which to lay their eggs. The larvae are

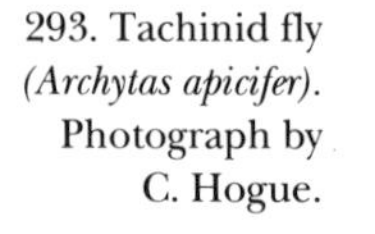

293. Tachinid fly *(Archytas apicifer).* Photograph by C. Hogue.

parasitoids that feed internally on the host; as the hosts are caterpillars, including cutworms and loopers, tachinid flies are considered beneficial to the gardener and farmer.

A common local species is *Archytas apicifer* (Figure 293); it is about 1/2 inch (13 mm) long and recognizable by its contrasting olive thorax and shiny purplish-black abdomen. It parasitizes cutworms.

## DOMESTIC FLIES

THIS GROUP INCLUDES SEVERAL FLY SPECIES that are more or less adapted to conditions produced by civilization, especially our organic wastes, which serve as the principal food of the larvae. Adults frequently enter houses and are pesky with their buzzing and crawling habits. Moreover, since the adults frequent filthy places—walking over feces, garbage, and other substances commonly contaminated with disease organisms—some may pick up and transmit germs in a mechanical way on their hairy feet, mouthparts, and bodies.

Some of these flies are known to accumulate around natural gas leaks. They are probably attracted by ethyl mercaptan, a smelly substance added to gas to make leaks detectable to the human nose. The odor of ethyl mercaptan is similar to that of volatile substances released during the decomposition of carrion, upon which many domestic flies oviposit and their larvae feed.

Many species also develop in dead animals, including humans. The insect species present in a corpse, and their stages of development, are regularly used in fixing the time of death of a person and are useful tools of forensic medicine.

Occasionally an adult fly may lay eggs in an untreated wound of an otherwise healthy person; the subsequent appearance of maggots in the wound is a condition called myiasis. The feeding of the larvae is usually confined to dead tissues, however, and they do little direct harm.

The domestic flies described below—those most likely to be seen by residents of the Los Angeles Basin—are representative of three families: the Muscidae (the House, Stable, False Stable, Little House, and Canyon flies), the Sarcophagidae or flesh flies, and the Calliphoridae or blow flies.

REFERENCES. Ecke, D. H., editor. 1963. A field guide to the domestic flies of California. California Mosquito Control Association, Field Guide Series, Publication no. 3.

Smith, K. G. V. 1986. *A manual of forensic entomology*. London: British Museum (Natural History).

■ HOUSE FLY
*(Musca domestica)*
Figures 294, 295

The House Fly is a truly domestic insect, so closely adapted to life in manmade environments that it is rarely found away from human habitations. The species, which is found throughout the basin and the world, is our worst pest among the flies.

The fly's color is generally dull grayish brown; there are four dark longitudinal stripes on the back of its otherwise solid pale gray thorax, and these are most clearly defined toward the front. It is small, with a body length of slightly less than 1/4 inch (6 mm). At rest, the wings are held out at a slight angle from the body, giving the fly a triangular shape when viewed from above. The larva is a typical cream-colored maggot. The spiracles located at the rear end of the body (Figure 295), which are closely spaced and nearly touch each other, are distinctively D-shaped, with sinuous slits.

All kinds of decaying and fermenting organic material—commonly decomposing lawn clippings, garbage, and feces of dogs, cats, horses, rabbits, and poultry—provide breeding places for the larvae. Pupation occurs away from the food in drier areas in the soil or under objects on the ground. Adult flies feed on liquids containing high concentrations of organic substances, such as syrups, milk, sputum, and other animal secretions. Dry materials may also be used as food; to do this, the fly disgorges a bit of its own saliva to liquify the solid and then drinks the dissolved substance. The insect's frequent regurgitation on almost every surface upon which it rests results in the small brown opaque spots we call "fly specks."

House Flies are vigorous insects and may breed throughout most of the year in the basin. They are most common, however, during the warmer months.

Because its mouthparts are soft and adapted only for sponging up free liquids, the House Fly cannot bite. Suspected instances of House Fly bites are usually attributable to a similar species, the Stable Fly.

REFERENCES. Colwell, A. E., and H. H. Shorey. 1975. The courtship behavior of the House Fly, *Musca domestica* (Diptera: Muscidae). *Annals of the Entomological Society of America,* vol. 68, pp. 152-156.

Eldridge, B. F., and M. T. James. 1957. The typical muscid flies of California. *Bulletin of the California Insect Survey,* vol. 6, no. 1, pp. 1-17.

West, L. S. 1951. The Housefly: Its natural history, medical importance, and control. Ithaca, N.Y.: Comstock.

294. House Fly. Photograph by C. Hogue.

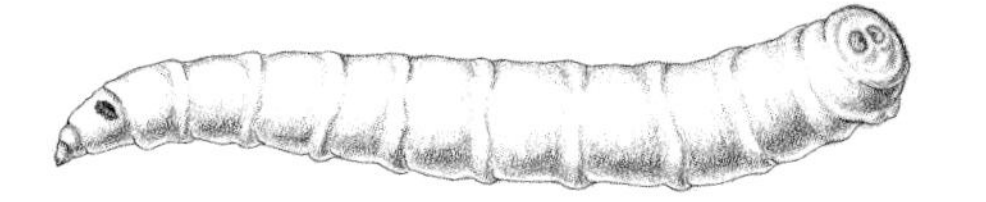

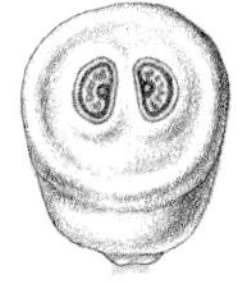

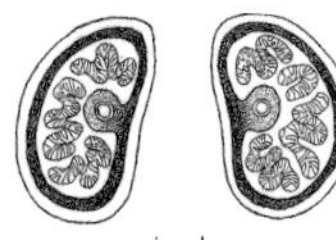


295. House Fly larva, side view (top) and view from rear (bottom left) with closeup of spiracles (bottom right). Drawing by T. Ross after various sources.

296. Stable Fly. Illustration from B. Greenberg, 1971, *Flies and disease,* vol. 1 (Princeton, N.J.: Princeton University). Courtesy Princeton University Press.

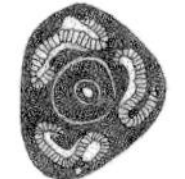

297. Stable Fly larval spiracles. Drawing by T. Ross after K. Smith, editor, 1973, *Insects and other arthropods of medical importance,* fig. 149E (London: British Museum of Natural History).

■ **Stable Fly**
*(Stomoxys calcitrans)*
Figures 296, 297

This muscid species resembles the House Fly and is about the same size (its body length is just under 1/4 in., or 6 mm). But the Stable Fly can be distinguished most easily by its heavy hard shiny black nonretractable proboscis. These mouthparts, unlike those of the House Fly, are adapted for blood-sucking and contain barbs that readily pierce the skin of people and animals. Also, this species has distinct rounded spots on its abdomen that are lacking in the House Fly. Both male and female Stable Flies are vicious biters, especially during warm humid weather. They are most common in late summer and fall and are frequently seen resting on the sunny surfaces of houses, cars, and fence posts.

The breeding habits of this species are similar to those of the House Fly. The maggots develop in substrata rich in cellulose—in animal manure and waste materials around farms, and in compost and piles of lawn clippings in urban areas. They are also found in piles of seaweed along beaches and may give rise to concentrations of adult flies that are extremely annoying to bathers; they are often called "beach flies," and their bites are also sometimes blamed on putative "sand fleas."

The maggot is similar to that of the House Fly, but the slits in the posterior spiracles are strongly sinuous, and the surrounding plates themselves are nearly triangular and widely separated (by a distance twice their width).

■ **False Stable Fly**
*(Muscina stabulans)*
Figure 298

The False Stable Fly is another muscid species that resembles the House Fly, but this fly is larger (about 3/4 in., or 20 mm, long), and the tip of its scutellum is pale yellow. The abdomen is almost black but is covered with gray in places, giving it a blotched appearance. In the outer portion of the wing, all of the veins are smoothly curved (in the House Fly, the central vein bends forward at a sharp angle near the tip of the wing).

The spiracles of the maggot are raised and on a rounded rather than D-shaped plate as in the House Fly.

This species' adult and larval habits are similar to those of the House Fly.

This is the fly you see on hot summer days, in the garage, under trees, in doorways, and in other shaded places, hovering in the air, seeming never to land nor to have a place to go. This aimless flight immediately distinguishes this species from our other domestic flies, which fly with purposeful direction and frequently come to rest. Swarms of Little House Flies consist mainly of males; the females usually remain at rest nearby.

■ LITTLE HOUSE FLY
*(Fannia canicularis)*
Figure 299, 300

The Little House Fly is 3/16 inch (5 mm) long, smaller than the House Fly, and it has a slightly more

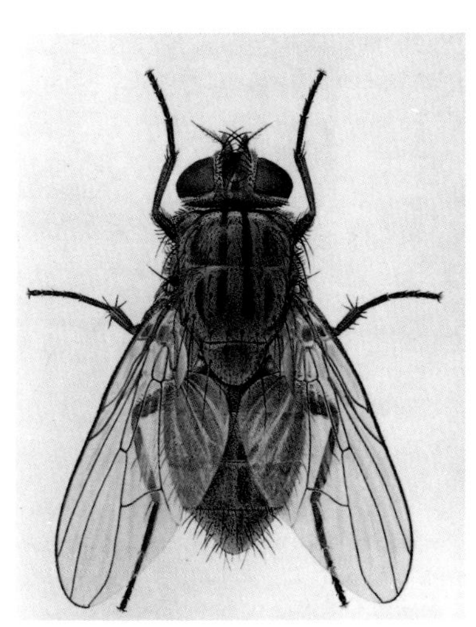

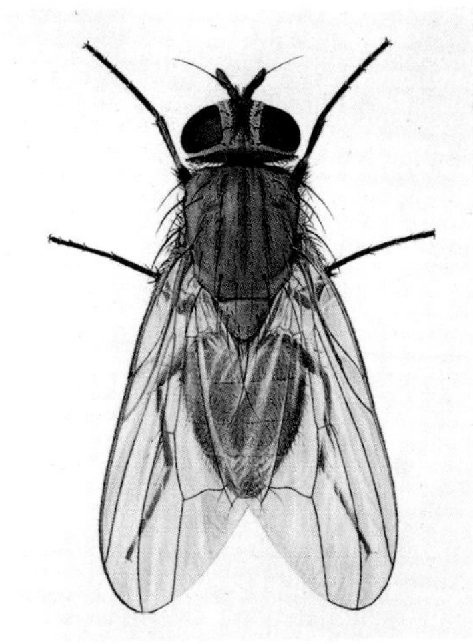

298 (above, left). False Stable Fly. Illustration from B. Greenberg, 1971, *Flies and disease,* vol. 1 (Princeton, N.J.: Princeton University). Courtesy Princeton University Press.

299 (above, right). Little House Fly. Illustration from B. Greenberg, 1971, *Flies and disease,* vol. 1 (Princeton, N.J.: Princeton University). Courtesy Princeton University Press.

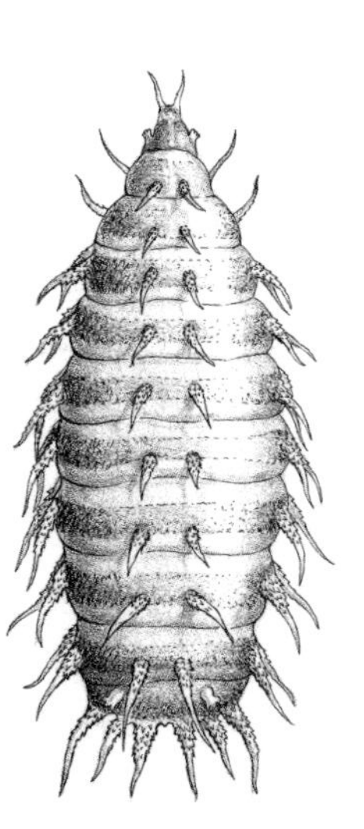

300 (left). Little House Fly larva. Drawing by T. Ross after various sources.

slender body. At rest its wings overlap and are held together in parallel fashion. Although a maggot, the larva is atypical in that it is flat, oval in shape, and possesses numerous branched projections radiating from edges of the body.

Breeding occurs in a wide variety of rotting organic materials. The larvae in our area show some preference for chicken manure, and the species abounds around poultry ranches. In urban areas breeding sites are similar to those of the other domestic flies.

■ CANYON FLY (*Fannia benjamini* complex) Figure 301

These muscid flies are common to dry chaparral and oak woodland areas throughout California. Seeking moisture, they have the very annoying habit of flying into the faces and especially the eyes of humans and animals. Because of such intimate contacts, the Canyon Fly is suspected host of the Eyeworm, *Thelazia californiensis,* a nematode parasite of human beings.

The several species belonging to this complex are small (body length about 1/8 in., or 3 mm) and mottled gray, and all have short legs and clear wings. *Fannia benjamini* is the most common local species, but others closely related to it may be present. The larva is similar to that of the Little House Fly but has only been seen in laboratory cultures because its natural habitat remains unknown.

REFERENCES. Poorbaugh, J. H. 1969. Laboratory colonization of the canyon fly, *Fannia benjamini* Malloch, and speculation on the larval habitat. *California Vector Views,* vol. 16, no. 3, pp. 21-24.

Turner, W. J. 1976. *Fannia thelaziae,* a new species of eye-frequenting fly of the *banjamini* group from California and description of *F. conspicua* female. *Pan-Pacific Entomologist,* vol. 52, pp. 234-241.

## ■ FLESH FLIES (Family Sarcophagidae)

RESEMBLING OVERGROWN HOUSE FLIES, most individuals of this family are pale gray overall, with the thorax longitudinally striped with black and the abdomen marked by a checkerboard pattern of light and dark spots that seem to change positions according to the angle of reflected light. Flesh flies have rather broad pads on the feet; males of many species have a red-tipped abdomen (Figure 302).

Most of the members of this family are wild flies, but many species accidentally enter dwellings that are near their breeding sites. The larvae live in fish and animal carcasses and other decomposing organic

301. Canyon Fly. Drawing by T. Ross.

302. Flesh fly male. Photograph by C. Hogue.

303. Larva of flesh fly *(Parasarcophaga* species) from side (top) and rear (bottom left) with close-up of spiracles (bottom right). Drawing by T. Ross after M. James, 1947, *The flies that cause myiasis in man,* fig. 14 (United States Department of Agriculture, Miscellaneous Publications, no. 631).

matter, particularly discarded meat. They frequently infest the remains of poisoned garden snails.

Unlike the other flies mentioned here, many flesh flies do not lay eggs but instead give birth to living, active larvae. The adult feeds like a House Fly and does not bite.

The maggots (Figure 303) grow to larger sizes than those of House Flies and other muscids (they are up to $^{3}/_{4}$ to 1 in., or 20 to 25 mm, long). The spiracles are found in the bottom of a deep pocket or concavity in the blunt rear end of the larval body.

REFERENCE. Aldrich, J. M. 1916. Sarcophaga *and allies in North America.* LaFayette, Ind.: Entomological Society of America (Thomas Say Foundation).

## ■ BLOW FLIES (Family Calliphoridae)

THIS FAMILY CONTAINS shiny blue-black or greenish flies. In most species, the thorax is dull blue to black, the abdomen deep metallic blue.

Blow fly biologies are all somewhat similar. Eggs are usually laid on the meat of animals that have just died. The egg-laying process is called "blowing," and this is the root of the common name for the family. The term "bottle flies" is also applied to these insects (bottle meaning "little bot") because the larvae are confused with those of true bot flies (insects of the family Cuterebridae and their relatives), which infest living animal tissue.

Adult blow flies sometimes hibernate in unheated buildings. During warm months they may seek shelter in cool places such as cellars, garages, and basements. The adults are somewhat variable in size (length of 3/16 to 3/8 in., or 5 to 12 mm), depending on the nutrition of the larvae.

Several species occur locally; those described below are the most common.

■ Green Bottle Flies *(Phaenicia* species; Figures 304, 305). In the two species of green bottle fly *(Phaenicia sericata* and *P. cuprina),* the body is light metallic green or bronzy throughout. These are the principal garbage-infesting flies in the basin. The maggots of *P. sericata* occasionally infest untended wounds in humans, and the adults are the flies most frequently seen feeding on fresh dog feces. The larvae of both species develop on meat or fresh decaying vegetative matter (especially in household garbage cans or trash containing garbage).

■ Common Blow Fly *(Eucalliphora lilaea;* Figures 306, 308 top). Adults of this species are variable in size and coloration; typical specimens have a dull black thorax that is without lines and is usually noticeably narrower than the abdomen, which is deep metallic blue. The upper bristles alongside the dark mouthpart cavity are stout (Figure 308 top). The Common Blow Fly breeds in freshly dead carcasses of small animals. Adults readily enter houses and may also occur there when they emerge from the bodies of poisoned or trap-killed mice and rats.

■ European Blue Bottle Fly *(Calliphora vicina;* Figures 307, 308 bottom). This species is superficially identical in behavior and appearance to the Common Blow Fly; its habits are nearly the same, and its abdomen is also much broader than its thorax. But this species has

304. Green bottle flies *(Phaenicia sericata)*. Photograph by C. Hogue.

305. Larva of green bottle fly *(Phaenicia sericata)*. Drawing by T. Ross after various sources.

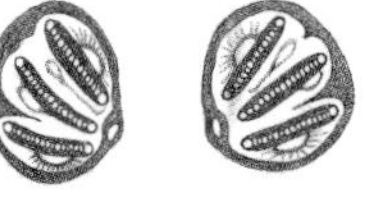

306. Common Blow Fly. Photograph by C. Hogue.

fewer and finer upper bristles along the side of its mouthpart cavity (Figure 308 bottom), which is pale, rather than dark.

■ Black Blow Fly *(Phormia regina)*. The body of this species is a dark dull metallic blue throughout (it has no lines on its thorax); its abdomen is as narrow as the thorax.

■ Wheeler's Blue Bottle Fly *(Paralucilia wheeleri)*. In this species, the thorax and abdomen are deep metallic blue, and the thorax has three strong dark longitudinal lines.

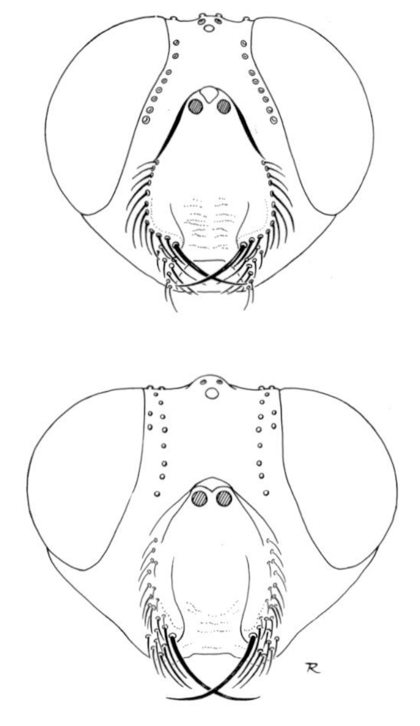

307 (left). European Blue Bottle Fly. Illustration from B. Greenberg, 1971, *Flies and disease,* vol. 1 (Princeton, N.J.: Princeton University). Courtesy Princeton University Press.

308 (right). Front view of heads of Common Blow Fly (top) and European Blue Bottle Fly; note heavier upper facial bristles in Common Blow Fly. Drawing by T. Ross.

■ Cluster Fly *(Pollenia rudis;* Figure 309). This fly does not have a metallic surface. Instead, the abdomen has a checkered pattern, and the thorax is lead gray with three diffuse black bars; long crinkly golden hairs are present over most of the body. The species is generally scarce in our region. The larvae are internal parasites of earthworms.

■ *Chrysomya* species. In 1988 specimens of blow flies of the genus *Chrysomya (C. megacephala* (Figure 310) and *C. rufifacies)* were found for the first time in the Los Angeles area. In both species, the body is shiny metallic green throughout, and the posterior margins of the abdominal segments are black. They may also be recognized by the swelling on the thorax near the base of the wing and the lower basal wing folds (calypters), both of which are hairy.

These two species are native to the Orient and Australia, where they are potentially serious pests because they breed in carrion, garbage, and open privies and also cause myiasis in domestic animals. Locally they are fond of dead fish. They seem to have become permanently established here.

REFERENCES. De Coursey, R. M. 1927. A bionomical study of the cluster fly, *Pollenia rudis* (Fabr.) (Diptera: Calliphoridae). *Annals of the Entomological Society of America,* vol. 20, pp. 368-384.

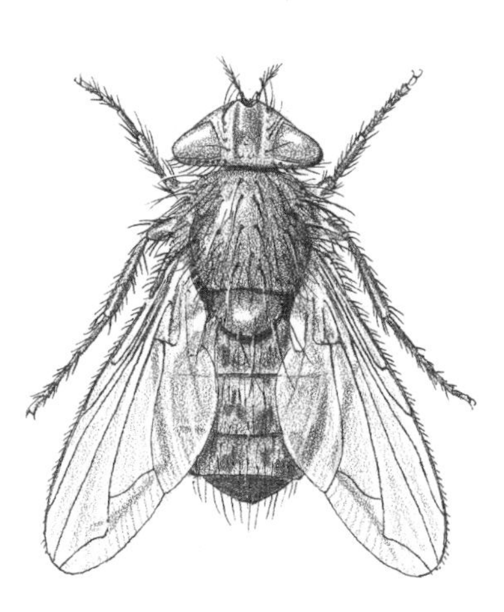

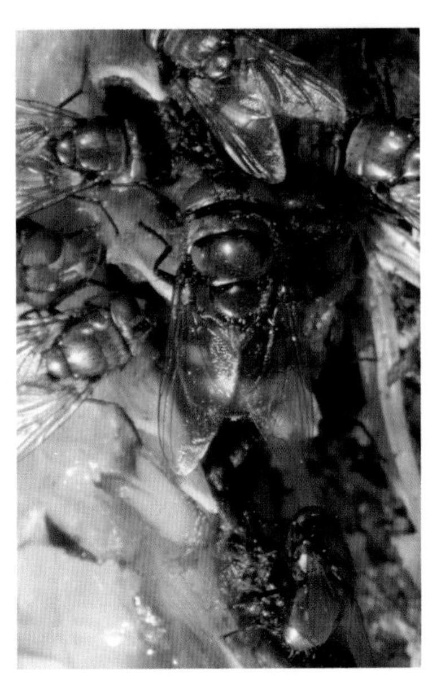

309 (left). Cluster Fly. Drawing by C. Hogue from J. A. Powell and C. L. Hogue, 1979, *California insects* (Berkeley, Calif.: University of California).

310 (right). Introduced blow fly *(Chrysomya megacephala;* center) among green bottle flies. Photograph by C. Hogue.


James, M.T. 1955. The blowflies of California. *Bulletin of the California Insect Survey,* vol. 4, no. 1, pp. 1-34.

Merritt, R. W. 1969. A severe case of human cutaneous myiasis by *Phaenicia sericata* (Diptera: Calliphoridae). *California Vector Views,* vol. 16, pp. 24-26.

Morse, A. P. 1911. *Lucilia sericata* as a household pest. *Psyche,* vol. 18, pp. 89-92.

Olsen, A. R., and T. H. Sidebottom. 1990. Biological observations on *Chrysomya megacephala* (Fabr.) (Diptera: Caliphoridae) in Los Angeles, California, and the Palau Islands. *Pan-Pacific Entomologist,* vol. 66, pp. 126-130.

Poorbaugh, J. H., and E. M. Fisher. 1988. *Chrysomya* blow flies. California Department of Food and Agriculture Detection Advisory Publications, no. PD61-88.

Richards, P. G., and F. O. Morrison. 1972. A summary of published information on the Cluster Fly, *Pollenia rudis* (Fabricius) (Diptera: Calliphoridae). *Phytoprotection,* vol. 53, pp. 103-111.

Rognes, K. 1987. The taxonomy of the *Pollenia rudis* species-group in the Holarctic Region (Diptera: Calliphoridae). *Systematic Entomology,* vol. 12, pp. 475-502.

Thomson, A. J. 1975. The biology of *Pollenia rudis,* the Cluster Fly (Diptera: Calliphoridae), part 4: A preliminary model of the Cluster Fly life cycle, and some possible methods of biological control. *Canadian Entomologist,* vol. 107, pp. 855-863.

■ WOOD RAT BOT FLY
*(Cuterebra latifrons)*
Figure 311

This large fly (body often exceeding 3/4 in., or 20 mm, in length) is seldom seen because it is a wild species with fleeting habits. The body is bulky, looking very much like that of a bumble bee, and black in color, with the exception of an area of dense white velvet over the lower half of the head and thorax; the abdomen is somewhat mottled on the sides. The wings are dusky brown.

The full-grown larva is a very large (to 1 in., or 25 mm, in length) spiny skinned grub ("bot" or "warble"), which has developed within the body of a wood rat *(Neotoma fuscipes)*. Female flies lay their eggs on the sticks that make up the rat's house, usually those near the entrance to the tunnel leading to the inner nest. The larvae develop within the egg shell but do not hatch until stimulated by the motion and body heat of a passing rat. Once hatched, they attach to the host, quickly burrow into its body through mucous membranes, and then migrate deeply through internal tissues and begin to develop. The mature (third instar) larva lies just beneath the skin, often at the neck or shoulder (Figure 311). It forms a swollen nodule with a hole in the center through which it breathes (the posterior breathing pores may be seen as the larva presses them against the hole to take in air from the outside). When the larva is ready to pupate, it emerges through this hole and drops to the ground, where it forms the protective puparium usual in species of flies.

The feeding and burrowing of these endoparasites seem to do little harm to the host, and the wounds produced by the warbles heal quickly once the larvae leave to pupate. All adult bot flies lack mouthparts and do not feed.

There are two additional species of bot flies in this genus that parasitize wild hosts in the basin: *Cuterebra tenebrosa* (entirely black-haired and also on wood rats) and *C. lapivora* (on cottontail rabbits).

REFERENCES. Catts, E. P. 1967. Biology of a California rodent bot fly *Cuterebra latifrons* Coquillett (Diptera: Cuterebridae). *Journal of Medical Entomology,* vol. 4, pp. 87-101.

_______. 1964. Laboratory colonization of rodent bot flies (Diptera, Cuterebridae). *Journal of Medical Entomology,* vol. 1, pp. 195-196.

Meyer, R. P., and M. E. Bock. 1980. Aggregation and territoriality of *Cuterebra lepivora* (Diptera: Cuterebridae). *Journal of Medical Entomology,* vol. 17, pp. 489-493.

Ryckman, R. E. 1953. *Cuterebra latifrons* reared from *Neotoma fuscipes macrotis* (Diptera: Cuterebridae). *Pan-Pacific Entomologist,* vol. 29, pp. 155-156.

Ryckman, R. E., and C. C. Lindt. 1954. *Cuterebra lepivora* reared from *Sylvilagus andubonii sanctidiegi* in San Bernardino County, California. *Journal of Economic Entomology,* vol. 47, pp. 1146-1148.

Sabrosky, C. W. 1986. *North American species of* Cuterebra, *the rabbit and rodent bot flies (Diptera: Cuterebridae).* Thomas Say Foundation Monograph. College Park, Md.: Entomological Society of America.

---

■ **Sheep Bot**
*(Oestris ovis)*
Figures 312, 313

The Sheep Bot is a large-sized fly (body length 1/2 in., or 13 mm) that is grayish-brown with dark spots on the thorax and abdomen; the head is yellowish and lacks mouthparts.

The female's eggs hatch within her body. She enters the eyes, nose, or throat of a host (a sheep, goat, horse, or other hoofed animal) and expels the live-borne larvae directly onto the surface tissues, into which they burrow.

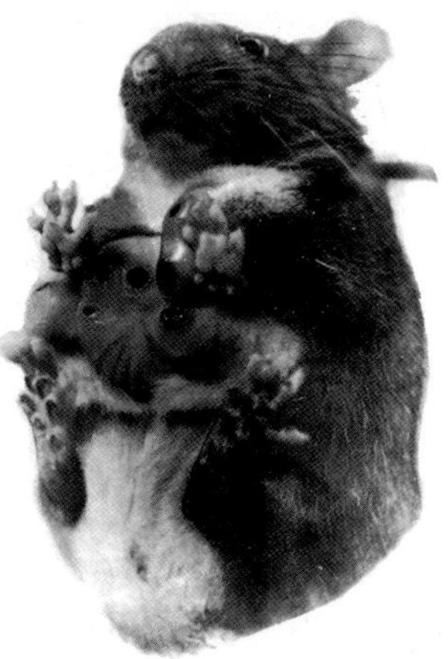

311. Wood Rat Bot Fly adult resting on leaf (top) and wood rat infested with fly's larvae; each of the holes in the wood rat's abdomen and leg is the breathing hole of a single larva. Photograph by E. Catts, Jr.

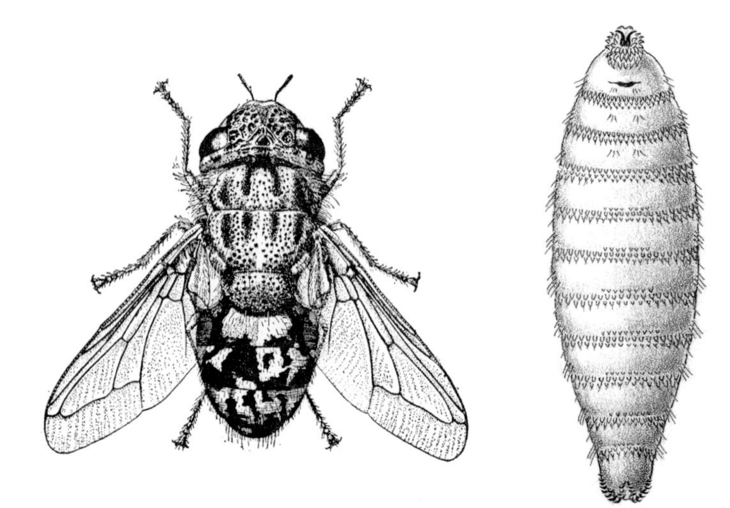

312 (left). Sheep Bot. Drawing from M. James, 1947, *The flies that cause myiasis in man,* fig. 62 (United States Department of Agriculture, Miscellaneous Publications, no. 631).

313 (right). Sheep Bot first instar larva. Drawing by T. Ross after various sources.

When first hatched, the grub is tiny (body length about 1/16 in., or 1 mm) and white, with rows of microscopic shark-toothed-shaped hooklets on its body. In its usual hosts the maggot works its way into the sinus cavities of the head (or tunnels into the skull or jaw bones) to mature. The full grown larva is nearly 1 inch (25 mm) long, a thick grub with its segments prominently marked with dark bands.

Sheep and goat herders occasionally suffer from infestations of the first instar larvae of this fly in the conjunctiva of the eye: some local cases have been recorded, and the affliction is well known on Santa Catalina Island.

REFERENCES. Brown, H. S., Jr., T. C. Hitchcock, Jr., and R. Y. Foos. 1969. Larval conjunctivitis in California caused by *Oestrus ovis. California Medicine,* vol. 111, no. 4, pp. 272-274.

Ryckman, R. E. 1981. Ophthalmomyiasis due to the sheep bot fly in southern California (Diptera: Oestridae). *Bulletin of the Southern California Academy of Sciences,* vol. 80, pp. 137-138.

■ FRUIT FLIES
*(Drosophila* species)
Figures 314, 315

"Fruit fly" is an unsatisfactory name for these insects because it also refers to another group, the "true" fruit flies (Family Tephritidae), whose larvae feed on growing fruit. Other names for *Drosophila* are "sour flies" and "vinegar flies." The small maggot larvae of our several common species live in fermenting fruit, feeding on the yeasts that flourish in such material.

The adults of the best known species, *Drosophila melanogaster,* are usually associated with man and are found in houses, in orchards and fields, and as habitual inhabitants of fruit racks in grocery stores. They are tiny (body length about 3/32 in., or 2 mm), red-eyed, and orange-bodied; these are the gnat-like flies that seem to appear out of nowhere when a lunch bag is opened or a banana is peeled. Their olfactory senses, which are located in the third antennal segment, are extremely keen, and they respond from a considerable distance to the odors of the various volatile organic compounds in fruits.

*Drosophila melanogaster* is the famous "Fruit Fly" of genetics research. Doubtless more is known about the heredity of this species than almost any other

314. "Vinegar" fly *(Drosophila melanogaster).* Photograph by M. Badgley.

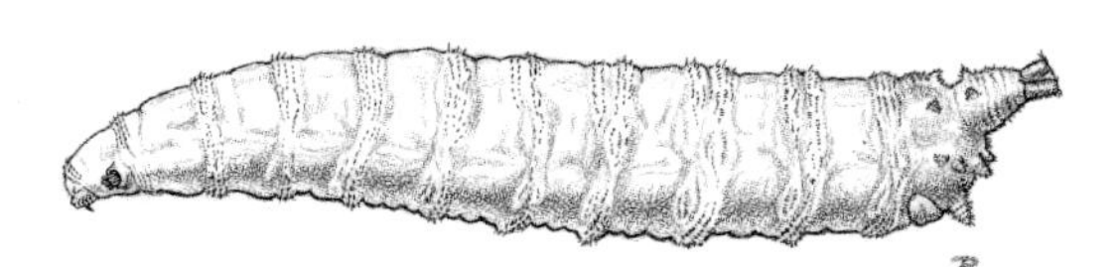

315. Larva of "vinegar" fly *(Drosophila melanogaster).* Drawing by T. Ross after J. McAlpine, 1987, *Manual of Nearctic Diptera,* vol 2, p. 1015, fig. 17 (Ottawa: Agriculture Canada, Monographs, no. 28).

organism on earth. Detailed maps have been drawn of the giant chromosomes found in the cells of the larva's salivary glands. The theory of heredity as it is now generally accepted was first derived mainly from studies of this fly. The species was chosen for this type of scientific work because of its short life cycle, its great ability to proliferate, and the ease with which it can be bred and manipulated in large quantities in the laboratory.

REFERENCES. Demerec, M., editor. 1950. *Biology of* Drosophila. New York: Wiley.

Demerec, M., and B. Kaufmann. 1965. Drosophila *guide: Introduction to the genetics and cytology of* Drosophila melanogaster, 7th edition. Washington, D.C.: Carnegie Institute of Washington.

Rubin, G. M. 1988. *Drosophila melanogaster* as an experimental organism. *Science,* vol. 240, pp. 1453-1459.

## TRUE FRUIT FLIES (Family Tephritidae)

ADULT MEMBERS OF THIS FAMILY are readily distinguished from the the drosophilid fruit flies by their wings, which are "pictured" (distinctly marked) with a pattern of black, brown, or multicolored spots or bars. They are also larger (1/8 to 1/4 in., or 3 to 6 mm).

The larvae typically feed in the interior of growing fruit, and many species are highly destructive to citrus, pome, and other fruit and vegetable crops. The Mediterranean Fruit Fly *(Ceratitis capitata;* Figure 316) belongs in this category and is the species of greatest economic importance. Apparently, it is not a permanent resident in the continental United States. But because there is always a likelihood of introduction, and because termporary infestations have frequently been detected, the species poses a constant threat to the country's agricultural future.

The "Medfly" has over 260 host plants, including avocado, citrus, pome fruits, peppers, and guavas. Females lay only a few eggs at a time, and the larvae feed, usually in fruit tissue, for just under two weeks. The life cycle averages twenty-one to thirty days under summer conditions, although adults can live up to a year.

Agricultural officials maintain constant vigilance to detect introductions of fruit flies. Through the use of cardboard traps containing synthetic pheromone lures that attract adults, several species have been found in the basin over the years. These include the Oriental Fruit Fly *(Dacus dorsalis),* Mexican Fruit

316. Mediterranean Fruit Fly, or "Medfly." Photograph by M. Badgley.

317. Walnut Husk Fly. Photograph by M. Badgley.

Fly *(Anastrepha ludens)*, and a few other rare intruders. Lasting infestations have fortunately been avoided by a combination of luck, natural failure of these tropical insects to survive our cold winters, and eradication programs. In the summer of 1989 and at intervals thereafter, numerous viable Medfly females were found in traps in the Los Angeles area. While I write these pages (May 1990), officials of the California State Department of Food and Agriculture are waging a major campaign against the flies, spraying regions of suspected infestation with Malathion droplets from helicopters and releasing sterile males in attempts to eradicate the invaders. The public has reacted very negatively to the Malathion spraying because of mistrust of its safety to people, pets, and wildlife, concern for the environment, the nuisance of having to protect property during spraying, and the annoyance of the noise of the repeated nighttime helicopter forays overhead. The uproar has resulted in attempts by local governments to curtail the spraying, which have so far been to no avail because the infestation is classified as a state of emergency. The threat of the fly and the future of the control campaign remain uncertain.

The most lastingly injurious species among the several fruit flies native to the basin is the Walnut Husk Fly *(Rhagoletis completa;* Figure 317), whose larvae primarily infest walnuts. It was introduced into southern California (specifically, into Chino) from the eastern United States in 1926. The maggots destroy the husk, but their feeding may also result in reduction of kernel quality. The species has severely depressed walnut cultivation in the Los Angeles area.

REFERENCES. The saga of the 1989-1990 Medfly war was reported almost daily in the pages of the *Los Angeles Times* for that period.

Anonymous. 1975. Walnut Husk Fly. *Diamond Walnut News,* vol. 57, no. 3, pp. 17-20.

Boyce, A. M. 1934. Bionomics of the Walnut Husk Fly, *Rhagoletis completa. Hilgardia,* vol. 8, pp. 363-579.

Carey, T. R., and R. V. Dowell. 1989. Exotic fruit fly pests and California agriculture. *California Agriculture,* vol. 43, pp. 38-40.

Foote, R. H., and F. L. Blanc. 1963. The fruit flies or Tephritidae of California. *Bulletin of the California Insect Survey,* vol. 7, pp. 1-117.

Rhode, R. H., J. Simon, A. Perdomo, J. Gutierrez, C. F. Dowling, Jr., and D. A. Linquist. 1971. Application of the sterile-insect-release technique in Mediterranean fruit fly suppression. *Journal of Economic Entomology,* vol. 64, pp. 708-713.

Wasbauer, M. S. 1972. An annotated host catalog of the fruit flies of America north of Mexico (Diptera: Tephritidae). Occasional Papers of the California Department of Agriculture, Bureau of Entomology, no. 19, pp. 1-172.

Weems, H.V., Jr. 1981. Mediterranean fruit fly, *Ceratitis capitata* (Wiedemann) (Diptera: Tephritidae). Entomology Circular, Florida Department of Agriculture and Consumer Services, no. 230, 8 pp.

## ANTHOMYIID AND KELP FLIES (Families Anthomyiidae and Coelopidae)

■ BEACH FLIES *(Fucellia* and *Coelopa* species) Figures 318, 319

Although they seldom are attracted to or land upon humans, two kinds of flies are often noticed by beachgoers: these are species of the genera *Fucellia* (family Anthomyiidae) and *Coelopa* (family Coelopidae). Adult *Coelopa* (body length 3/16 in., or 5 mm) are dark and have a very flat, naked thorax and stout legs. In winter enormous quantities mass together in recesses on rocky shores. *Fucellia* (body length 1/8 in., or 3 mm) have a rounded bristled thorax and are similar in appearance to House Flies. They often gather in large numbers on boats an-

318. Beach fly *(Fucellia* species). Drawing by T. Ross.

319. Beach fly *(Coelopa* species). Drawing by T. Ross.

chored just offshore from our beaches, where they become extremely bothersome, crawling over everything on board. When the craft gets underway, the flies are driven away by the wind.

In the late summer, adults of these species (also known as kelp flies) swarm over kelp that has washed ashore and may literally darken the sand with their numbers. The maggots of both types feed on stranded masses of seaweed that have begun to rot, and pupation occurs in the upper layers of sand beneath the algae. In the past, before human cleanup crews patrolled beaches, these flies were responsible for the degradation of piles of kelp. Now, from time to time, wrack is inadvertently buried by bulldozers, and the flies' breeding is encouraged.

REFERENCES. Aldrich, T. M. 1918. The kelp-flies of North America (genus *Fucellia,* family Anthomyidae). *Proceedings of the California Academy of Science* (series 4), vol. 8, pp. 157-179.

Huckett, H. C. 1971. The Anthomyiidae of California. *Bulletin of the California Insect Survey,* vol. 12, pp. 1-121.

Poinar, G. O., Jr. 1977. Observations on the kelp fly, *Coelopa vanduzeei* Cresson in southern California (Coelopidae: Diptera). *Pan-Pacific Entomologist,* vol. 53, pp. 81-86.

## SHORE FLIES (Family Ephydridae)

■ PETROLEUM FLY
*(Helaeomyia petrolei)*
Figure 320

Living in our area is one of the world's biological curiosities. This is the Petroleum Fly, so called because the larva lives its entire life submerged in the pools of crude oil that accumulate around natural oil seeps and in commercial oil fields. Oil pool maggots have been known to petroleum technologists for many years, but it was not until 1898 that the maggots were reared by entomologists and their nature defined. It is not certain what physiological adaptations permit this maggot to live in this normally inhospitable medium, although an exceptionally impermeable cuticle probably helps.

The evidence that the larvae cannot subsist on the oil alone is fairly conclusive. Experiments indicate that the maggots feed primarily on larger insects that fall into the oil, although more research along these lines is needed. Under natural conditions the larvae are constantly in motion, imparting a shimmering quality to the surface of the oil.

The mature larvae leave the oil to pupate on nearby dry debris and vegetation. The small black gnat-like flies (which are about 1/16 in., or 1.5 mm, long) stay in the immediate vicinity of the pools, hiding in soil cracks or walking about on oily surfaces.

The species is known throughout the oil fields of southern California and is also reported from petroleum fields and tarseeps in Cuba and Trinidad. It is well established at the oil seeps at the La Brea "Tar Pits" on Wilshire Boulevard.

REFERENCES. Easterly, C. O. 1913. The "Oil Fly" of southern California. *Bulletin of the Southern California Academy of Sciences,* vol. 12, pp. 9-11.

Thorpe, W. H. 1930. The biology of the petroleum fly *(Psilopa petrolii,* Coq.). *Transactions of the Entomological Society of London,* vol. 78, pp. 331-337.

_______. 1931. The biology of the petroleum fly. *Science* (new series), vol. 73, pp. 101-103.

## PHORID FLIES (Family Phoridae)

■ LABORATORY GNAT
*(Megaselia scalaris)*
Figure 321

This is a small gnat-like fly (body length 1/16 in., or 1 mm) that frequently appears on laboratory walls and benches and on desks in offices adjoining insectaries or in other places where animals are kept and studied. The larvae are known to infest food, dead insects, and accumulated debris in rearing containers.

Adults are apparently attracted to the odors of decomposition emanating from dead flesh and veg-

320. Petroleum Fly.
Drawing by T. Ross.

321. Laboratory Gnat.
Drawing by T. Ross.

etable matter. The species is also a pest in hospitals, where it is found in debris in waste disposal chutes.

The pale brown or brownish-yellow adults are recognized by their compressed shape, numerous body and leg spines, and wings that are held flat over the back. They are also characteristically jerky in their running movements and reluctant to fly even when disturbed.

A related species *(Conicera tibialis)* infests mortuaries and mausoleums and is sometimes called "coffin fly" because of the larva's ability to develop in cadavers, even those that are embalmed with formaldehyde.

REFERENCES. Colyer, C.N. 1954. The "coffin" fly, *Conicera tibialis* Schmitz (Dipt., Phoridae). *Journal of the Society of British Entomology*, vol. 4, pp. 203-206.

Kats, H. 1987. Managing mausoleum pests. *Pest Control Technology*, March 1987, pp. 72-74.

Robinson, W. H. 1975. *Megaselia (M.) scalaris* (Diptera: Phoridae) associated with laboratory cockroach colonies. *Proceedings of the Entomological Society of Washington*, vol. 77, pp. 384-390.

## LOUSE FLIES (Family Hippoboscidae)

OCCASIONALLY, A PERSON HIKING OR WORKING in mountainous areas will notice a small (1/8 in., or 3 mm, long) flattened brown fly (Figure 322) that has landed on his skin or clothing. Looking much like a winged tick, the fly clings tenaciously or crawls sluggishly for a moment before flying away. This is one of the two common local species of "deer louse flies" *(Neolipoptena ferrisi* or *Lipoptena depressa),* which normally live as ectoparasites on deer. Large numbers (as many as a thousand or more) may infest the hide of a single animal.

Upon emerging from the pupa, this fly—which possesses wings that are fully developed although fragile—flies among the trees or shrubs in search of the host (it can survive at this stage for only a few days in the absence of the normal host). Upon successfully finding a deer, it immediately crawls through the hair to the skin and begins to suck blood. Here it remains as a permanent parasite, soon losing its wings through wear.

Local sportsmen often see one or both of two additional hippoboscid species on quail they have bagged. These are the so-called "quail louse flies" *(Lynchia hirsuta* and *Stilbometopa impressa).* They are larger (3/16 to 1/4 in., or 5 to 6 mm, in body length) than the deer louse flies and much more rapid in their movements: they move with equal facility forward,

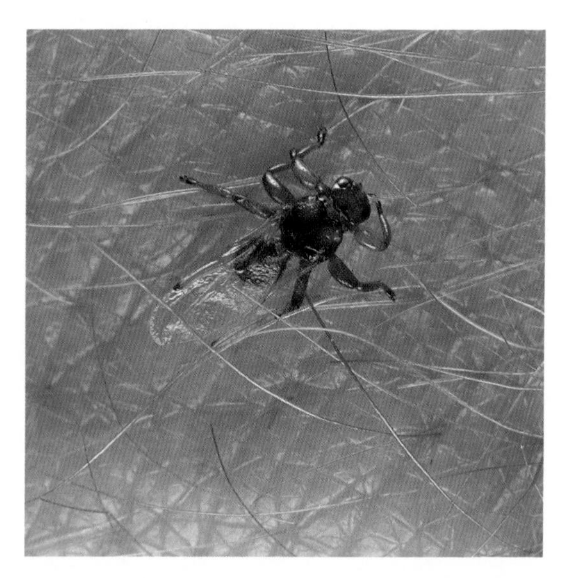

322. Deer louse fly *(Lipoptena depressa)* on human hand. Photograph by C. Hogue.

backward, or sideways through and over the feathers. Domestic pigeons carry a louse fly, *Lynchia americana,* although they are more likely to be infested with yet another species, *Pseudolynchia canariensis.* These and other species of bird louse flies may also be found on wild birds in the basin.

All louse flies are blood suckers, although none feeds regularly on humans. They may transmit disease between wild animals but not to and between people.

Development of these flies is of a special type: the larva is not free-living but matures within the body of the female parent. When the puparium is formed, the female deposits it on the host whence it soon falls off onto the ground.

REFERENCES. Hare, J. E. 1953. Bionomics of *Lipoptena depressa* (Say), the common louse fly of deer in Western North America (Diptera: Hippoboscidae). *Microentomology,* vol. 18, pp. 38-51.

Herman, C. 1945. Hippoboscid flies as parasites of game animals in California. *California Fish and Game,* vol. 31, pp. 16-25.

Tarshis, I. B. 1958. New data on the biology of *Stilbometopa impressa* (Bigot) and *Lynchia hirsuta,* Ferris (Diptera: Hippoboscidae). *Annals of the Entomological Society of America,* vol. 51, pp. 95-105.

# 13 FLEAS (Order Siphonaptera)

THE SCIENTIFIC NAME of this holometabolous order refers to the feeding habits of these insects (*siphon* means sucking) and their wingless condition (*a,* without; *pteron,* wing). Fleas possibly descended from primitive blood-sucking flies that lived in mammal or bird nests and had little use for wings in their parasitic way of life. All members of the order are small (most are about 1/16 in., or 2 mm, long), and all are ectoparasitic on warm-blooded vertebrates.

Fleas have peculiar bodily structures that make them suited to a life in fur or feathers. Especially notable are the combs of flattened spines arranged in linear rows on the lower edge of the head and on the rear of the prothorax in some species; these spines, in conjunction with a streamlined compressed body, facilitate the flea's progress through the host's pelt (their presence or absence is also very helpful in identifying flea species). Anyone who has tried to catch fleas on their dog or cat or to hold onto a flea has discovered another purpose of the spines—to make the flea "slippery" and hard to handle. Enlarged claws at the tips of flea's legs enable the insect to grasp and hold onto the host. Enlarged hind legs with powerful muscles give the flea the ability to hop great distances.

Fleas are important pests of people and their pets because of the annoyance caused by flea bites. Although some species are capable of transmitting diseases, allergic responses to flea bites are by far the most common local reason for complaints. The species most often involved in cases of flea bites is the Cat Flea. Cat Fleas may bite anywhere on a person's body but, because they are usually found on the floor or ground, they most often attack the lower leg.

Occasionally an infestation of fleas in the home may be traced to a stray animal that has had a litter or has died under the house, but infestations usually appear in homes with a resident cat or dog. People are seldom bitten until the animal has been away from the premises for several days. During the pet's absence, many new fleas will have matured, and adults that were already present will have been without their normal food source for a time (although the adults

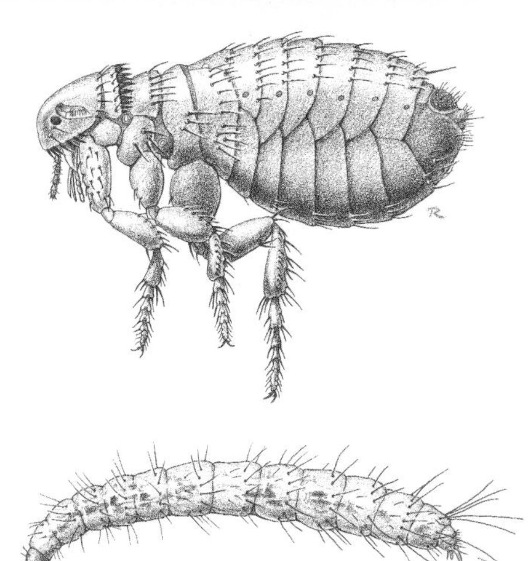

323. Cat Flea female (top) and larva. Drawings by T. Ross after various sources.

324. Oriental Rat Flea. Photograph of LACM specimen by C. Hogue.

can survive for long periods, up to 18 months, without food). Active and hungry, these fleas will begin to greedily feast on the nearest warm-blooded animals, i.e., people.

Fleas breed in the largest numbers under a pet's bedding. The larvae hatch from eggs that have been placed on the bedding of the host, in furniture, and outdoors in the soil or that have become lodged in floor cracks and crevices. They feed on bits of organic debris, including partially digested blood dropped by adult fleas. Pupation occurs in a loose silken cocoon to which particles of dust or sand adhere, providing a degree of camouflage.

Some species breed in sandy areas and possibly have contributed to the common term "sand fleas," which is used to explain insect bites experienced by bathers on the beach. Terrestrial amphipod crustaceans (species of the genera *Orchestia* and *Orchestoidea),* seen scurrying nearby, are often blamed for these bites by lay persons unaware that these animals do not bite but feed only on organic debris. This unjustified incrimination is reinforced by the fact that these "beach fleas" or "hoppers" have a compressed flea shape and hop about in flea-like fashion. Probably Stable Flies, punkies, or mosquitoes are the real culprits in most cases of seaside insect bite.

Fleas can transmit infectious diseases, most notoriously Bubonic Plague (caused by the bacterium, *Yersinia pestis).* Plague of rodent-flea origin occasionally crops up in hilly areas around the Los Angeles Basin and is always a potential threat to our health. The State Vector Surveillance and Control Section (under the California Department of Health Services) and Los Angeles County health authorities maintain a constant vigil on the distribution of fleas and plague organisms throughout the state and are ready to enact swift controls in the event of an outbreak. Human cases of plague are rare, but they do occur. The destructive epidemic of 1924, which resulted in the deaths of thirty-four people in central Los Angeles, was associated with domestic rats rather than wild rodents.

Murine Typhus, another disease that passes from host to host by means of fleas, occurs locally in certain areas. The chief vector is probably the Mouse Flea *(Leptopsylla segnis).*

The most common fleas and their hosts in the Los Angeles Basin are the following.

■ Cat Flea *(Ctenocephalides felis;* Figure 323). This is a truly domestic flea and our most abundant species; it is very common on cats and dogs (the Dog Flea, *C. canis,* also found on both cats and dogs, is rare in our area). Combs of spines are present on both the head and thorax of the Cat Flea.

■ Stick-tight Flea *(Echidnophaga gallinacea).* This is a wild species that preys on ground squirrels and wood rats but may also be found on domestic rats and fowl. When feeding, the adult buries a part of its head in the host's flesh. The thorax of this species is highly contracted, the head is blunt in the front, and both head and thorax lack combs.

■ Human Flea *(Pulex irritans).* This flea, which is so common in other metropolitan areas, is at present very rare or nonexistent in the basin. It lacks combs and has a smoothly rounded head and thorax. The closely related *Pulex simulans,* a flea of deer and other wild mammals, is sometimes mistaken for the Human Flea because it occasionally attacks people in wilderness areas.

■ Oriental Rat Flea *(Xenopsylla cheopis;* Figure 324). This is another domestic species that is common on rats and is a good plague vector. It has no combs, but large bristles are evident in front of the eye, and the front margin of the head is smoothly rounded.

■ Mouse Flea *(Leptopsylla segnis).* This flea is also found on local rats and mice. It has a head comb composed of only four spines on each side.

■ Ground Squirrel Flea *(Diamanus montanus).* This is a wild flea and plague transmitter that parasitizes ground squirrels. Only the thoracic comb is present, and the flea has very long mouthparts.

■ Northern Rat Flea *(Nosopsyllus fasciatus).* This is yet another domestic flea normally found on rats and mice. Like the Ground Squirrel Flea, this species has only the thoracic comb, but its mouthparts are short rather than long.

■ Wild Mouse Flea *(Malaraeus telchinus).* This flea is found in the wild on various kinds of field mice and wood rats. It is devoid of combs; the anterior margin of the head has small angular notches.

■ Rabbit Flea *(Cediopsylla inaequalis).* This is a sylvan species found on cottontail and jack rabbits. The comb is restricted to the head, is vertical, and has about eight spines.

REFERENCES. Bennet-Clark, H. C., and E. C. A. Lucey. 1967. The jump of the flea: A study of the energetics and model of the mechanism. *Journal of Experimental Biology,* vol. 47, pp. 59-76.

Hubbard, C. A. 1947. Fleas of western North America. Ames, Iowa: Iowa State College.

Keh, B., and A. M. Barnes. 1961. Fleas as household pests in California. *California Vector Views,* vol. 8, pp. 55-58.

Schwan, T. G. 1985. Fleas on roof rats in six areas of Los Angeles County, California: Their potential role in the transmission of plague. *American Journal of Tropical Medicine and Hygiene,* vol. 34, pp. 372-379.

# 14 BEETLES (Order Coleoptera)

THE MOST DISTINCTIVE FEATURE of this holometabolous order of insects are the thickened hard or leathery fore wings (called "elytra"), which cover and protect the hind wings and usually meet in a straight line down the back. Only the hind wings are used for flight, the elytra being useless for this purpose.

Weevils, thought by many people to make up a special category of insects, are actually beetles with the heads elongated into snouts.

The Coleoptera is the largest order of living organisms—nearly a quarter of a million kinds of beetles are known. They vary tremendously in size, structure, and biology and are found almost everywhere. Only a few of the most conspicuous species in the Los Angeles Basin are discussed here.

REFERENCES. Arnett, R. H., Jr. 1960-1962. *The beetles of the United States.* Washington, D.C.: Catholic University.

Jaques, H. E. 1951. *How to know the beetles.* Dubuque, Iowa: W. C. Brown.

White, R. E. 1983. *A field guide to the beetles of North America.* Peterson Field Guide Series. Boston: Houghton Mifflin.

## GROUND BEETLES (Family Carabidae)

SEVERAL SPECIES of long-legged, fast-running beetles, known collectively as the ground beetles, are common in the basin, especially under stones, boards, or other objects lying on damp ground. They typically hide by day where it is dark and damp and then venture forth at night in search of the small soft-bodied invertebrates that are their prey. Both adults and larvae are predaceous and live in the same habitats.

All ground beetles have smooth shiny bodies and wing covers that are finely grooved lengthwise; all are little known biologically. The more common species can be recognized by their size and color.

■ Small ground beetles *(Amara* (Figure 325) and *Bembidion* species). These are seldom more than 1/8 inch (3 mm) long and are shiny gray to black (sometimes with a greenish sheen).

■ Black Ground Beetle *(Pristonychus complanatus;* Figure 326). The largest of the ground beetles listed here

(⁵/₈ in., or 10 mm), this species is all black and has a narrow prothorax; the front of the head is flat and protruding. The Black Ground Beetle was introduced into the basin, but coleopterists (beetle specialists) are not sure of its origins.

■ Tule Beetle *(Tanystoma maculicolle;* Figure 327). The wing covers are bicolored—pale brown marginally and dark in the center. The species is about 3/8 inch (9 mm) long.

■ Rufous Ground Beetle *(Calathus ruficollis;* Figure 328). The wing covers of this species, which are a rich reddish brown, contrast with the dark wide prothorax; the beetle is about 3/8 inch (9 mm) long.

325. Small ground beetle (*Amara* species). Photograph by C. Hogue.

326. Black Ground Beetle. Photograph by C. Hogue.

327. Tule Beetle. Photograph by C. Hogue.

Other ground beetles are generally much larger and of some special biological interest; they are described in the following sections.

REFERENCES. Ball, G. E., and T. Negre. 1972 . The taxonomy of the Nearctic species of the genus *Calathus* Bonelli (Coleoptera: Carabidae: Agonini). *Transactions of the American Entomological Society,* vol. 98, pp. 412-539.

Liebherr, T. K. 1984. Description of the larval stages and bionomics of the tule beetle, *Tanystoma maculicolle* Coleoptera: Carabidae). *Annals of the Entomological Society of America,* vol. 77, pp. 531-538.

______. 1985. Revision of the Platynine carabid genus *Tanystoma* Motschulsky (Coleoptera). *Journal of the New York Entomological Society,* vol. 93, pp. 1182-1211.

---

■ SNAIL EATERS
*(Scaphinotus* species)
Figure 329

These beetles can be recognized by their inflated elliptical wing covers (which make the hind body region much broader than the prothorax) and their extra long mandibles and legs. Our two species, *Scaphinotus cristatus* and *S. punctatus,* are black and 3/8 to 7/8 inch (14 to 22 mm) in length; like all their close relatives, they prey on slugs, snails, and other insects. They are relatively uncommon and occur in moist canyons, usually near streams, in the San Gabriel and Santa Monica Mountains.

REFERENCE. Noonan, G. R. 1967. Observations on the ecology and feeding habits of adult *Scaphinotus punctatus* LeConte (Coleoptera: Carabidae). *Pan-Pacific Entomologist,* vol. 43, pp. 21-23.

---

■ COMMON CALOSOMA
*(Calosoma semilaeve)*
Figure 330

During the spring this beetle may be so common as to constitute a pest. The adults are large (about 1 in., or 25 mm, long) and run free during the day rather than being nocturnal and confined to burrows or cavities under objects on the ground, as are most ground beetles.

The Common Calosoma sometimes enters homes and, when disturbed, emits a disagreeable chemical that smells something like burnt rubber or electrical insulation. Because of its size, black color, and activeness, it is sometimes mistaken for the Oriental Cockroach.

Wireworms and caterpillars, especially cutworms, are the favorite prey of both adults and larvae. Consequently, the species is very beneficial to the gardener.

REFERENCE. Bidaspow, T. 1959. North American caterpillar hunters of the genera *Calosoma* and *Callisthenes* (Co-

328. Rufous Ground Beetle. Photograph by C. Hogue.

329. Snail eater *(Scaphinotus* species). Photograph by C. Hogue.

330. Common Calosoma *(Calosoma semilaeve)*. Photograph by J. Hogue.

leoptera, Carabidae). *Bulletin of the American Museum of Natural History,* vol. 116, pp. 225-344.

■ BOMBARDIER BEETLES *(Brachinus* species) Figure 331

These smallish beetles (less than 1/2 in., or 13 mm, long) are easily recognized among the ground-dwelling beetles by their dark blue or brown wing covers and contrasting red head and prothorax. They are found near water—by streams and lakes and in marshy areas, and they are sometimes abundant under stones and other objects on moist ground.

These beetles are best known for their ability, when alarmed, to forcibly eject a liquid from glands at the rear end of the body. The liquid, which is composed partly of nitrogen and benzoquinones, volatilizes instantly upon contact with the air, generating a small puff of "smoke" accompanied by an audible popping or crackling sound (crepitation). If the vapor touches human skin, the spot will burn for an instant and then turn dark brown for several days before returning to its normal color. Enemies of these beetles (birds, skunks, and other insects) no doubt are affected by the vapor in much the same manner as people and do not soon forget the uncomfortable experience of being "popped."

Larval development in these beetles can be extremely rapid, with only one or two days passing between the hatching of the egg to maturity of this stage. The larvae of some *Brachinus* bombardiers have been observed preying on the pupae of certain water scavenger beetles *(Berosus* and *Tropisternus* species of the family Hydrophilidae), and this habit is probably general for all species in the genus.

REFERENCES. Aneshansley, D. J., and T. Eisner. 1969. Biochemistry at 100°C: Explosive secretory discharge of bombardier beetles *(Brachinus)*. *Science,* vol. 165, pp. 61-63.

Erwin, T. L. 1965. A revision of *Brachinus* of North America, part 1: The California species. *Coleopterists' Bulletin,* vol. 19, pp. 1-19.

_______. 1967. Bombardier beetles (Coleoptera, Carabidae) of North America, part 2: Biology and behavior of *Brachinus pallidus* Erwin in California. *Coleopterists' Bulletin,* vol. 21, pp. 41-55.

---

■ **FALSE BOMBARDIERS**
*(Chlaenius* species)
Figure 332

These are flattened medium-sized beetles (a little over 1/2 in., or 13 mm, long) that are usually dark steely blue in color with reddish-brown legs (one species—*C. sericeus*—is a bright metallic green). Their mandibles are asymmetrical; the left one is long and straight, the right short and curved. When disturbed they give off a disagreeable odor resembling that of fresh leather.

Members of the genus are common along the margins of ponds and streams. They are principally nocturnal and feed on caterpillars, beetle larvae, and other ground-dwelling insects.

REFERENCES. Bell, R. T. 1960. A revision of the genus *Chlaenius* Bonelli (Coleoptera: Carabidae) in North America. *Miscellaneous Papers of the Entomological Society of America,* vol. 1, pp. 98-166.

Larochelle, A. 1974. A world list of prey of *Chlaenius* (Coleoptera: Carabidae). *Great Lakes Entomologist,* vol. 7, pp. 137-142.

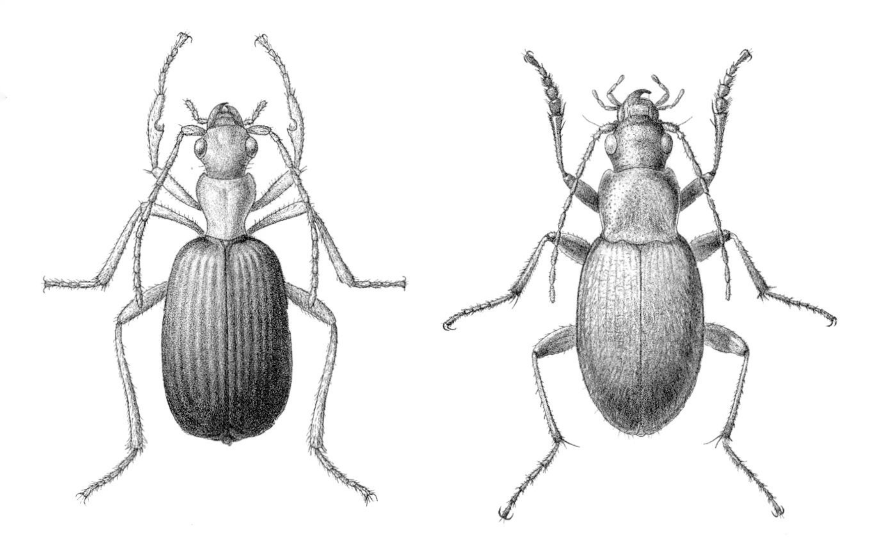

331 (above, left). Bombardier Beetle (*Brachinus* species) Drawing by T. Ross.

332 (above, right). False bombardier (*Chlaenius* species). Drawing by T. Ross.

333 (right). Oregon Tiger Beetle. Photograph by P. Nordin.

## TIGER BEETLES (Family Cicindelidae)

■ OREGON TIGER BEETLE
*(Cicindela oregona)*
Figure 333

This 1/2 inch (13 mm) long common beetle occurs almost everywhere that there are sandy flats or sand bars near permanent water because the larvae live in small vertical holes in moist compact sandy soil. These conditions are found in arroyos, by our large streams, ponds, and rivers (especially where they empty into the sea), and at the seashore itself.

The larva has a large flat head, which it uses to block the entrance to its burrow. It may also be recognized by a conspicuous hump on the fifth abdominal segment.

Both larvae and adults are predaceous. The

former snatch small prey that happen to stray close to their burrows. The agile fast-flying beetles hunt other small creatures, which they capture and kill with their powerful sickle-shaped mandibles. Although they may be found year round, they are most active in the hot sunshine during the middle hours of late spring and summer days. Their general appearance and behavior in flight is like that of flies, for which they may be mistaken.

Although some parts of the undersurface of the Oregon Tiger Beetle reflect iridescent green and bronzy hues in the proper light, the species is overall a rather somber-colored member of an otherwise brilliantly marked group: its upper surfaces are brown, and there are irregular yellow splotches on the wing covers.

REFERENCES. Misumi, D. H. 1967. An experimental investigation of the life history of *Cicindela oregona.* Master's thesis, California State University, Los Angeles, 41 pp.

Nagano, C. D. 1980. Tiger beetles of the genus *Cicindela* (Coleoptera: Cicindelidae) inhabiting the marine shoreline of southern California. *Atala,* vol. 8, no. 2, pp. 33-42.

## AQUATIC AND SEMIAQUATIC BEETLES

The aquatic and semiaquatic bugs have already been introduced (order Heteroptera; see Chapter 7). A corresponding group of beetles belonging to several families has members living in, on, or near water—mostly in ponds, in slow-moving streams, and in the vegetation at lake margins.

### ■ WATER SCAVENGER BEETLES (Family Hydrophilidae)

The water-loving species of this family most commonly inhabit marshy places—weedy shallow ponds and stream pools or flooded areas with grass growing out of the water. The larvae are found in similar places, usually among algae and weeds or in marginal plant debris. Adults are herbivorous, but the larvae are mainly carnivorous, feeding on one another and other small aquatic invertebrates.

■ ELLIPTIC WATER SCAVENGER *(Tropisternus ellipticus)* Figure 334

Of the several species of water scavenger beetles in the basin, the Elliptic Water Scavenger is the most common. It is greenish black and broadly oval, with an arched back; its length is about 3/8 to 1/2 inch (10 to 13 mm). It stridulates (makes a shrill chirping noise) when seized.

334. Elliptic Water Scavenger; its air supply is visible as a silvery sheen on the beetle's underside. Photograph by P. Bryant.

335. A school of whirligig beetles *(Dineutes* species). LACM photograph.

These beetles live in ponds and other quiet-water habitats in the basin. Females fasten their silken egg cases onto objects below the water surface. The larvae, which have short finger-like projections on the sides and back of the abdomen, prey on small aquatic insects. Pupation occurs in moist soil.

Adults graze and scavenge under water, consuming algae, detritus, and animal remains, often chirping as they feed. They fly well and disperse to new habitats that become available after the winter rains.

REFERENCES. Ryker, L. C. 1975. Calling chirps in *Tropisternus natator* (D'Orchymont) and *T. lateralis nimbatus* (Say) (Coleoptera: Hydrophilidae). *Entomological News,* vol. 86, pp. 179-186.

______. 1975. Observations on the life cycle and flight dispersal of a water beetle, *Tropisternus ellipticus* LeConte, in western Oregon (Coleoptera: Hydrophilidae). *Pan-Pacific Entomologist,* vol. 51, pp. 184-194.

## ■ WHIRLIGIG BEETLES (Family Gyrinidae)

THE SEMIAQUATIC WHIRLIGIGS live up to their name as they wildly whirl and gyrate in schools (Figure 335) on the surfaces of small ponds and quiet stream pools. Structurally, they are separated from other beetles found in association with water by their front legs, which are extra long, and their unusual compound eyes: they appear to have four eyes because each eye is divided into two widely separated parts, allowing vision simultaneously above and below the water surface.

Though they normally live on the surface film, the adults dive readily and can take flight after crawling out of the water. Their food consists of organic material scavenged from the water surface. The larvae are long and slender and bear several pairs of featherlike gills on their sides; they live on the bottom and prey on other small aquatic insects.

All of our species are small (slightly less than 1/4 in., or 5 mm, in length) and shiny black and belong to the genera *Dineutes* and *Gyrinus*.

## ROVE BEETLES (Family Staphylinidae)

■ Devil's Coach Horse *(Staphylinus olens)* Figure 336

Members of this beetle's family are characterized by short wing covers; most of the abdomen is exposed and freely movable. The Devil's Coach Horse is large (body length 1 1/3 in., or 33 mm) and solid black; when alarmed it opens its formidable jaws and rears its abdomen over its back (Figure 336), moving it threateningly as if attempting to sting. Although the beetle has no sting, the action is effective in scaring off enemies. There are two cream-colored eversible glands at the tip of the abdomen from which the beetle can emit a malodorous yellowish fluid, a habit giving it its scientific name *olens,* meaning "stinking."

The larva is elongate (just under 1 in., or 25 mm, long), with long legs and a large round head that is wider than the thorax. Pupation occurs in a cell in the soil.

Both the adults, which appear in September and October, and the larvae, which are present in the spring, are voracious predators, searching on the ground for slugs and snails, which they consume greedily. They are probably effective in controlling populations of the pestiferous and all too common Brown Garden Snail *(Helix aspersa).* They will also eat carrion.

The species is native to Europe but was apparently introduced into California around or before 1931.

OTHER SCIENTIFIC NAMES. *Ocypus olens.*

REFERENCES. Nield, C. W. 1976. Aspects of the biology of *Staphylinus olens* (Müller), Britain's largest staphylinid beetle. *Ecological Entomology,* vol. 1, pp. 117-126.

336. Devil's Coach Horse in threat posture. Photograph by C. Hogue.

337. Click beetle on its back preparing to attempt to flip itself over. Photograph by C. Hogue.

338. Larva of click beetle ("wireworm") in lettuce root. Photograph by C. Hogue.

Orth, R. E., I. Moore, T. W. Fisher, and E. F. Legner. 1975. A rove beetle, *Ocypus olens,* with potential for biological control of the Brown Garden Snail, *Helix aspersa,* in California, including a key to the nearctic species of *Ocypus. Canadian Entomologist,* vol. 107, pp. 1111-1116.

_______. 1975. Biological notes on *Ocypus olens,* a predator of Brown Garden Snail, with descriptions of the larva and pupa (Coleoptera: Staphylinidae). *Psyche,* vol. 82, pp. 292-298.

## CLICK BEETLES (Family Elateridae)

THERE ARE MANY SPECIES of these slender drably colored beetles living in the Los Angeles Basin. All have a special mechanism on the underside of the body between the prothorax and mesothorax that, when sharply snapped closed, makes an audible clicking sound and arches the body violently. A click beetle stranded on its back (Figure 337) employs this device to flip itself into the air. With luck, it lands on its feet after the first try and ambles off; more often several attempts are necessary for the insect to right itself.

In addition to their acrobatic behavior, click beetles can be recognized by their elongate shape and the sharp angles at the rear corners of the prothorax. They are often seen around porch lights at night. The larvae, called "wireworms" (Figure 338), live primarily in the soil, where they feed on or within herbaceous plant roots, tubers, and stems.

REFERENCE. McClure, N. E. 1933. The click beetle's click (Coleop.: Elateridae). *Entomological News,* vol. 44, pp. 145-147.

## PANTRY BEETLES AND BOOKWORMS

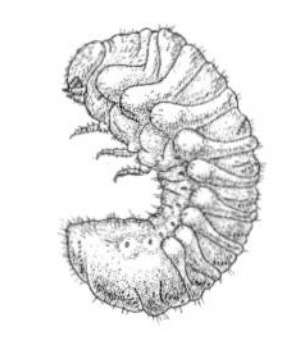

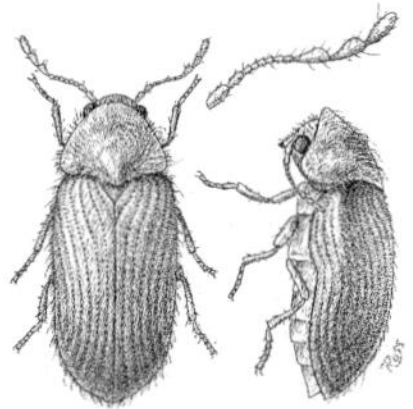

339. Drug Store Beetle larva and adult (top and side views and detail of antenna). Drawing by T. Ross after E. Back and R. Cotton, 1938, U.S. Department of Agriculture, Farmer's Bulletins, no. 1260.

SEVERAL SPECIES OF SMALL BEETLES are counterparts to the pantry moths in that they infest dried food products. Though the larvae rarely consume an appreciable quantity of the food, their presence alone is sufficient to render it unpalatable.

Pantry beetles are likely to be found in all kinds of dry organic material used by people as food. They may infest such common foods as dry breakfast cereal, rice, oats, wheat, peas, candy, spices, dried fruit, noodles and spaghetti, nuts, pet food, and beans as well as materials not usually thought of as food, including tobacco, red pepper, drugs, herbs, and even certain types of upholstery stuffing.

Several species act as intermediate hosts and vectors of the human tapeworms *Hymenolepis nana* and *H. diminuta.* People acquire infections by ingesting beetles containing the larval (or cysticercoid) stages of the tapeworm, which will often remain viable in infested corn meal and wheat flour that is undercooked.

The principal local offenders in this category are listed and characterized below. The worm-like larvae of certain of these species, notably the Drug Store Beetle and Cigarette Beetle, include paper in their diets. These larvae (and those of a few additional species not described here) may burrow themselves into the hearts of books, consuming the pages and ruining them. This habit gave rise to the satirical comparison of a learned or studious person to a "bookworm." Fortunately, bookworms in the United States generally confine their ravages to the leather and cardboard of the binding and cover; damage to pages deep within a book from bookworms is common only in damp tropical climates.

The appearance of these pests in a tightly sealed package of dried food is a source of wonder to housekeepers. Entry is commonly by way of minute imperfections in the seal, but some species may bore through paper and cardboard containers to get at the contents. In other cases, infestations occur when the foods are stored in bulk in railroad cars, warehouses, and at other stops along the processing line.

Although the layman tends to refers to all these forms as "weevils," the only true weevils on the following list are those described last; the remainder belong to other beetle families.

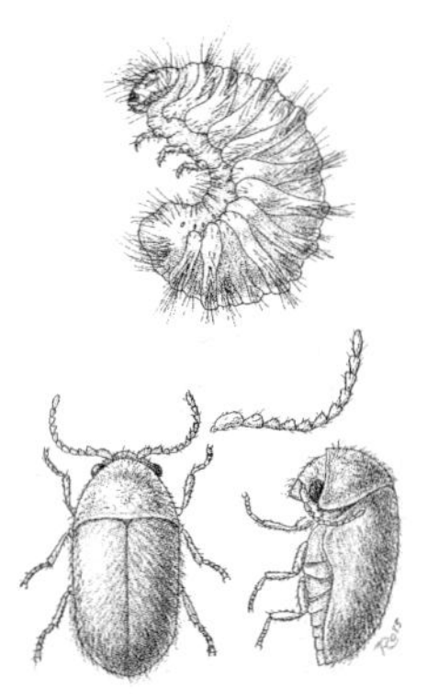

340. Cigarette Beetle larva and adult (top and side views and detail of antenna). Drawing by T. Ross after E. Back and R. Cotton, 1938, U.S. Department of Agriculture, Farmer's Bulletins, no. 1260.

■ Drug Store Beetle *(Stegobium panaceum,* family Anobiidae; Figure 339). An enlarged prothoracic shield covers much of the head dorsally. The beetle is elongate and oval in shape in side view, with clubbed antenna; its body length is up to 1/8 inch (3 mm). The larva has few hairs and is C-shaped.

■ Cigarette Beetle *(Lasioderma serricorne,* family Anobiidae; Figure 340). Like the Drug Store Beetle, this species has an enlarged prothoracic shield. But the Cigarette Beetle has a shorter body (body length 1/16 in., or 1 1/2 mm) and is triangular in shape in side view; the antennae are serrate. The larva is hairy and C-shaped.

■ Saw-toothed Grain Beetle *(Oryzaephilus surinamensis,* family Sylvanidae). This is a very small beetle (1/8 in., or 3 mm, or less). It is elongate and slender and has teeth on the margins of its prothorax. The larva is elongate, straight, and pale in color.

The Merchant Grain Beetle *(O. mercator;* Figure 341) is very similar to the Saw-toothed Grain Beetle in appearance. Its distinguishing feature is a small swell-

341. Merchant Grain Beetle among dry breakfast cereal flakes. Photograph by M. Badgley.

ing on the head behind the eye which is lacking in *O. surinamensis.*

■ Flour Beetles *(Tribolium* species, family Tenebrionidae; Figure 342). These are moderately small insects (1/8 to 3/16 in., or 3 to 4 mm, long). They have an elongate body with smooth margins to the prothorax. The larvae are worm-like and well pigmented. Some species have been used in classic laboratory studies of population biology.

■ Mealworm *(Tenebrio molitor,* Family Tenebrionidae; Figures 343, 344). This is the largest of the pantry beetles (1/2 to 5/8 in., or 13 to 15 mm, long). It is elongate and shiny brown, with smooth margins on the prothorax. The larvae are worm-like, pigmented, and very smooth; these are the "Li-Cut" worms sold as fishing bait and as food for pets. (A close relative, *T. obscurus,* is very similar in appearance and habits; it is dull brown and is found in grain storehouses.)

■ Grain Weevils *(Sitophilus* species, family Curculionidae; Figure 345). These are very small beetles (1/16 in., or 2 mm, long). They are true weevils in that the head is elongated into a snout. The larvae are obese pale grubs without obvious legs.

REFERENCES. Back, E. 1939. *Bookworms.* Annual report for 1939, Smithsonian Institution.

Dunkley, L. C., and D. F. Mettrick. 1971. Factors affecting the susceptibility of the beetle *Tribolium confusum* to infection by *Hymenolepis diminuta. Journal of the New York Entomological Society,* vol. 70, pp. 133-138.

Hinton, H. E. 1945. *Monograph of beetles affecting stored products.* London: British Museum of Natural History.

King, C. E., and P. S. Dawson. 1972. Population biology and the *Tribolium* model. *Evolutionary Biology,* vol. 5, pp. 133-227.

Sokoloff, A. 1972, 1974. The biology of *Tribolium;* with special emphasis on genetic aspects, 2 volumes. London: Oxford University.

## LADYBIRD BEETLES (Family Coccinellidae)

IMMORTALIZED IN THE NURSERY RHYME, "Ladybird, ladybird, fly away home," these handsome round beetles are known to everyone. In most species, both the adults and larvae are voracious predators of other small soft-bodied insects, particularly aphids and scale insects; their appetites for these pests make them beneficial in the garden and on the farm.

Our commonest local species are listed below; the first five feed primarily on aphids:

■ Convergent Lady Beetle *(Hippodamia convergens;* Figures 346-348). This is the species most often seen in the garden. It is 3/16 to 1/4 inch (4 to 6 mm) long and is either solid red or red with several small black spots

342. Flour beetle (*Tribolium* species) in dry breakfast cereal. Photograph by M. Badgley.

343. Mealworm adults. Photograph by C. Hogue.

344. Two pupae (left) and two larvae of the Mealworm. Photograph by C. Hogue.

345. Grain weevils (*Sitophilus* species) among rice grains. Photograph by C. Hogue.

(the black spots are usually present, but variants that are very lightly spotted or solid blood-red are common, and these are hardly distinguishable from the California Ladybird described below). The common name refers to a pair of converging pale dash-like marks that are present on the black prothorax. The Convergent Lady is the most important ladybird used in the biological control of aphids.

During the late summer and fall, the adults congregate in great masses in mountain canyons and other cool protected places (Figure 347). Here they hibernate for up to nine months, frequently buried beneath the snow, until the first warm spring days, when they move back to the valleys. While still massed, they are collected by entrepreneurs, who sell them in nurseries for release in home gardens.

Specimens sometimes accumulate on beach driftage after having been carried out to sea by Santa Ana winds and drowned while making their translocation flights.

The larva is elongate and black or gray in general color, with indistinct orange spots on its back. Because of its shape and voracity, it has been likened to a tiny alligator. The pupa is red and black and attached to the substrate at the posterior end.

■ Two-spotted Ladybird Beetle *(Adalia bipunctata;* Figure 349). This species is similar to the foregoing in appearance but is practically always deep red with a single black spot in the center of each wing cover. Its body length is 1/8 to 3/16 inch, or 4 to 5 mm.

■ California Ladybird Beetle *(Coccinella californica).* The adults of this species are superficially almost identical to solid-colored individuals of the preceding two. But this species is slightly larger (body length 3/16 to 1/4 in., or 5 to 7 mm) and more oval in shape, and its prothoracic shield is all black except for broad white borders at the sides near the front.

■ Ashy Gray Ladybird *(Olla v-nigrum,* formerly *O. abdominalis;* Figure 350). This beetle is variable in size but is usually small (body length 1/8 to 1/4 in., or 3 to 6 mm). It has two strikingly different color variants: black with a pair of large pale-yellow spots on its back

346. Convergent Lady Beetles; note the variation in size and number of spots on the prothorax and wing covers. Photograph by C. Hogue.

347. Overwintering aggregation of Convergent Lady Beetles in San Gabriel Mountains. Photograph by C. Hogue.

348. Convergent Lady Beetle larva. Photograph by C. Hogue.

349. Two-spotted Ladybird Beetle. Photograph by C. Hogue.

350. Ashy Gray Ladybird. Photograph by P. Bryant.

or pale gray with numerous small black spots (Figure 350).

■ Two-stabbed Ladybird *(Chilocorus orbus;* Figure 351). This is a small beetle (body length 1/8 in., or 3 mm) that is all black with two large blood-red spots.

■ Mealybug Destroyer *(Cryptolaemus montrouzieri;* Figures 352, 353). This useful ladybird was introduced from Australia in 1892 to control mealybugs. The prothoracic shield and the tips of the wing covers are reddish, the rest of the wing is black. The larva is white and covered with waxy excrescences, giving it an appearance similar to that of the mealybugs upon which it feeds. It is 1/8 inch (3 mm) in length.

■ Vedalia *(Rodolia cardinalis;* Figure 354). This ladybird is black with large rounded E-shaped red markings back to back on the wing covers; it is 3/32 inch (2.5 to 3.5 mm) long. This is the famous beetle imported from Australia at the end of the nineteenth century in a successful attempt to control the Cottony Cushion Scale, which threatened the southern California citrus industry.

REFERENCES. Booth, R. G., and R. D. Pope. 1986. A review of the genus *Cryptolaemus* (Coleoptera: Coccinellidae) with particular reference to the species resembling *C. montrouzieri* Mulsant. *Bulletin of Entomological Research,* vol. 76, pp. 701-717.

Caltagirone, L. E., and R. L. Doutt. 1989. The history of the Vedalia beetle importation to California and its impact on the development of biological control. *Annual Review of Entomology,* vol. 34, pp. 1-16.

Davis, J. R., and R. L. Kirkland. 1982. Physiological and environmental factors related to the dispersal flight of the Convergent Lady Beetle, *Hippodamia convergens* (Guerin-Meneville). *Journal of the Kansas Entomological Society,* vol. 55, pp. 187-196.

Doutt, R. L. 1958. Vice, virtue and the Vedalia. *Bulletin of the Entomological Society of America,* vol. 4, pp. 119-123.

Gordon, R. D. 1985. The Coccinellidae (Coleoptera) of America north of Mexico. *Journal of the New York Entomological Society,* vol. 93, fasc. 1, pp. 1-912.

Grossman, J. 1990. L.A.'s the place for biological pest control: How a little lady beetle saved California's citrus industry. *Terra,* vol 28, no. 3, pp. 38-43.

Hagen, K. S. 1970. Following the ladybug home. *National Geographic,* vol. 137, pp. 543-553.

______. 1962. Biology and ecology of predaceous Coccinellidae. *Annual Review of Entomology,* vol. 7, pp. 289-326.

Hubbell, S. 1991. Ladybugs. *The New Yorker,* 7 October 1991, pp. 103-111.

## IRONCLAD BEETLES (Family Zopheridae)

■ IRONCLAD BEETLE *(Phloeodes pustulosus)* Figure 355

This beetle derives its name from its extremely hard body wall, which may be difficult to pierce even with a sharp pin. The adult is about 3/4 to 1 inch (20 to 25 mm) long and dull gray-black in color. The surface texture of the

351. Two-stabbed Ladybird. Photograph by P. Bryant.

352. Mealybug Destroyer on mealybug colony. Photograph by M. Badgley.

353. Mealybug Destroyer larva. Photograph courtesy of Los Angeles County Agricultural Commissioner's Office, J. Wallen.

354. Vedalia feeding on Cottony Cushion Scale. Photograph by P. Bryant.

wing covers and prothorax is very rough. Adult Ironclads are fairly abundant locally under the loose bark of dead trees, especially oaks. They are thought to feed on punky fungus-ridden wood.

## DARKLING BEETLES (Family Tenebrionidae)

■ STINK BEETLES
*(Eleodes* species)
Figure 356

Several species of smooth shiny black beetles belonging to this genus are common in the basin, primarily in hilly areas. They are medium to large (1 to 1 1/4 in., or 25 to 32 mm, long), and their wing covers are fused along the midline making it impossible for them to fly. These conspicuous beetles are usually encountered as they amble along the ground. Individuals may also be found under stones and loose tree bark, where the long cylindrical larvae also often live. The larvae, which are nut-brown in color and prominently jointed, feed on the seeds, roots, and underground parts of many kinds of herbaceous plants. The adults are general feeders. Eggs are laid in the soil in the spring.

When a stink beetle is disturbed or its wandering is interrupted, it stands on its head and points its rear end into the air (Figure 356). For this headstanding habit, these insects are sometimes called "acrobat beetles." Adults may emit a disagreeable though weak odor when handled.

REFERENCE. Tschinkel, W. R. 1975. A comparative study of the chemical defensive system of tenenbrionid beetles: Defensive behavior and ancillary features. *Annals of the Entomological Society of America,* vol. 68, pp. 439-453.

■ WOOLY DARKLING
*(Cratidus osculans)*
Figure 357

This moderate-sized darkling beetle ( 1/2 to 3/8 in., or 12 to 16 mm) is immediately recognizable by its wooly pelage of reddish-brown hair; it also has a boxy shape. It is often seen in vacant lots and along paths through brushy or wooded areas. John Doyen, of the University of California at Berkeley, tells me that the larva lives in dry soil and ground litter, feeding on organic detritus.

## SOLDIER BEETLES (Family Cantharidae)

■ BROWN LEATHERWING
*(Cantharis consors)*
Figure 358

Like all members of its family, the Brown Leatherwing has soft wing covers. In this species these structures are medium brown, while the prothorax and legs are orange. The total length of the beetle is about 5/8 to 3/4 inch (15 to 20 mm).

Adults frequently come to porch lights in the late spring (April to May). They give off a strong unpleasant musty odor when handled or crushed and may also exude a yellow fluid. Little else is known of the habits of the adults, and the early stages remain undescribed. Both are probably ground dwellers that live in plant litter and prey on other insects.

355. Ironclad Beetle. Photograph by C. Hogue.

356. Stink beetle *(Eleodes* species) in threat posture. Photograph by C. Hogue.

357. Wooly Darkling. Photograph by C. Hogue.

358. Brown Leatherwing. Photograph by C. Hogue.

## FIREFLIES (Family Lampyridae)

■ PINK GLOWWORM
*(Microphotus angustus)*
Figure 359

People from the eastern half of the United States who have moved to Los Angeles often remark on the absence of "fireflies." There are actually four local beetles of the firefly family, but none of these communicate by bright flashing light like their eastern relatives. However, the female of the Pink Glowworm (which is 1/2 in., or 13 mm, long) communicates her location to the male (1/4 in., or 6 mm, long) by emitting a continuous uniform luminescent glow. The adult male has the usual firefly beetle form, but the female is "larviform" (wingless and elongate like the larva; Figure 359).

The males are not seen as often as the females because they give light only when disturbed, and the light is weak and not used in communication. The female is fairly common in late spring to early summer in the foothill canyons (a colony was reported from Griffith Park, near the Greek Theater, in 1989). Found at night by its glow and in the daytime under stones lying on leaf mold in grassy areas, the adult Pink Glowworm is easily recognized by the pink color of the flattened segments; the terminal segments are yellowish. The segments of the larvae of both male and female are blackish with pink margins.

REFERENCE. Green, J. W. 1959. Revision of the species of *Microphotus,* with an emendation of the Lampyrini (Lampyridae). *Coleopterists' Bulletin,* vol. 13, pp. 80-96.

■ CALIFORNIA GLOWWORM
*(Ellychnia californica)*
Figure 360

Adults of this medium-sized lampyrid (which is 5/8 in., or 15 mm, long) are often seen flying during spring days near streams lined with willows in foothill canyons (Santa Monica and San Gabriel Mountains). They are nonluminescent in all stages. The body is flat and mostly black, but there are conspicuous bright rosy red marks on the sides of the prothoracic shield. Although nothing is known about the larva, it probably lives as a predator under bark.

REFERENCE. Fender, K. M. 1970. *Ellychnia* of western North America (Coleoptera: Lampyridae). *Northwest Science,* vol. 44, pp. 31-43.

## SCARAB BEETLES (Family Scarabaeidae)

THIS BEETLE FAMILY contains the famous sacred scarabs *(Scarabeus sacer* and other species) that were venerated in the religion of ancient Egypt. Scarabs were depicted in the hieroglyphic script of this civilization,

359. Pink Glowworm adult females. Photograph by C. Hogue.

360. California Glowworm adult male. Photograph by C. Hogue.

and effigies were common in jewelry. Images of the beetle replaced the heart in mummies of high-ranking persons.

The scarab family is a large one, with many representatives in the Los Angeles Basin. Its varied members, including june beetles, chafers, and dung rollers, have in common antennae with a terminal club composed of leaf-like plates, which may be opened and closed. The larvae of most of our species are grub-like root feeders.

REFERENCES. Ritcher, P. O. 1966. White grubs and their allies, a study of North American scarabaeoid larvae. *Oregon State University Studies in Entomology,* vol. 4, pp. 1-212.

_______. 1967. Keys for identifying larvae of Scarabaeoidea to the family and subfamily (Coleoptera). Occasional Papers of the California Department of Agriculture, no. 10, 8 pp.

---

■ Rain Beetles
*(Pleocoma* species)
Figure 361

The first measurable late fall or winter rains stimulate the adult males of this genus of scarab beetle to emerge from their subterranean burrows, in which they have lain as pupae for more than a month. In some species a soaking rain of several inches is needed to initiate activity, but the species found at higher elevations will often fly a few minutes after the onset of the year's first shower. In the foothill and canyon areas, the males seem to emerge in response to the rain and may be attracted to light, congregating around store fronts or dwellings during a drizzle or downpour.

When searching for the burrows of the flightless female beetles, the males fly in slow sweeping patterns low over the ground and brush. The two major periods of flight and mating activity for most species occur shortly before dark and again before daybreak, but the occasional male may be seen flying throughout the night or at midday. Other species will fly actively as long as it is raining, no matter what the time of day.

Between flights the males may hide beneath leaves and debris or reenter the ground, digging shallow burrows (about 4 in., or 10 cm, deep) with their shovel-like heads and strong fore legs. The females are seldom found outside their burrows, which may extend from 4 to 6 feet (120 to 180 cm) into the soil.

The male apparently locates the burrow of the female by detecting a strong scent that she puts into the air. He enters her burrow to copulate, after which the female plugs the burrow's opening with pulverized soil and digs deep in the ground. The eggs, which are laid in a spiral at the base of the burrow and packed tightly in fine soil, do not mature until the following spring or early summer.

Female adults and larvae produce a musky odor. The newly hatched larvae shed their skin and then move through the soil in search of food. The life cycle may last as long as ten or twelve years, with the larvae feeding on the roots of varied kinds of plants, usually hardwood shrubs or trees and in particular oaks and conifers. The mouthparts of the adults are atrophied and useless for feeding.

There are three species in the Los Angeles area: *Pleocoma puncticollis, P. badia* (Figure 361), and *P. australis.* These are fairly large beetles: the males are about 3/4 inch (20 mm), and the females are up to 1 1/2 inches (40 mm). The first species is shiny black, with dense black hair on the undersides; it is found primarily in the Santa Monica Mountains, where it is strongly attracted to light at night. *Pleocoma australis* is dark brown with light-reddish hair; it occurs in the foothills of the San Gabriel Mountains. *Pleocoma puncticollis* and *P. australis* have four to five plates in the antennae; *P. badia,* which is dark to light brown, has seven.

REFERENCES. Hovore, F. T. 1971. A new *Pleocoma* from southern California with notes on additional species. *Pan-Pacific Entomologist,* vol. 47, pp. 193-201.

______. 1977. New synonymy and status changes in the genus *Pleocoma* Leconte (Coleoptera: Scarabaeidae). *Coleopterists' Bulletin*, vol. 31, pp. 229-238.

______. 1979. Rain beetles. Small things wet and wonderful. *Terra*, vol. 17, no. 4, pp. 10-14.

Smith, R. F., and W. L. Potts. 1945. Biological notes on *Pleocoma hirticollis vandykei* Linsley. *Pan-Pacific Entomologist*, vol. 21, pp. 115-118.

Stein, W. I. 1963. *Pleocoma* larvae, root feeders in western forests. *Northwest Science*, vol. 37, pp. 126-143.

---

■ GREEN FRUIT BEETLE
*(Cotinus mutabilis)*
Figures 362, 363

This scarab occurs only sporadically in the basin, although it may be abundant in our few remaining fruit orchards or in suburban neighborhoods where backyard fruit trees grow. Originally native to Arizona and New Mexico, it gradually spread westward and became noticeable in the Los Angeles area after the 1960s. The adults feed on a wide variety of fruits, including cactus fruit (which may be its wild host), figs, peaches, apricots, nectarines, and grapes.

The beetle is easily recognized by its large size (its length is 3/4 to 1 1/8 in., or 20 to 30 mm) and the general velvety olive-green color of the top of the thorax and wing covers (the latter are marked with contrasting dull brownish-orange marginal bands). The underside of the body is shiny metallic green. It is equipped with a short shovel-like horn on the front of the head for enlarging openings in the skin of fruit. On hot days in August and September, individuals may be seen in flight, buzzing clumsily in circles or zigzagging near the ground.

The adults are active from late summer to early fall and, during this period, lay their eggs in compost piles and other accumulations of decomposing plant litter. The larvae are fairly large (2 in., or 50 mm, long) and C-shaped; the body is pale translucent white, and

361. Copulating rain beetles *(Pleocoma badia)*. Photograph by F. Hovore.

the head is dark brown. The first two molts are completed in the fall, the third in the following spring. Larvae move forward on their backs with an undulating motion of the entire body. They obtain purchase on the substratum with transverse rows of stiff short stout bristles on the back of the thorax. Because of this peculiar manner of locomotion, they are known as "crawly-backs."

Larvae tunnel into compost piles and soil under accumulations of decomposing plant litter, emerging periodically to feed on surface material. In local populations, development also commonly occurs beneath manure piles and haystacks in horse stables (the larvae do not affect turf).

OTHER SCIENTIFIC NAMES. An earlier scientific name for this beetle was *Cotinus texanus.* It is also known by the common names "Peach Beetle" and "Fig Beetle."

REFERENCES. Chittenden, F. H., and D. E. Fink. 1922. The green june beetle. Professional Paper, U.S. Department of Agriculture, Bulletin 891, pp. 1-52. [Treats a related species of the eastern United States.]

Nichol, A. A. 1935. A study of the fig beetle, *Cotinus texana. Technical Bulletin of the Arizona Agricultural Experiment Station,* vol. 55, pp. 154-198.

Stone, M. W. 1982. The Peach Beetle, *Cotinus mutabilis* (Gory and Percheron), in California (Coleoptera: Scarabaeidae). *Pan-Pacific Entomologist,* vol. 58, pp. 159-161.

---

■ TEN-LINED JUNE BEETLES *(Polyphylla decemlineata* and *crinita)* Figure 364

These are large brown scarabs with conspicuous longitudinal white stripes on their wing covers. They are attracted to light at night and are seen frequently in midsummer in the San Gabriel Mountains and, occasionally, in nearby foothill communities. *P. crinita* is 7/8 to 1 inch (23 to 25 mm) long and has conspicuous erect hairs on the back of its prothoracic plate. In contrast, *P. decemlineata* (Figure 364) is larger (1 to 1 1/8 in., or 25 to 29 mm, long) and does not have the prothoracic hairs.

The larvae are large (1 to 2 in., or 25 to 50 mm, long) and feed on the roots of a wide variety of plants. The life cycle takes three to four years. Adults feed at night on needles of coniferous trees. They make loud squeaking noises when handled.

REFERENCES. Young, R. M. 1967. *Polyphylla* Harris in America north of Mexico, part 1: The *diffracta* complex (Coleoptera: Scarabaeidae: Melolonthinae). *Transactions of the American Entomological Society,* vol. 93, pp. 279-318.

362. Green Fruit Beetle. Photograph by C. Hogue.

363. Green Fruit Beetle larva. Photograph by C. Hogue.

364. Ten-lined june beetle *(Polyphylla decemlineata)*. Photograph by C. Hogue.

______. 1988. A monograph of the genus *Polyphylla* Harris in America north of Mexico (Coleoptera: Scarabaeidae: Melolonthinae). *Bulletin of the University of Nebraska State Museum,* vol. 11, no. 2, pp. 1-115.

■ COMMON JUNE BEETLES
Figures 365-369

Most of these scarab beetles are monotonously dark to light brown, although some bear vivid reddish or greenish hues; they range in size from 1/4 to 3/4 inch (6 to 19 mm) long. They are most often encountered on warm summer evenings when they come to outdoor

365. Serica *(Serica* species). Photograph by C. Hogue.

lights. The larvae, called "white grubs," are root feeders and are sometimes injurious to turf, ornamentals, and other valuable plants.

There are many species of june beetles—the four types most commonly found in the Los Angeles area are those listed below. (Fortunately, the very destructive "true" june beetles of the genus *Phyllophaga,* subfamily Melolonthinae, are not found here.)

■ Sericas *(Serica* species, subfamily Melolonthinae; Figure 365). These beetles are small to medium sized (1/4 to 3/8 in., or 6 to 10 mm), cylindrical, and usually smooth and hairless. The wing covers are tan to black with faint longitudinal lines. The most abundant species seems to be *Serica alternata.*

■ Hairy June Beetle (*Phobetus comatus,* subfamily Melolonthinae; Figures 366, 367). This species is hairy and medium sized (its length is 9/16 in., or 15 mm); the body is generally frail and soft, and the wing covers are pliable and translucent. Larvae feed on the tree roots that are just below the surface. Adults are present during the spring (March and April).

■ Dusty June Beetle (*Parathyce palpalis,* subfamily Melolonthinae; Figure 368). This medium-brown beetle is the largest of the five species described here (its body length is 3/4 to 7/8 in., or 20 to 23 mm). Its upper surface is coated with minute elongate white scales, its undersides with white hairs. If the beetle is handled, the scales are easily dislodged in patches, giving it a smudged or blotchy appearance. The heart-shaped scutellum is noticeably paler than the rest of the back.

■ Cyclocephalas (*Cyclocephala* species, subfamily Dynastinae). Species in this genus, such as *C. hirta* (Figure 369), are very common locally. They are medium-sized (the body length is 1/2 in., or 13 mm) and pale tan, some with fine specks or lines along the back.

REFERENCE. Hardy, A. R. 1974. Revisions of *Thyce* LeConte and related genera (Coleoptera: Scarabaeidae). Occasional Papers of the California Department of Agriculture, no. 20, 47 pp.

366. Hairy June Beetle. Photograph by C. Hogue.

367. Hairy June Beetle larva. Photograph by C. Hogue.

368. Dusty June Beetle. Photograph by P. Bryant.

369. Hairy Cyclocephala. Photograph by C. Hogue.

■ CARPET BEETLES
Figures 370-373

The larvae of these beetles are more often seen than are the adults. They are small brownish elongate or oval creatures with rows of stiff short hairs on the body segments and longer tufts projecting from the rear and sides. Frequently only the cast skins are found.

The larvae prefer dark undisturbed places; because they feed on all sorts of natural textiles, they can do great damage in the home.

The adults, which feed on flower pollen and nectar, are small (body length of 1/8 in., or 3 mm, or less) and oval, with short legs; they may be entirely black in color or variegated black, brownish, and white.

The following three genera are the most common:

■ *Attagenus* (Figures 370, 371). Adults of this genus are shiny and solid dark brown or black. The larvae are slender—about four times as long as wide, with short stiff reclining hairs.

■ *Anthrenus* (Figures 372, 373). These are the common carpet beetles. The adults have conspicuous variegated patterns of red, white, or black splotches on the back. The larvae are stout and less than three times as long as they are wide; they have hairs along the body that are generally stiff and erect, and there are two tufts at the rear end. These beetles are hated by entomologists because they often destroy collections of dry insect specimens.

■ *Trogoderma*. Adults of this genus are solid black, with patches of a dull pubescence (downy fuzz) forming irregular light markings. The larvae are stout, about three times as long as wide, with stiff erect hairs but without the terminal tufts common among species of *Anthrenus*. They are hardy: individuals of some species have been shown in experiments to survive up to five years without food. This genus includes the very important Khapra Beetle *(T. granarium)*, which is responsible for great losses of stored grain and flour. Fortunately, it has been eradicated from California, but its reentry is a constant threat. It does not fly and so its despersal is dependent on the transport of materials that it infests. The Warehouse Beetle *(T. variable)*, a flying species, is a pest in flour mills and granaries in the basin.

REFERENCE. Beal, R. S., Jr. 1954. Biology and taxonomy of the Nearctic species of *Trogoderma* (Coleoptera: Dermestidae). *University of California Publications in Entomology*, vol. 10, pp. 35-102.

370. *Attagenus* carpet beetle. Photograph by M. Badgley.

371. *Attagenus* carpet beetle larva. Photograph by M. Badgley.

372. *Anthrenus* carpet beetle. Photograph by P. Bryant.

373. *Anthrenus* carpet beetle larva. Photograph by P. Bryant.

## COMMERCIAL WOOD-BORING BEETLES

WOOD USED IN THE MANUFACTURE of furniture, cabinets, picture frames, tool handles, and boxes and in the construction of homes is attacked by several types of beetles that, under natural conditions, breed in dead tree branches, fallen twigs, and other cellulose material. The larvae cause the most damage as they bore through the interior of the wood, but emerging adults also make unsightly holes in the surface. None normally attack or reinfest finished wood that has been lacquered, shellacked, varnished, or painted; oviposition takes place only on bare wood or on freshly milled lumber or timber, often while it is resting in storage yards.

Ironically, furniture made of "wormwood" (wood that has been riddled by wood-boring beetles) is considered fashionable by some people and commands a high price. By damaging the wood, these beetles are actually creating an economically desirable product.

The principal types of wood-boring beetles in the basin are described in the following sections.

### ■ BARK BEETLES (Family Scolytidae)

■ FALSE FIVE-SPINED IPS
*(Ips paraconfusus)*
Figure 374

The adults of this species are very small (1/8 in., or 3 mm, long) and dark brown. The prothorax is large and partly conceals the back of the head; the wing covers are finely haired and have linear series of punctures; the antennae are clubbed.

The species develops under the bark of pines—in our area, primarily Monterey Pine. Usually only unhealthy or cut trees are attacked, but healthy trees are sometimes infested. The larvae make fine tunnels through the growth layer beneath the bark, and these tunnels may connect, girdling and killing the tree.

The species resembles and has been confused with a close relative, the Pinyon Pine Engraver *(Ips confusus)*.

REFERENCES. Bright, D. E., and R. W. Stark. 1973. The bark and ambrosia beetles of California. Coleoptera: Scolytidae and Platypodidae. *Bulletin of the California Insect Survey,* vol. 16, pp. 1-169.

Lanier, G. N. 1970. Biosystematics of North America *Ips* (Coleoptera: Scolytidae): Hopping's Group IX. *Canadian Entomologist,* vol. 102, pp. 1139-1163.

Strubble, G. R., and R. C. Hall. 1955. The California five-spined engraver, its biology and control. U.S. Department of Agriculture Circulars, no 964, 21 pp.

Wood, D. L. 1963. Studies on host selection by *Ips confusus* (LeConte) (Coleoptera: Scolytidae), with special reference to Hopkins host selection principle. *University of California Publications in Entomology,* vol. 27, pp. 241-282.

## ■ BRANCH BORERS (Family Bostrichidae)

■ STOUT'S HARDWOOD BORER
*(Polycaon stouti)*
Figures 375, 376

Adults of this 3/4 inch (19 mm) long black beetle may appear in the fall (September) in unlikely places, such as in the hallways and rooms of new buildings, in warehouses, and in homes. Their occurrence is explained by their breeding habits. The larvae are wood

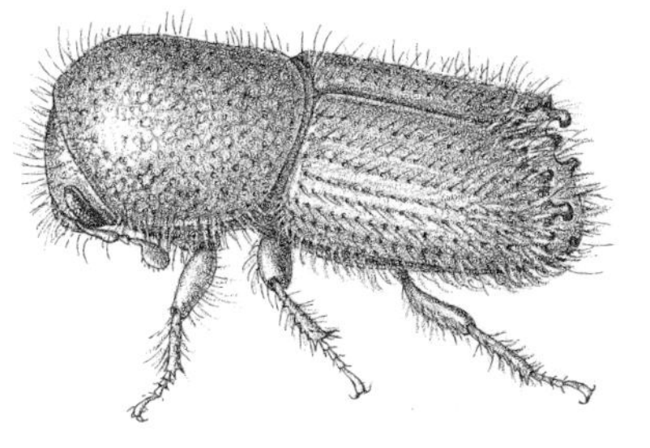

374. False Five-spined Ips. Drawing by T. Ross.

375. Stout's Hardwood Borer. Photograph by C. Hogue.

376. Stout's Hardwood Borer larva. Photograph by C. Hogue.

borers that feed within various hardwoods such as oak, California Laurel, alder, maple, and eucalyptus—construction woods that are often used in building boxes, shipping crates, storage racks, and the slats used behind acoustic ceiling tiles; the larvae will also infest finished wood products such as cupboards, cabinets, and furniture. The adult Stout's beetles may emerge from these products after the construction is completed and even after the product has been finished.

Adults normally attack dead, dying, or weakened trees but may occasionally infest apparently healthy ones. Females bore into timber and lay their eggs on the walls of their tunnels. Larval development may be prolonged, and emergence may be delayed for many years. There is no evidence that the species reinfests lumber or manufactured wood products once the adults have emerged from them.

REFERENCE. Middlekauff, W. W. 1974. Delayed emergence of *Polycaon stouti* Lec. from furniture and interior woodwork (Coleoptera: Bostrichidae). *Pan-Pacific Entomologist,* vol. 50, pp. 416-417.

## ■ HARDWOOD BORERS (Family Lyctidae)

■ POWDER-POST BEETLES
*(Lyctus* species)
Figure 377

Powder-post beetles are so named because their borings reduce the interior of seasoned hardwoods to a fine powder. Oak, ash, hickory, maple, and walnut are their preferred hosts.

The adults are small (1/4 in., or 6 mm, long) elongate brownish-black beetles. The females often lay their eggs in depressions on the surface of the wood from which they emerged. Hence the same wood, if left unfinished, may be infested repeatedly.

REFERENCES. Gerberg, E. 1957. A revision of the New World species of powder-post beetles belonging to the family Lyctidae. U.S. Department of Agriculture, Technical Bulletin, no. 1157.

Wright, C. G. 1969. Biology of the southern Lyctus Beetle, *Lyctus planicollis. Annals of the Entomological Society of America,* vol. 53, pp. 285-292.

## LONG-HORNED BEETLES (Family Cerambycidae)

VERY LONG ANTENNAE typify these beetles and give them their common name. The larvae, called roundheaded borers, feed on wood. Most of our species are found only in dead wood out of doors, usually logs in mountain areas, and do not become pests of houses or furniture.

There are many species of long-horned beetles in our area. The first two of the five types described in the following sections, the Pine Sawyer and the California Prionus, are conspicuous mainly because of their large size. They are mountain dwellers and not regular residents of the basin, but they are included here because they occasionally enter our area on the northern fringes or make themselves conspicuous to campers visiting the San Gabriel or San Bernardino Mountains. Both are dark brown beetles with long antennae whose larvae burrow in the dead and decomposing wood of hardwood and coniferous trees.

■ PINE SAWYER
*(Ergates spiculatus)*
Figure 378

This is the largest local beetle, a shiny brown species that is 2¼ inches (57 mm) long. The lateral margins of the prothorax are armed with many small sharp teeth or spines. The larvae, which are also very large (up to 2½ in., of 63 mm, long), are pale grubs that burrow in the wood of dead or felled coniferous trees; they also feed in fallen logs, stumps, and telephone poles. Though similar to those of the California Prionus, Pine Sawyer larvae are characterized by four blunt teeth on the front of the head.

The females are attracted to lights, and camp-

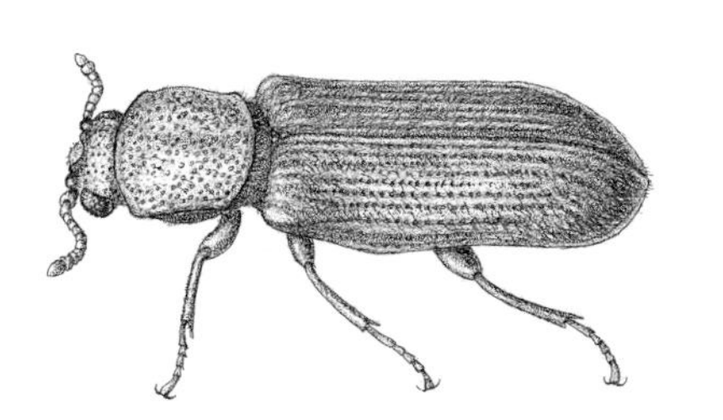

377. Powder-post beetle *(Lyctus* species). Drawing by T. Ross.

378. Pine Sawyer. Photograph by C. Hogue.

ers in pine flats in the neighboring mountains are frequently startled when these beetles loudly buzz into their lanterns on warm summer evenings. Adults are active from July to September.

■ **California Prionus**
*(Prionus californicus)*
Figure 379

Adults of this species are superficially similar to those of the Pine Sawyer in size and coloring. But the California Prionus beetle has a slightly broader and shorter body (its length is 2 in., or 51 mm), and the teeth of the lateral margins of the prothorax are coarse and only three in number. The antennal segments of the male (Figure 379) are conspicuously lobed.

The mature larva is about 3 inches long and 3/4 inch in diameter (70 by 20 mm). It is white with a dark brown head; there are no teeth on the front of the head. Because they bore into the roots of living native trees, the immatures frequently are injurious to oaks, madrone, and cottonwoods as well as to various fruit trees. They also develop in old eucalyptus stumps. Pupation occurs in an underground cell near the roots of the host or in the wood itself.

Adults emerge in early summer. They are nocturnal or crepuscular fliers and are often attracted to lights.

■ **Nautical Borer**
*(Xylotrechus nauticus)*
Figure 380

This beetle frequently hitchhikes on firewood, and this explains its mysterious appearance indoors. The larva, which is about 3/4 inch (20 mm) long when mature, is pale dirty brown and enlarged just behind the head. It bores into the heartwood of dead oak and other hardwoods. The adult is 3/8 to 5/8 inch (10 to 16 mm) long, grayish brown to nearly black, with three irregular transverse lines across its wing covers. For a long-horned beetle, the Nautical Borer has relatively short antennae.

■ **Milkweed Borer**
*(Tetraopes basalis)*
Figure 381

Wherever milkweed *(Asclepias* species) grows, this long-horned beetle may be found congregating on the leaves and flowers. Individuals are 3/8 to 1/2 inch (10 to 13 mm) long and entirely red, except for a pattern of black spots: there is usually one behind each eye, four in a square on the prothorax, and two placed obliquely in the center of each wing cover; there is an additional spot at the side of the front corner of each wing cover.

When abundant, these beetles are somewhat gregarious, and numerous individuals may occupy the same plant. They are not usually active and are not apt to fly even when alarmed, though at such times (especially when held between the fingers), they may make a squeaking or purring noise.

Adults occur from April to August, feeding on the foliage and flowers of the host plant. The white grub-like larvae burrow within the roots of the living plant. Pupation occurs in the soil adjacent to the plant 1 or 2 inches (2 to 5 cm) below the surface.

REFERENCE. Chemsak, J. A. 1963. Taxonomy and bionomics of the genus *Tetraopes* (Cerambycidae: Coleoptera). *University of California Publications in Entomology*, vol. 30, pp. 1-90.

379. California Prionus male. Photograph by C. Hogue.

380. Nautical Borer. Photograph by C. Hogue.

381. Milkweed Borer. Photograph by F. Hovore.

■ **Banded Alder Borer** *(Rosalia funebris)* Figure 382

This is a handsome beetle: the fairly large adults (1 to 1½ in., or 25 to 40 mm, long) are pale bluish white with contrasting black transverse bands across the wing covers. Although the species is also popularly known as the California Laurel Borer, it does not use California Laurel *(Umbellaria* species) as a primary host. It is normally noticed resting on the dead trunks of its host trees, which are alder, ash, and other hardwoods. In the basin it sometimes infests Coast Live Oak and eucalyptus.

Adults may be attracted to the volatilizing esters in fresh paint.

REFERENCE. Chemsak, J. A., and E. G. Linsley. 1971. Some aspects of adult assembly and sexual behavior of *Rosalia funebris* Motschulsky under artificial conditions (Coleoptera: Cerambycidae). *Pan-Pacific Entomologist,* vol. 47, pp. 149-154.

■ **Eucalyptus Long-horn Borer** *(Phoracantha semipunctata)* Figure 383

Since the introduction of this Australian beetle to southern California, probably in 1982 (the first recorded sighting was near El Toro in Orange County in 1984), it has steadily moved into and through the Los Angeles Basin. Until this borer came on the scene, its host, eucalyptus, had been virtually free of major pests since its own arrival here after the 1860s. It is uncertain, however, how serious a threat the Eucalyptus Long-horn Borer poses, because the beetles probably only attack trees weakened by lack of moisture, disease, or other stresses. They also infest freshly cut wood.

Damage by larvae is characteristic and may be extensive because of their large size (length of up to 1½ in., or 40 mm). They form deep broad galleries under the bark and, as they reach maturity, they girdle the tree and may kill it. Larvae are present from spring to fall, and there may be several generations per year. During the warm summer, the life cycle may only require three months.

Adults vary in length from ½ to 1¼ inches (13 to 32 mm) and bear conspicuous orange to cream-colored markings on their shiny reddish-brown wing covers and body—zigzag splotches on the basal half of the wing covers and a spot on the tip. The legs are medium brown; segments 3 to 7 of the antennae are spined.

REFERENCES. Anonymous. 1986. An Australian insect is killing southern California's eucalyptus. *Sunset Magazine,* October 1986, pp. 76-77.

Penrose, R. L. 1985. The eucalyptus borer, a pest new to

382. Banded Alder Borer. Photograph by F. Hovore.

383. Eucalyptus Longhorn Borer. Photograph by C. Hogue.

California. *California Plant Pest and Disease Report,* vol. 4, no. 3, pp. 80-84.

## LEAF BEETLES (Family Chrysomelidae)

THESE BEETLES ARE all plant feeders found primarily on flowers and foliage. Many are very small and can hop (they are nicknamed "flea beetles"). Many species are brightly colored, metallic blue being a common hue. The family is large, and a number of its member species occurs in the basin; those described in the following sections attract the most attention.

■ DICHONDRA FLEA BEETLE
*(Chaetocnema repens)*
Figure 384

The larva of this species is a serious pest of dichondra lawns because of its abundance and habit of feeding on the roots of this plant. The larvae are active during the warmer times of the year and are found to depths of 5 inches (13 cm) in the soil.

The adults are very small ($^{1}/_{16}$ in., or 1.5 mm, long) and bronzy reddish black. The wing covers have rows of minute punctures. The femur of the hind leg is greatly enlarged and contains strong muscles that enable the beetle to jump like a flea. Adults overwinter among the leaves of the host plant and may emerge to feed on warm days.

REFERENCE. McCrea, R. 1973. A new species of the flea beetle genus *Chaetocnema* found on dichondra in California. *Pan-Pacific Entomologist,* vol. 49, pp. 61-66.

---

■ **Western Spotted Cucumber Beetle** *(Diabrotica undecimpunctata undecimpunctata)* Figure 385

This is a small (1/4 in., or 6 mm, long) pale-green beetle with several black spots on its wing covers. The legs, antennae, prothorax, and abdomen are black.

The immature stages develop in the soil. In the spring, females lay their eggs at the base of the host grasses and legumes. (Later generations of the year develop on agricultural crops such as alfalfa, potatoes, curcurbits, and orchard fruits.) Larvae feed on the exterior of small roots and burrow within larger roots and tubers; they pupate in an earthen chamber about 1 inch (25 mm) below the surface. The pupae have been observed to vibrate the tip of the abdomen violently when disturbed.

Although they sometimes feed on flowers and fruits, adults mainly eat the epidermis of the leaves, which as a result may be reduced to a translucent framework. The species passes the winter in the adult stage.

REFERENCE. Smith, R. F., and A. E. Michelbacher. 1949. The development and behavior of populations of *Diabrotica 11-punctata* in foothill areas of California. *Annals of the Entomological Society of America,* vol. 42, pp. 497-510.

---

■ **Elm Leaf Beetle** *(Xanthogaleruca luteola)* Figures 386, 387

This serious pest of elms, primarily the Siberian Elm, entered the Los Angeles Basin via the San Fernando Valley in the 1970s. It is still extending its range, but it is not expected to become common in the basin because winter temperatures are mild, inducing the adults to emerge and attempt to feed before new leaves are available. Local infestations may be severe, however, with the larvae literally stripping trees of their leaves.

The yellow elongate eggs are laid in clusters on the undersides of leaves. Larvae are yellow grubs about 1/2 inch (13 mm) in length, with a row of black tubercles down each side of the back. They eat the surface tissues from the leaves, leaving only a skeleton of veins. When mature they crawl down the trunk to pupate in ground litter at the base of the host.

The adults are small beetles (body length 1/4 in., or 6 mm) that are generally yellow with a black stripe around the edge of the wing cover and some black

384. Dichondra Flea Beetle; note enlarged femur of hind leg. Photograph by C. Hogue.

385. Western Spotted Cucumber Beetle. Photograph by P. Bryant.

386. Elm Leaf Beetle. Photograph by C. Hogue.

387. Elm Leaf Beetle larva. Photograph by C. Hogue.

spots behind the head. Like the larvae, adults also feed on elm, cutting small round holes in the leaves.

The species goes through three to four generations in a summer. Adults overwinter, to become active again in the spring.

OTHER SCIENTIFIC NAMES. *Pyrrhalta luteola.*

REFERENCES. Calman, J. 1984. Elm leaf beetle control. Integrated Pest Management Newsletter (Agriculture Committee/Weights Measurement Department, County of Los Angeles), vol. 3, no. 1, pp. 1-2.

Weber, R. G. 1976. Sexing the Elm Leaf Beetle, *Pyrrhalta luteola* (Coleoptera: Chrysomelidae). *Annals of the Entomological Society of America,* vol. 69, pp. 217-218.

---

■ BLUE MILKWOOD BEETLE
*(Chrysochus auratus cobaltinus)*
Figure 388

This species is medium sized (3/8 in., or 10 mm, long) and a beautiful metallic green, blue-green, or blue. At times, congregations of these beetles are seen on the leaves of milkwood plants. The larvae are not known, but they probably burrow in the roots.

## WEEVILS (Family Curculionidae)

WEEVILS ARE A FAMILY of beetles in which the front part of the head is usually elongated into a snout or proboscis. The mouthparts are located at the tip of this apparatus and are generally modified for gnawing through tough-skinned buds, fruits, seeds, or nuts.

The weevil family is the largest family of animals—there are some 35,000 species now known throughout the world. Many forms live in the basin; the following are those most frequently noticed, aside from those found in the pantry (see Pantry Beetles).

---

■ YUCCA WEEVIL
*(Scyphophorus yuccae)*
Figure 389

This is a large weevil (5/8 in., or 17 mm, long) that is found from March to July, sometimes in considerable numbers, crawling among the sword-like leaves forming the basal rosette of the common yucca *(Yucca whipplei),* which grows on canyon slopes of our foothills. The adults feed on the sap of the living plant; they are smooth, black, and fully winged, with flattened and grooved wing covers and a long downcurved beak. The larvae bore into the wood of the flower stalks and into the basal stems of the nonflowering plants.

REFERENCE. Vaurie, P. 1971. Review of *Scyphophorus* (Curculionidae: Rhynchophorinae). *Coleopterist's Bulletin,* vol. 25, pp. 1-8.

388. Blue Milkweed Beetle. Photograph by C. Hogue.

389. Yucca Weevil. Photograph by C. Hogue.

390. Fuller's Rose Weevil. Photograph by C. Hogue.

■ FULLER'S ROSE WEEVIL
*(Asynonychus godmanni)*
Figure 390

This weevil was originally introduced from South America at the end of the last century. It is known to attack many cultivated plants and is especially troublesome in citrus groves, where the adults trim new growth from young trees. The eggs are cemented to citrus fruit under the green calyx, but the legless larvae live underground, feeding on roots and making furrows in the bark.

The adults are active at night from July to November. Around the garden the beetles may be seen on roses, berry veins, gardenias, and other ornamentals or vegetables, but the species also develops on weeds. Adults sometimes rest on the walls of buildings or hide in crevices or under leaves during the day; they will feign death if knocked off onto the ground.

The adult is about 1/3 inch (8 mm) in length and has a short snout for a weevil. It is generally gray-brown in color, but there is an oblique white area on the side of each wing cover. No males have ever been found, and the females produce viable eggs without mating (a phenomenon known as parthenogenesis). Adults cannot fly.

OTHER SCIENTIFIC NAMES. This weevil has been referred to by a number of names, including *Pantomorus godmani, Aramigus fulleri,* and *Strophomorphus canariensis.*

REFERENCES. Chadwick, C. E. 1965. A review of Fuller's rose weevil *(Pantomorus cervinus* (Boh.)). (Col., Curculionidae). *Journal of the Entomological Society of Australia (N.S.W.),* vol. 2, pp. 1-11.

Dickson, R. C. 1950. The Fuller rose beetle: A pest of citrus. Bulletin of the University of California Agricultural Experimental Station, no. 718, 8 pp.

Lanteri, A. A. 1986. Revisión del género *Asynonychus* Crotch (Coleoptera: Curculionidae). *Revista de la Asociación de Ciencias Naturales del Litoral,* vol. 17, pp. 161-174.

■ STRAWBERRY ROOT WEEVIL
*(Otiorhynchus cribricollis)*
Figure 391

Like Fuller's Rose Weevil, this species is small (1/4 in., or 6 mm) and has a short snout; but the Strawberry Root Weevil is a dark reddish-brown and has conspicuous punctures in its wing covers. The larvae are serious pests of strawberries: they live underground and eat the fine roots and crown of the plant. The adults feed on the foliage of this and other plants (including privet), usually at night, and they may be fairly common in gardens. Adults are flightless, and the species is apparently wholly parthenogenetic (males are not known in North America).

The Strawberry Root Weevil is sometimes mis-

391. Strawberry Root Weevil. Photograph by C. Hogue.

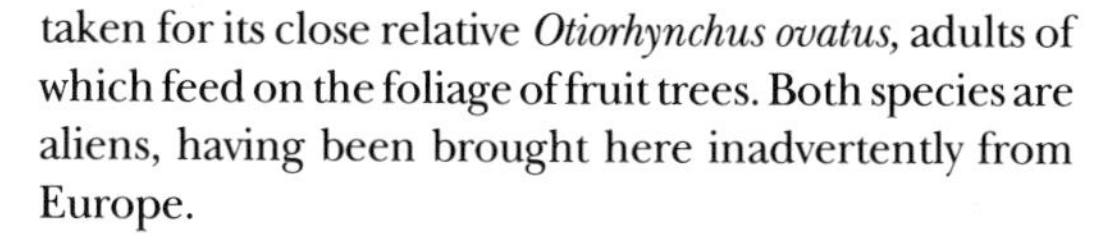

392. Vegetable weevil. Drawing by T. Ross.

taken for its close relative *Otiorhynchus ovatus,* adults of which feed on the foliage of fruit trees. Both species are aliens, having been brought here inadvertently from Europe.

REFERENCE. Warner, R. E., and F. B. Negley. 1976. The genus *Otiorhynchus* in America north of Mexico (Coleoptera: Curculionidae). *Proceedings of the Entomological Society of Washington,* vol. 78, pp. 240-262. [Contains useful key to species, but distribution information is in error.]

■ VEGETABLE WEEVILS *(Listroderes costirostris* and *difficilis)*
Figures 392, 393

These are small weevils (3/8 in., or 10 mm, long), both the larvae and adults of which eat the foliage of various garden or field crops. The larvae also feed extensively on the roots, especially tuberous types; the species is a bad pest of dichondra lawns. I have seen the larvae in the basin in dense growths of the common aster family weed *Cotula australis.*

The pale-green grubs are small (the body length is about 1/4 in., or 6 mm) and fat. They feed in the spring and pupate in April.

The brown adults, which are active in the summer, are characterized by a V-shaped marking across the wing covers. The species are native to Brazil and

393. Vegetable Weevil larvae. Photograph by C. Hogue.

were introduced into this country in the early 1900s and into California in the early 1920s.

OTHER SCIENTIFIC NAMES. Two very similar, often indistinguishable species are confused under the name Vegetable Weevil; the name *L. obliquus* is also sometimes mistakenly used for these species. Both were introduced separately from parthenogenetic populations in South America.

REFERENCES. High, M. M. 1939. The vegetable weevil. United States Department of Agriculture Circulars, no. 530, 25 pp.

Lovell, O. H. 1932. The vegetable weevil *Listroderes obliquus*. University of California Agricultural Experiment Station Bulletin, no. 546, 19 pp.

# 15 ANTS, WASPS, AND BEES (Order Hymenoptera)

The Hymenoptera is the third largest order of insects and undoubtedly the most beneficial from our standpoint because its contains a great many valuable predators and parasites of insect pests as well as the important pollinators (especially the bees) of plants.

Members of this holometabolous order are familiar to most of us, and little description is necessary to define the three major groups: ants, bees, and wasps. Ants are always distinguishable by the knobs or nodes on the segments of the waist. Bees always have finely branched (feather-like) hairs on the body (although the branches are visible only with a strong lens). And all Hymenoptera without these features are wasps, most of which are narrow waisted (wood wasps and sawflies have the abdomen broadly joined to the thorax).

The well-developed ovipositor of females of most species of Hymenoptera is often modified to form a stinger, a dominant characteristic and one of which many of us are painfully aware.

## ANTS (Family Formicidae)

The ants are among the most fascinating of insects. Because of the interest of biologists and laymen alike, there exists a great deal of literature describing their classification and social habits and giving almost unbelievable accounts of their strange anatomy and habits.

The ants make up but a single family in the large order Hymenoptera. They are descended from and remain very similar to wasps, but they are structurally and physiologically specialized to a highly developed social way of life. Within their social order there are typically three castes—male reproductives, female (queen) reproductives, and female "nonproductives" or "spinster sisters" (workers). A fourth caste, the soldiers, are found only in a certain few species.

The process of founding new ant colonies and the life within the average established colony are similar in some ways to those among the termites. Periodically, when conditions are right within the colony and in the general environment, winged fertile males and females disperse on nuptial or "wedding" flights. Mating takes place in the air during these flights, and the supply of sperm received by the female remains viable in her body for her lifetime.

Soon thereafter the male dies, and the female (now properly called the queen) establishes a new colony. After breaking off her wings by rubbing them against the ground or rocks, she seals herself in a suitable nesting site, lays a batch of eggs, and rears the resulting young to maturity. This first mature generation then assumes responsibility for feeding and maintaining the colony.

When they are in their nuptial swarms, ants are sometimes mistaken for termites (which are called "white ants"). The ant is, however, structurally very different from the termite and is distinctive in three features: the ant's hind wings are much smaller than its fore wings; its antennae are elbowed; and its abdomen is separated from the thorax by a narrow waist from which a strong knob or node usually projects upward. In contrast, the termite's wings, fore and hind, are equal in size and shape; its antennae are straight or gently curving; and it possesses a broad simple waist.

Because of their well-organized societies, their ability to communicate with others of their kind, and their often human-like behavior, ants have been considered intelligent animals. Although they are capable of simple types of learning, they cannot reason and must therefore carry on all their activities according to inherited instinctive patterns. Communication is achieved by means of touch (antennae tapping), or by the passing of odors of glandular secretions from one ant to another; no complicated language or voice exists. An example of chemical communication is the "odor trail." An individual, having located a food source, will mark the path between this food and the nest with a chemical substance it secretes from the abdomen. Other members of the colony then follow the path back to the food, aided by sensitive smelling organs located on the antennae and mouth.

When people disturb ant nests they sometimes see workers carrying larvae and pupae to safety, and they mistakenly assume that these white forms are "ant eggs." Although the eggs are to be found in the nest, they are not moved in this way and are in any case almost too small to see.

Quite a few species of ants live in the basin; those described in the following sections are the most familiar.

REFERENCES. Creighton, W. S. 1950. The ants of North America. *Bulletin of the Museum of Comparative Zoology, Harvard University,* vol. 104, pp. 1-585.

Hölldobler, B., and E. O. Wilson. 1990. *The ants.* Cambridge, Mass.: Belknap Press (Harvard University).

Mallis, A. 1941. A list of the ants of California with notes on their habits and distribution. *Bulletin of the Southern California Academy of Science,* vol. 40, pp. 61-100.

Smith, M. R. 1965. House-infesting ants of the eastern United States. U.S. Department of Agriculture, Technical Bulletin, no. 1326, pp. 1-105.

Wheeler, W. M. 1913. *Ants, their structure, development and behavior.* New York: Columbia University.

Wilson, E. O. 1971. *The insect societies.* Cambridge, Mass.: Belknap (Harvard University).

## ■ KEEPING ANTS IN CAPTIVITY

MOST GROUND NESTING ANTS may be cultured nicely for observation in an ant house, which may be made at home with very little effort or purchased—complete with ants—from a toy store. Basically, all that is required is a rectangular narrow wooden frame, to either side of which as been laminated a piece of window glass or thick transparent plastic. Access holes should be left at the top, and a stand affixed to the bottom.

Establishing and maintaining an ant colony is very instructive. Any ground-nesting species can be kept, but the California Harvester Ant is a good species to use because of its large size and availability (it is best to collect individuals of this species during the winter when they are not active and are less likely to sting).

Although many observations may be made on a group of confined workers, it is desirable to keep the queen as well. To obtain a queen, the entire nest will probably need to be excavated and carefully searched. A queen can be recognized by her larger size and wing stubs; colonies of some species have more than one queen.

After a supply of ants has been obtained, they should be introduced into the ant house, which has been previously filled with soil similar to that in which the ants are found in nature (dry beach sand should be avoided; the ants cannot excavate it successfully because it keeps caving in). A small amount of water should be added to the soil from time to time to keep it slightly damp.

The ants will soon dig a system of galleries and set up housekeeping. Their normal food should be kept available, of course: insects caught at your porch light or mealworms are suitable for carnivorous types, and wheat germ meal or dog food is good for the vegetarians. A drop or two of honey once in a while will be most welcome.

■ ARGENTINE ANT *(Iridomyrmex humilis)* Figure 394

This is our most common ant, the little blackish species (its length is 1/8 in., or 3 mm) that invades our homes and yards in search of food and water. Abundant in urban areas, it develops to prodigious numbers, and single colonies may harbor thousands of workers. It often becomes particularly noxious at the onset of cool weather in the fall, when colonies converge and move to sheltered, warmer quarters under homes, and foraging columns begin to seek food indoors.

The Argentine Ant is, as its name suggests, native to South America (Argentina and Brazil), and it is an undesirable alien in our country. It was apparently introduced into New Orleans before 1891 in coffee shipments from Brazil, and it has since spread rapidly over much of the United States.

The species is one of the most persistent and troublesome of all our house-infesting ants. Argentine Ant workers seek out and feed on almost every type of food, although they are especially fond of sweets. Making themselves most objectionable, the ants invade the house through minute crevices and cracks—filing along baseboards, across sinks, and over walls and tables in endless trails. They also have another undesirable habit: by protecting and tending scale insects and aphids, worker ants foster these injurious garden pests.

Shallow nests are made in the ground, often under rocks or wood; the galleries extend only to depths of 6 to 7 inches (15 to 18 cm) below the surface. There may be a number of queens in a single colony.

The Argentine Ant is a highly competitive species and is quick to exterminate other species of ants, including natives, in territory that it has just invaded. This ant has no sting; its bite is feeble but can be felt.

REFERENCE. Ribble, D. W. 1978. Visible trails of the Argentine Ant. *Journal of the Kansas Entomological Society,* vol. 51, pp. 796-797.

■ CALIFORNIA HARVESTER ANT *(Pogonomyrmex californicus)* Figure 395

This is the big "red ant" with the bad sting that mothers warn their children to avoid. Apart from its large size (1/4 in., or 6 mm) and red color, the species may be recognized by its large head, which is finely grooved and "bearded" (the long hairs on the chin function to carry sand or soil when the ant is excavating its nest).

The nest is constructed in dry sandy soil and is evident as a low flat crater encircled by a small area devoid of plant growth. The workers harvest seeds for food and litter the periphery of the nest with the husks. Each evening they close the entrances to the nest, reopening it in the morning. The nest is kept permanently closed during the winter, when there is no outside activity.

The California Harvester Ant is the special prey of horned toads.

REFERENCES. Michener, C. D. 1942. The history and behavior of a colony of harvester ants. *Scientific Monthly,* vol. 55, pp. 248-258.

Mintzer, A. C. 1982. Copulatory behavior and mate selection in the harvester ant, *Pogonomyrmex californicus* (Hymenoptera: Formicidae). *Annals of the Entomological Society of America,* vol. 75, pp. 323-326.

Spangler, H. G., and C. W. Rettenmeyer. 1966. The function

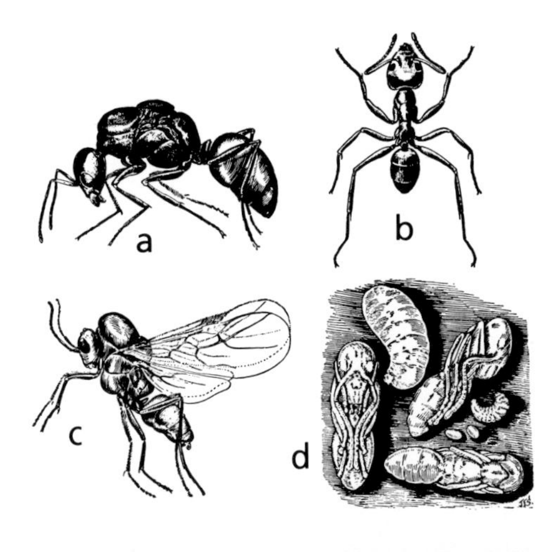


394. Argentine Ant queen (a), worker (b), winged male (c), and eggs, larvae, and pupae (d). Drawings from E. Back, 1937, U.S. Department of Agriculture Leaflets, no. 147, fig. 4.

395. California Harvester Ant. Photograph by P. Bryant.

of the ammochaetae or psammophores of harvester ants, *Pogonomyrmex* spp. *Journal of the Kansas Entomological Society,* vol. 39, pp. 739-745.

Weber, N. A. 1959. The stings of the harvesting ant, *Pogonomyrniex occidentalis* (Cresson), with a note on populations (Hymenoptera). *Entomological News,* vol. 70, pp. 85-90.

---

■ SOUTHERN FIRE ANT
*(Solenopsis xyloni)*
Figure 396

A close relative of the infamous imported Fire Ant *(Solenopsis invicta)* of the eastern United States, our species shows similar pestiferous habits and can inflict a painful sting. Workers are especially fond of fatty or oily foods and may attack and kill young or newly hatched poultry, quail, and other wild birds. They also build ugly mounds on lawns (evidence of their ground nests), gnaw vegetable and other crops, and make themselves obnoxious in many other ways.

The workers are distinguished by the characteristic way they raise and vibrate their abdomens when disturbed, and also by their color—a yellowish-red on the head and thorax and black on the abdomen. They vary in length from 1/32 to 1/4 inch (1 to 6 mm).

Locally, these ants are known to damage electrical equipment. They enter "pull boxes," which contain the wiring for traffic signals, and remove insulation from wires, causing shorts and signal failure.

REFERENCE. MacKay, W. P., D. Sparks, and S. B. Vinson. 1990. Destruction of electrical equipment by *Solenopsis xyloni* McCook (Hymenoptera: Formicidae). *Pan-Pacific Entomologist,* vol. 66, pp. 174-175.

---

■ THIEF ANT
*(Solenopsis molesta)*
Figure 397

These are tiny yellowish ants (1/16 in., or 1.5 mm, long) that commonly enter houses in search of food. They prefer items rich in fat or protein and will often feed on grease, meat, and cheese. Because of their small size, they are not easily excluded from food containers. They nest in the soil and normally feed on dead insects and the like.

The name "Thief Ant" alludes to their habit of establishing nests in or near those of other ants, which they then rob of food and brood.

---

■ PHARAOH ANT
*(Monomorium pharaonis)*
Figure 398

This species is sometimes mistaken for the common Argentine Ant. But the Pharaoh Ant is slightly smaller (3/32 in., or 2 mm), and its waist is composed of two segments instead of one. In addition, although its sting is ineffectual, the Pharaoh Ant does have a stinger, which the Argentine Ant lacks. The body color ranges from yellowish to light brown or reddish.

The species normally establishes its nests in

hideaways in buildings. It is especially common in places where food is handled—in hotels, boarding houses, and markets, for example. Omnivorous, it feeds on every manner of human foods, but it has a distinct preference for grease, fat, and meats.

The Pharaoh Ant was introduced from its native Africa.

REFERENCE. Peacock, A. D., D. W. Hall, I. C. Smith, and A. Goodfellow. 1950. The biology and control of the ant pest *Monomorium pharaonis* (L.). Miscellaneous Publications of the Scottish Department of Agriculture, no. 17, 50 pp.

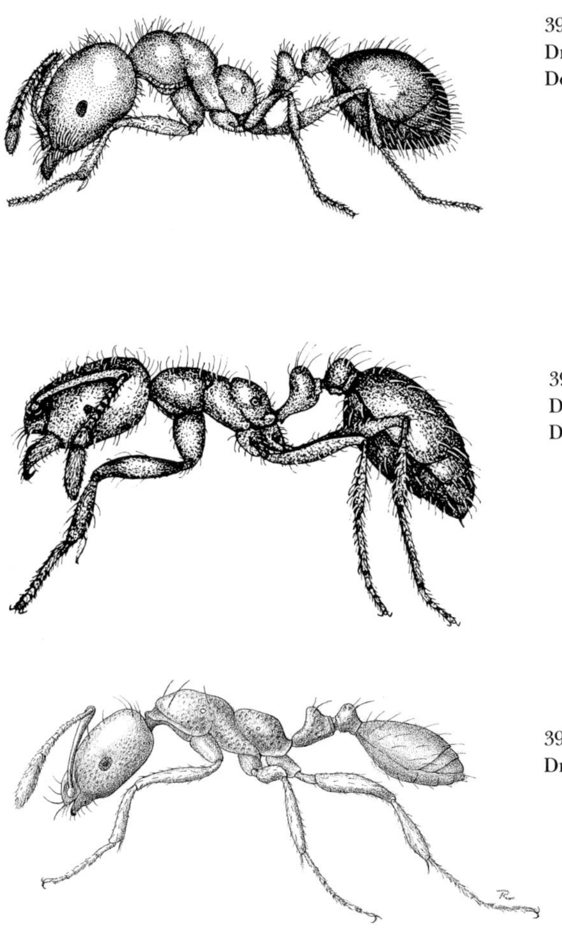

396. Southern Fire Ant. Drawing by R. DeNicola.

397. Thief Ant. Drawing by R. DeNicola.

398. Pharaoh Ant. Drawing by T. Ross.

**■ Velvety Tree Ant**
*(Liometopum occidentale)*
Figure 399

Although it sometimes nests in houses, the typical nesting areas of this species are cavities in and beneath the bark of trees growing along stream banks. Workers range up and down the tree trunks and branches, tending aphids from which they extract honeydew.

About 1/4 to 1/2 inch (6 to 13 mm) in length, this ant is easily recognized by its red thorax, glistening velvety black abdomen, and brownish black head. It has a strongly pungent odor when crushed or disturbed.

**■ Carpenter Ants**
*(Camponotus* species)
Figure 400

Carpenter ants are so called because many species build their nests in wood. Usually they inhabit preexisting cavities, such as abandoned termite galleries and rot holes, or burrow into rotten wood. But certain species (fortunately none of our local ones) excavate sound wood and are destructive to house timbers, telephone poles, and the like. They do not eat the wood but merely tunnel through it to form the nest. Their natural food consists of both dead and live insects, honeydew, and other sugary substances (rotting fruit, plant sap, etc.).

These are large ants (workers are up to 3/8 in., or 10 mm, long). Our three local species—*Camponotus vicinus, C. semites taceus,* and *C. clarithorax*—usually exhibit a reddish or yellowish thorax and black abdomen. In southern California, the ants most often locate their nests in the soil and only occasionally inhabit wooden structures. A fourth species, *Camponotus laevigatus,* which is all black, is an extremely common wood nesting carpenter ant found in fallen coniferous logs at higher elevations in the San Gabriel Mountains.

## WASPS

All members of the order Hymenoptera, excepting bees and ants, are technically considered to be wasps. Of great variety, they range from the large stingless horntails to the minute chalcidoids. Most familiarly associated with the name, however, are the stinging social and semisocial nest-building forms, including paper wasps, yellow jackets, and hornets.

A word about the stings of wasps is appropriate here. It is true that the females of most species are capable of stinging, but, with few exceptions, they must be roughly handled to be induced to do so. The effects of a sting—temporary pain and swelling—are usually not serious. The barbs on the wasp's stinger,

399. Velvety Tree Ant (right) dragging carpenter ant carcass to nest; larvae of a parasitic fly are feeding on the carcass. Photograph by C. Hogue.

400. Carpenter ant *(Camponotus clarithorax)*. LACM photograph.

unlike those of the Honey Bee, are very fine and do not catch in the flesh. However, the stinger may be repeatedly reinserted into the skin, and multiple stings, or even a single sting on a sensitive person, may cause general trauma or more serious consequences.

REFERENCES. Bohart, R. M., and R. C. Bechtel. 1957. The social wasps of California. *Bulletin of the California Insect Survey,* vol. 4, no. 3, pp. 73-101.

Evans, H. E. 1963. *Wasp farm.* New York: Natural History.

## ■ PAPER WASPS (Family Vespidae)

■ YELLOW JACKET
*(Vespula pensylvanica)**
Figure 401

With an abdomen that is broad to the base and marked on the back with black bars and spots, this very common wasp is unlike the narrow-waisted Golden Polistes and Mud Dauber. It is also a little smaller; its length is 7/16 to 9/16 inch (11 to 15 mm).

Nests consist of multiple layers of cells enclosed in a globular outer shell. The nest material is formed by

*When first used the scientific name of this wasp was inadvertently misspelled with a single letter "n" in the first syllable; the rules of scientific nomenclature require that the original spelling, even though erroneous, be preserved.

mixing saliva with fibers rasped from dead wood until a pulp not unlike papier maché is formed. Nests range in size from 2 to 12 inches (5 to 30 cm) in diameter and are usually situated in underground chambers and, rarely, in hollows in trees and attics of houses.

Colonies consist of as many as 15,000 individuals, which are differentiated into males, queen, and female workers; the latter are somewhat smaller than the queen and sexually impotent.

The Yellow Jacket is attracted to fruit and meat (it is sometimes called "meat bee") and at times may be very abundant; because of its inquisitiveness and painful sting, it can be a great nuisance to picnickers and campers.

REFERENCES. Akre, R. D., J. F. MacDonald, and W. B. Hill. 1974. Yellow jacket literature (Hymenoptera: Vespidae). *Melanderia,* vol. 18, pp. 67-93.

Bohart, R. M., and R. C. Bechtel. 1957. The social wasps of California. *Bulletin of the California Insect Survey,* vol. 4, pp. 73-102.

Duncan, C. D. 1939. A contribution to the biology of North American vespine wasps. *Stanford University Publications in Biological Science,* vol. 8, pp. 1-272.

MacDonald, J. F., R. D. Akre, and W. B. Hill. 1974. Comparative biology and behavior of *Vespula atropilosa* and *V. pensylvanica* (Hymenoptera: Vespidae). Melanderia, vol. 18, pp. 1-93.

Wells, H., and P. H. Wells. 1988. Foraging patterns of yellow jackets, *Vespula pensylvanica,* in an artificial flower patch. *Bulletin of the Southern California Academy of Sciences,* vol. 87, pp. 12-18.

---

■ GOLDEN POLISTES
*(Polistes fuscatus aurifer)*
Figure 402

This is another very common local wasp, which can be distinguished from the Yellow Jacket by its narrowed waist and nearly solid yellow abdomen. It is moderate-sized (body length 3/4 in., or 20 mm).

The umbrella-shaped nests, which are made of a paper-like substance similar to that produced by the Yellow Jacket, are composed of a single layer of cells and attached by a short stem to the underside of overhanging surfaces (eaves or fence rails, for example). Adult wasps gather caterpillars, which they skin and chew before feeding them to the grub-like larvae developing in the cells.

A new nest is constructed each spring by an over-wintering female that mated the previous autumn. Colonies consist of males, females, and workers, all of the same general form. Workers are fairly easily provoked and sting forcibly.

401. Yellow Jacket on fruit. Photograph by C. Hogue.

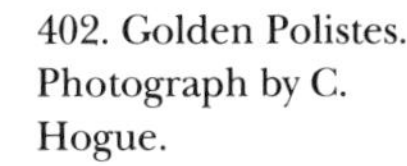

402. Golden Polistes. Photograph by C. Hogue.

## ■ SPHECID WASPS (Family Sphecidae)

■ MUD DAUBER
*(Sceliphron caementarium)*
Figures 403, 404

Like the Golden Polistes, this species has a narrow waist; it is a much longer one, however, and its abdomen is usually predominantly black instead of yellow. The body length is 1 inch (25 mm). Adult Mud Daubers twitch their wings while walking or standing.

The mud nests (Figure 404) built by the females are stuck under eaves, beneath rocks, to walls and rafters of attics and garages, or wherever there is protection from the elements. They consist of two to six or more elongate cells placed in series side by side. Females provision these cells with paralyzed spiders upon which the larvae feed.

Nest building and egg laying take place in the late summer and fall, after which the female dies. Larval development proceeds rapidly, and the mature larva passes the winter nestled in its brown cocoon within the mud cell. Pupation occurs in the spring, and the adult emerges soon after.

Mud Daubers are not aggressive and seldom sting. Even when they do so, the effects are generally mild.

REFERENCE. Shafer, G. D. 1949. *The ways of a mud-dauber.* Stanford, Calif.: Stanford University.

---

■ **Blue Mud Wasp** *(Chalybion californicum)* Figure 405

This is a strikingly beautiful wasp, with its shiny blue, blue-green, or blackish color. The female is often seen crawling into cracks and corners of walls or over objects on the ground, looking for its prey, spiders (it is particularly fond of combfooted spiders of the family Theridiidae). These wasps have been observed deliberately tapping on or vibrating webs to tempt the spider occupant out of hiding so that it can then be captured.

When it is successful in catching its prey, the Blue Mud Wasp paralyzes the spider with its sting and carries the body to its nest, placing it there as food for its larvae. It does not construct nests of its own but uses instead abandoned Mud Dauber nests or simple crevasses.

The males are rare and, at 3/8 to 1/2 inch (9 to 13 mm), considerably smaller than the females, which are 3/4 to 7/8 inch (20 to 23 mm). Both sexes have a short narrow waist and a somewhat hairy body. The wings are opaque and colored like the body.

REFERENCES. Coville, R. E. 1976. Predatory behavior of the spider wasp, *Chalybion californicum* (Hymenoptera: Sphecidae). *Pan-Pacific Entomologist,* vol. 52, pp. 229-233.

Landes, D. A., M. S. Obin, A. B. Cady, and J. H. Hunt. 1987. Seasonal and latitudinal variation in spider prey of the mud dauber *Chalybion californicum* (Hymenoptera, Sphecidae). *Journal of Arachnology,* vol. 15, pp. 249-256.

---

■ **Sand Wasps** *(Bembix* species) Figure 406

Also known as digger wasps, these insects are recognizable by their stout shape and greenish-white or bluish-white abdominal markings. A close look at the head reveals that they have unusually long mouthparts (including a triangular labrum). The fore legs are fringed with long hairs, which are used in digging.

Sand wasps are characteristic inhabitants of dry sandy areas such as beach bluffs and mesas, sand dunes, and arroyos; I have seen them working in the long jump pit on the track at the University of Southern California. They fly low and rapidly over the ground seeking prey and tending their burrow nests.

The nests are shallow tubes running obliquely into the soil; each contains a single larva, which the female keeps supplied with a diet of fresh flies and other insects. In practicing this form of continuous provisioning of the larvae, sand wasps differ from spider wasps, mud daubers, and many other digging wasps, which provide only a single cache of food that

403. Mud Dauber on mud nest. Photograph by C. Hogue.

404. Cutaway view of Mud Dauber nest, showing cells with larvae (white grubs) and cocoons. LACM photograph.

405. Blue Mud Wasp. Photograph by C. Hogue.

406. Sand wasp *(Bembix comata)* at entrance to its burrow nest. Photograph by C. Hogue.

must last throughout the larva's development. Sand wasps are not social insects, as are hornets and yellow jackets; yet, as a result of the tendency of individuals to nest in the same area, a type of colony develops.

Several genera and species of these interesting insects live in the basin. The most common is the Western Sand Wasp *(Bembix comata;* Figure 406), which is found on and in compact sand near the beaches.

REFERENCES. Bohart, R. M., and D. S. Horning, Jr. 1971. The California bembicine sand wasps. *Bulletin of the California Insect Survey,* vol. 13, pp. 1-49.

Evans, H. E. 1957. *Studies on the comparative ethology of digger wasps of the genus* Bembix. New York: Comstock.

## ■ SPIDER WASPS (Family Pompilidae)

■ TARANTULA HAWKS
*(Pepsis* species)
Figure 407

Tarantula hawks are giants, indeed among the world's largest wasps; some female individuals have a body length of nearly 2 inches (50 mm). In addition to their great size, these insects are easily recognized by the dark steely-blue body and contrasting orange-red wings. Differences in the antennae distinguish the sexes: in the females they are curved; in the males they are straight or only gently curving. But be careful! The females—which have powerful stingers—may now and then straighten their antennae.

These wasps are no longer common in our area, but an occasional individual may be seen in the hilly parts of the basin, bumbling over milkweed in the spring and summer or over scale broom *(Lepidospartum)* in the fall, or buzzing low over the ground in search of the tarantulas *(Aphonopelma* species) on which it preys.

When a female wasp finds a tarantula, she alights and engages it in battle. The wasp then stings the spider on the underside between the legs (Figure 407) and usually succeeds in paralyzing but not killing it. She has previously dug a shallow burrow, using her mandibles and legs as pick and shovel, or selected an earth crack, rodent burrow, or even the burrow of a tarantula for a nest, and she now drags the paralyzed prey into this hole, lays an egg on the victim, and then seals the tunnel with soil. A supply of fresh food is thus insured for the developing larva.

The sting of the female tarantula hawk is described as extraordinarily painful by those who have experienced it.

REFERENCES. Hurd, P. D., Jr. 1948. Systematics of the California species of the genus *Pepsis* Fabricius (Hymenoptera: Pompilidae). *University of California Publications in Entomology,* vol. 8, pp. 123-150.

Petrunkevitch, A. 1926. Tarantula versus tarantula-hawk: A study in instinct. *Journal of Experimental Zoology,* vol. 45, pp. 367-394.

Williams, F. X. 1956. Life history studies of *Pepsis* and *Hemipepsis* wasps in California (Hymenoptera, Pompilidae). *Annals of the Entomological Society of America,* vol. 49, pp. 447-466.

## ■ VELVET-ANTS
(Family Mutillidae; *Dasymutilla* species)

VELVET-ANTS ARE NOT REALLY ANTS at all but are wingless female wasps (the males are winged like other wasps). With their conspicuous brightly colored vestiture of long fine hairs (Figure 408), both sexes always attract attention in the arid sandy areas they prefer. Although most frequent in our deserts and foothills, they live also on the coastal dunes and bluffs.

We know very little about the life cycles of these wasps. Available information suggests that they are parasitic on other ground-nesting wasps and bees.

The female stings unhesitatingly, causing a sharp pain that quickly subsides and produces little or no swelling or other side effects. Females are capable of emitting a faint squeaking sound when confined.

Several species of velvet-ants occur in the basin; all exhibit contrasting black and yellow, red, orange, or white color patterns and are about 1/4 to 7/8 inch (6 to 20 mm) long. Unfortunately for the layman, the principal colors are not consistent according to the species (a species may exhibit two or more color phases in different geographic areas), and it is nearly impossible to distinguish the species with a superficial examination.

REFERENCE. Hurd, P. D. 1951. The California velvet-ants of the genus *Dasymutilla* Ashmead. *Bulletin of the California Insect Survey,* vol. 1, pp. 89-118.

407. Tarantula hawk *(Pepsis chrysothymus)* stinging prey. LACM photograph.

## ■ CUCKOO WASPS (Family Chrysididae)

THESE ARE SMALL BEE-LIKE WASPS that are usually deep metallic blue or green in color and sparsely haired (Figure 409). They can roll themselves into a ball by burying the head in the ventrally concave abdomen.

Like the bird of the same common name, the Cuckoo Wasp is parasitic—it raises its young in the nests of bees and other wasps. The larva kills the rightful occupant of the nest and develops on the provisions left in the cell by the nestbuilder.

## ■ GALL WASPS (Family Cynipidae)

WE FREQUENTLY NOTICE GALLS because of their curious shapes and because they appear as abnormal growths on leaves, stems (Figure 411), flowers, and roots of plants and often mar the beauty of ornamentals. Galls take a variety of grotesque forms—they can resemble fleshy mushrooms, gnarled hard knots, star-shaped or spiny projections, or large smooth reddish spheres (the spheres are especially common on oak trees and are called oak "apples" because of their similarity to that fruit).

The adult gall wasp is a small insect (usually less than 1/4 in., or 6 mm, in length) and dark or somber in color. The thorax has a somewhat humped appearance, causing the head to be oriented downwards, and the abdomen is frequently compressed from side to side.

The mechanism by which the female wasp or feeding larvae cause the galls to be formed is not well understood. Some chemical stimulates the plant to grow excess tissue in a specific shape. This tissue is utilized by the wasp larva as food. Upon maturing, the adults burrow out of the gall, leaving a small hole on the surface.

Some gall wasps exhibit an alternation of generation in their life cycles: during one phase of development males as well as females are produced and reproduction is sexual, whereas in the other phase only females are present and reproduction is asexual. The insects of the two generations may be so different structurally as to appear to be distinct species.

Oaks seem to be a favored host for gall wasps. Many species infest our white, black, and scrub oaks, and the gall made by each is typical of the species. Two of the most common are the following:

■ Live Oak Gallfly *(Callirhytis quercuspomiformis;* Figure 410). This species forms on the Coast Live Oak *(Quercus agrifolia).* The apple galls (Figure 411) are

408. Velvet-ant female *(Dasymutilla coccinea)*. Photograph by C. Hogue.

409. Cuckoo wasp *(Chrysura pacifica)*. Photograph by C. Hogue.

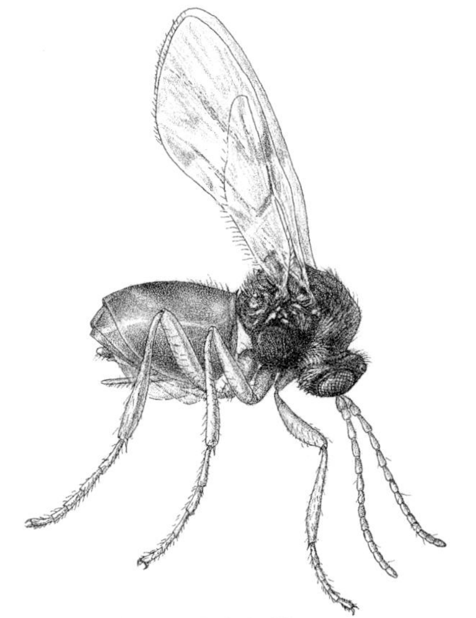

410. Live Oak Gallfly. Drawing by T. Ross.

411. Apple galls of Live Oak Gallfly. Photograph by F. Hovore.

caused by the asexual generation, the mushroom leaf galls by the sexual generation.

■ California Gallfly *(Andricus quercuscalifornicus)*. White Oaks *(Quercus lobata, dumosa,* etc.) are the hosts of this gall wasp. No alternative phase and host are known. Adults emerge from the apple galls in October.

REFERENCES. Brown, L. R., and C. O. Eads. 1965. Oak gall insects. Chapter 5 in *A technical study of insects affecting the oak tree in southern California,* edited by L.R. Brown and C.O. Eads. Bulletin of the University of California Agricultural Experiment Station, no. 810.

Russo, R. A. 1979. *Plant galls of the California Region.* Pacific Grove, Calif.: Boxwood.

Weld, L. H. 1957. Cynipid galls of the Pacific Slope (Hymenoptera, Cynipoidea). Published by the author, Ann Arbor, Michigan.

## ■ ICHNEUMONS (Family Ichneumonidae)

THESE SOLITARY WASPS are, for the most part, small and have more slender bodies and legs than do the familiar social and semisocial types. Often the abdomen is strongly compressed from side to side (Figure 412). They vary great in size, from small gnat-like insects to large wasps over 1 inch (25 mm) long. All ichneumons are parasitic on other insects; our most common local species belong to the genus *Ophion* and attack caterpillars.

The eggs are inserted into the body of the host by means of the female's short sharp ovipositor (which, incidentally, can penetrate human skin). The larvae feed on the internal tissues and, when mature, pupate within the host. *Ophion* adults frequently come to light at night.

## ■ APHID KILLERS (Family Aphidiidae)

FEMALES OF THESE SMALL PARASITOID WASPS insert their eggs in the abdomen of aphids. The larva consumes the host from within. When the adult emerges, it leaves behind a dry hollow shell with a round hole in the back of the abdomen (Figure 413). Large numbers of these "aphid mummies" on plants indicate successful wasp reproduction. Because of their efficacy in killing aphids, these wasps are considered good biological control agents, and they are often intentionally introduced to farms and orchards.

Species of the genus *Aphidius* are the main members of the family; *Aphidius ervi* is a common local representative. It is very small (its length is about

412. Ichneumon wasp (*Ophion* species). Photograph by C. Hogue.

413. Aphid "mummies," the remains of aphids killed by parasitism by an aphid killer wasp (*Aphidius* species). Photograph by C. Hogue.

3/32 in., or 2 mm) and shining black, with wings that are clear except for an elongate triangular spot midway along the front margin. The abdomen is long and slender and sharp-tipped.

REFERENCE. Schlinger, E. I., and J. C. Hall. 1960. Biological notes on Pacific Coast aphid parasites (Aphidiinae) and their aphid hosts (Hymenoptera: Braconidae). *Annals of the Entomological Society of America,* vol. 53, pp. 404-415.

## ■ CHALCIDOID WASPS
(Superfamily Chalcidoidea)

MOST OF THESE WASPS are very tiny (1/32 to 1/8 in., or 1 to 3 mm) and thus are seldom noticed. The presence of many species is nonetheless abundantly apparent because of their importance as parasitoids of injurious insects. There are several families represented in the basin by scores of species, and these attack a variety of hosts, especially the larvae of beetles, butterflies and moths, homopterans (cicadas, leafhoppers, and the like), and other hymenopterans—including other parasitoids! The larvae of a few feed on seeds and other plant parts; an example of such species are the well-known fig wasps *(Blastophaga* species), whose life inside the fruit of certain figs is a prerequisite for the fruit to ripen *(caprifigs, caprification).*

For the most part, chalcidoids are metallic and black, blue-black, or greenish. Almost all are winged. The wings have very few veins and are often fringed; they are held flat over the abdomen when the insect is at rest. The antennae are elbowed.

The following local chalcidoids have reputations as beneficial agents in biological control of insect pests.

■ Fly killers *(Muscidifurax* species; family Pteromalidae). These wasps destroy flies and so are welcome in feed lots, poultry ranches, dairies, animals pens, and backyards. Different species attack different fly species, and each of the common pestiferous flies—including the House Fly and Stable Fly—is prey to a certain fly killer. The eggs are inserted into the fly maggots and puparia, where the wasp's own immature stages feed and develop, eventually killing the host.

■ Scale parasitoids *(Aphytis* species, family Encyrtidae). The Red Scale Aphytis *(A. melinus;* Figure 414), which was introduced from India, is the notable local representative of this very beneficial genus of chalcidoids. It is a prime enemy of California Red Scale but also attacks other armored scales on crops. The females place their eggs directly into the body of the scale insect under the scale cover. The wasps mature rapidly on the blood and tissues of the host, destroying it.

■ Egg wasps *(Trichogramma* species, family Trichogrammatidae; Figure 415). These are miniscule creatures, among the smallest insects known, measuring less than 1/32 inch (1 mm) in length. They are so small that they can pass their developmental stages within the eggs of other insects, principally those of butterflies and moths. Large numbers are released in agricultural areas for control of lepidopterous pests of field crops, ornamentals, and orchards.

REFERENCES. DeBach, P. 1959. New species and strains of *Aphytis* (Hymenoptera, Eulophidae) parasitic on the California red scale, *Aonidiella aurantii* (Mask.), in the Orient. *Annals of the Entomological Society of America,* vol. 52, pp. 354-362.

DeBach, P., D. Rosen, and C. E. Kennett. 1971. Biological control of coccids by introduced natural enemies. Pages 165-194 in *Biological control,* edited by C.B. Huffaker. New York: Plenum.

Flanders, S. E. 1938. Identity of the common species of American *Trichogramma. Journal of Economic Entomology,* vol. 31, pp. 456-457.

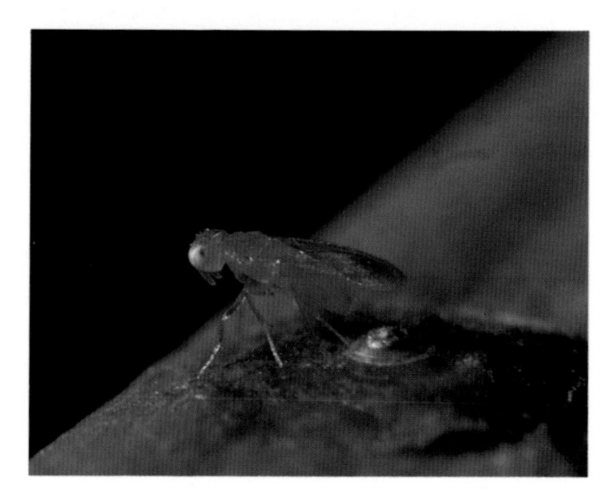

414. Scale parasitoid *(Aphytis melinus)* with scale insect. Photograph by M. Badgley.

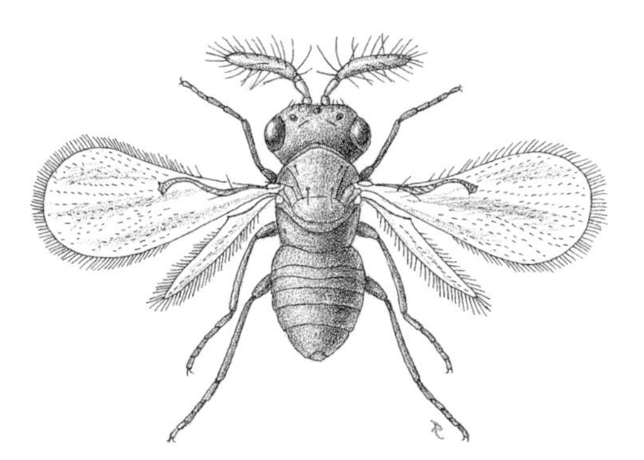

415. Egg wasp *(Trichogramma* species). Drawing by T. Ross after Z. Boucek, 1988, *Australian Chalcidoidea,* p. 30, fig. 20 (Wallingford, Oxon, U.K.: CAB International).


Kogan, M., and E. F. Legner. 1970. A biosystematic revision of the genus *Mucidifurax* (Hymenoptera: Pteromalidae) with descriptions of four new species. *Canadian Entomologist,* vol. 102, pp. 1268-1290.

Legner, D. R., and H. W. Brydon. 1966. Suppression of dung-inhabiting fly populations by pupal parasites. *Annals of the Entomological Society of America,* vol. 59, pp. 638-651.

Legner, E. F., and M. E. Badgley. 1982. Improved parasites for filth fly control. *California Agriculture,* vol. 36, nos. 9-10, p. 27.


## ■ HORNTAILS OR WOOD WASPS
(Family Siricidae)

THESE WASPS, which are just over an inch in length (25 to 30 mm), are so named because of a horn-like projection at the tip of the abdomen (Figure 416). Below this process, the females carry a very long rigid

416. Wood wasp male *(Sirex aureolatus).* Photograph by C. Hogue.

417. Wood wasp larva *(Tremex columba).* Drawing by T. Ross after various sources.

ovipositor used to insert the eggs deep into dead or dying coniferous wood, which is the food of the larvae (Figure 417). As the larvae develop, they burrow through the wood, forming tunnels that reduce its value as lumber.

These insects are widely disseminated in shipments of lumber or timber, and the adults may not emerge until several years have elapsed. In the basin, I have encountered instances of adults, especially those of the Long-tailed Wood Wasp *(Sirex longicauda),* exiting from wall studs through the plaster facing into the interior of homes. When a load of infested lumber has been used in a housing development, such emergence may be extensive and cause alarm among the new residents.

REFERENCE. Middlekauff, W. W. 1960. The siricid wood wasps of California. *Bulletin of the California Insect Survey,* vol. 6, pp. 59-78.

## BEES

EVOLUTIONARILY, BEES represent a branch off the more primitive wasps. Although some bees are parasitoids, the majority nourish their young on pollen and nectar, which they collect from flowers. The bees' thick body covering of plumose hairs entraps pollen, and

418. Leaf cutter bee *(Megachile* species). Photograph by C. Hogue.

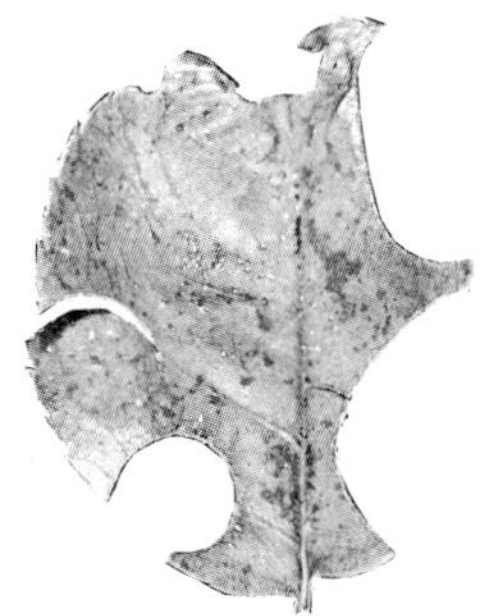

their elongate multifunctional mouthparts enable them to reach deep into flowers to the nectaries.

Although some bees, such as the Honey Bee, have highly developed social instincts, most are nonsocial or semisocial insects that lead independent lives without the development of castes for the division of labor in the family.

Bees' bodies are generally more robust than those of wasps. Females, like female wasps, do have stingers, which they are quick to use in defense. The stinger of the Honey Bee is barbed and clings to the wound in human skin.

## ■ MEGACHILID BEES (Family Megachilidae)

■ LEAF CUTTER BEES *(Megachile* species) Figure 418

Neatly cut semicircular notches in the leaf edges of one's rose bushes indicate the presence of these solitary bees in the neighborhood. The females of the family Megachilidae cut the leaf fragments with their mandibles and carry them to a natural tubular cavity or burrow, to form a lining for the nest. Provisions for the larvae consist of pollen and honey.

The few species of leaf cutter bees of the genus *Megachile* in the basin are medium sized (1/4 to 1/2 in., or 6 to 13 mm, long) and black in color; they have abundant hair on the head, thorax, and legs.

## ■ HALICTID BEES (Family Halictidae)

■ METALLIC SWEAT BEES (*Agapostemon* and *Augochlorella* species) Figure 419

These are solitary bees of medium size (3/8 in., or 10 mm) with brilliant metallic green bodies; the males of *Agapostemon* have yellow abdomens with black rings. They nest in tubular burrows dug in the ground, often in clay banks.

REFERENCE. Roberts, R. B. 1969. Biology of the bee genus *Agapostemon* (Hymenoptera: Halictidae). *University of Kansas Science Bulletin,* vol. 48, pp. 689-719.

## ■ ANTHOPHORID BEES (Family Anthophoridae)

■ CARPENTER BEES (*Xylocopa* species) Figures 420-422

These very large bees (3/4 to 1 in., or 20 to 25 mm) are commonly mistaken for bumble bees, to which they are not related. Carpenter bees are solid in color, usually black or steely blue (the males of one species are light brown). Bumble bees, on the other hand, are marked with yellow and black hair patterns.

Carpenter bees are so named because they bore into wood, forming tunnel-like nests for the rearing of the young. Many kinds of wood are used; although fence posts, building timbers, and telephone poles often are attacked, the effects are seldom damaging. In 1989 a female attracted attention by choosing to make her nest in the cross-section of a giant redwood that is on display at the Figueroa Street office of the Southern California Automobile Club.

Upon completion of the nest tunnel, the female places pollen and nectar at the far end, lays an egg, and closes off this part of the tunnel with a partition of wood pulp, creating a cell. Five or six such cells, each in front of the last, are formed and sealed in succession (Figure 422). In about a week the eggs hatch, and the larvae feed on the pollen and nectar until they mature. Adult bees emerge after about forty to forty-five days and leave through the entrance to the tunnel; since all mature at about the same time, the bee at the deep end must wait for those nearer to the single exit to leave before it can escape. Female bees hibernate during the winter and start new nests the following spring.

Carpenter bees are commonly found about flowers in the less-developed parts of the basin, and their formidable size and habit of "buzzing" people earns them a bad reputation. The females can sting but do so very reluctantly, causing only mild pain.

Three kinds of carpenter bees live in the basin:

■ Valley Carpenter Bee (*Xylocopa varipuncta;* Figures 420, 421). This is the species found in the lowlands

419. Metallic sweat bee male (*Agapostemon* species). Drawing by C. Hogue.

420. Valley Carpenter Bee male. Photograph by C. Hogue.

421. Valley Carpenter Bee female. Photograph by C. Hogue.

422. Cutaway view of nest of California Carpenter Bee and close-up of two nest cells (right); the bottom cell contains a larva that has partially consumed its pollen provision. Photograph by L. R. Brown.

and the most common. The male is light brown with yellow to golden hairs; the female is shiny black with bronzy reflections on the wings. The nest is bored into lumber and logs (often eucalyptus).

■ California Carpenter Bee *(Xylocopa californica diamesa;* nest shown in Figure 422). Both males and females are dark blue; in certain light the body reflects a greenish cast. Burrows are usually constructed in the dry dead stems of the Common Yucca *(Yucca whipplei).* The species is most common at higher elevations.

■ Mountain Carpenter Bee *(Xylocopa tabaniformis orpifex).* Both male and female are largely black; the male has some white hair on the thorax. Native softwoods are usually used for the nests. Like the California Carpenter Bee, this species is most common at higher elevations; it is also the smallest of the three forms described here.

REFERENCES. Cruden, R. W. 1966. Observations on the behavior of *Xylocopa c. californica* and *X. tabaniformis orpifex. Pan-Pacific Entomologist,* vol. 42, pp. 111-119.

Hurd, P. D., Jr. 1955. The carpenter bees of California. *Bulletin of the California Insect Survey,* vol. 4, pp. 35-72.

_______. 1958. Observations on the nesting habits of some New World carpenter bees with remarks on their importance in the problem of species formation. *Annals of the Entomological Society of America,* vol. 51, pp. 365-375.

## ■ TRUE BEES (Family Apidae)

■ BUMBLE BEES
*(Bombus* species)
Figure 423

Familiar to most everyone, bumble bees may be easily recognized by their large size (1/2 to 1 in., or 13 to 25 mm, long) and dense furry covering of black and yellow hairs. They are sometimes confused with the carpenter bees, which are also large but are solid black in color with no yellow patterns. Equally as characteristic of bumble bees as their size and coloration is their loud buzzing flight, to which the name "bumble" refers; the word comes from the Middle English *bumblen,* which means "to hum." The modern German word for bumble bee is *Hummel;* in Britain they are sometimes called "humble bees."

Bumble bees are annually social—that is, the colony dies out each winter. A new colony is formed in the spring by a fertile queen who has hibernated in a crevice in a trees or under bark or in another protected place; once she has laid the eggs and established the colony, she usually dies. The nest, which is constructed in a cavity in the ground, consists of a cluster of irregular, rounded wax cells sheltered by a matting of dry grass or twigs. The young are reared in these cells on a diet of pollen and honey. There is normally only one brood per year.

The females and workers sting severely, but our species, at least, are not easily provoked.

Four species occur sporadically throughout the basin:

- Sonoran Bumble Bee *(Bombus sonorus;* Figure 423). The body is largely yellow, but the face and the tip of abdomen are black and a thin black band runs across the thorax. This is our most common species, but it is no longer abundant. It nests at the base of shrubs and rushes in marshy flats near the coast.
- California Bumble Bee *(Bombus californicus).* This bumble bee is yellow on abdominal segment 2 only; the rest of the abdomen and the face are black.
- Vosnesenski's Bumble Bee *(Bombus vosnesenskii).* The abdomen of this species has a banded pattern like that of the California Bumble Bee, but the yellow is farther back, on segment 4. The face is yellow.
- Crotch's Bumble Bee *(Bombus crotchii).* This species is yellow on abdominal segments 2 and 3 only; the apex of the abdomen is reddish, and the face is black.

REFERENCES. Free, J., and C. Butler. 1959. *Bumblebees.* London: Collins.

Ryckman, R. E. 1953. Notes on the ecology of *Bombus sonorus* in Orange County, California and new parasite records. *Pan-Pacific Entomologist,* vol. 29, pp. 144-146.

Thorp, R. W., D. S. Horning, Jr., and L. L. Dunning. 1983. Bumble bees and cuckoo bumble bees of California (Hymenoptera: Apidae). *Bulletin of the California Insect Survey,* vol. 23, pp. 1-79.

423. Sonoran Bumble Bee worker. Photograph by P. Bryant.

■ HONEY BEE
*(Apis mellifera)*
Figure 424

Because its anatomy is unmodified by selective breeding practices and it can live on its own in nature, the Honey Bee is not a truly domesticated insect, although it is often referred to as such. The practice of apiculture, our keeping of bees for their products, is nearly as old as agriculture itself. Today we provide the bees with specially constructed hives, which replace the hollow trees and caves that were the original homes of the species. Even today, many bees escape their manmade homes and return to the wild, where they often survive quite well. Unfortunately for the homeowner, they frequently choose an attic for their new abode.

The Honey Bee originated in southern Asia and was not introduced into America from Europe until the middle of the seventeenth century. It is not known how the bees were first successfully transported on the long journey from the Old World to the New, but there is no doubt that colonies made the trip safely more than once.

There are at least five different races of the Honey Bee, and a multitude of strains within each of these. Each has its own structural and behavioral characteristics, some desirable from the apicultural standpoint, others undesirable. Much hybridization occurs between these types, and artificial insemination techniques are part of modern breeding technology.

The Honey Bee is an typical social insect: it lives in a colony that is made up of the offspring of a single reproductive pair, and all of the colony members exhibit a high degree of cooperation in their activities. Thus the colony is a large complex family. Members of this family are of three types, each with a specific body form and set of duties in the colony:

■ The queen, who may live four to five years as the sole reproductive female, lays eggs (2,000 or more a day) and acts as the control center of the colony. She takes no part in caring for the young.

■ The workers, which are nonreproductive females, carry out most of the hive and field activities—they gather nectar, pollen, and water; build the honeycomb; feed the queen; form the honey; and clean and ventilate the hive.

■ The drones are males whose only function is to fertilize the virgin queen at swarming time; they usually die soon afterwards.

The swarming of Honey Bees, the species' method of colony reproduction, is a spectacular phe-

424. Honey Bee worker. Photograph by C. Hogue.

nomenon. There are two types of swarms. In a dispersal swarm, an impregnated queen leaves the hive with a group of workers to seek a new home. In a mating swarm, one virgin queen leaves the hive with a few drones and many workers. The queen mates with a drone from another colony; a new home may be sought or the swarm may return to the old one just vacated.

Swarms are a frequent source of concern to local residents. Swarming bees are usually rather docile (as are bees most of the time) and will not sting unless provoked. However, the bees may temporarily settle in trees or shrubs near places of habitation, and they sometimes establish hives in chimneys or in the walls of houses. Such situations are best handled by experienced apiarists (beekeepers).

Stings are usually of little consequence, but hypersensitive persons may suffer extreme reactions or even death. The stinger of the worker is coarsely barbed and usually remains in the wound, pulling out of the bee with vital organs attached and thus killing it. (The queen can also sting, but the drones cannot, as they lack stingers.) The stinger itself should be removed from the wound as soon as possible to prevent the entrance of an unnecessary amount of venom, which continues to flow into the wound under pressure from the attached, pulsating poison sac. The stinger should be scraped out of the skin with a knife or fingernail or lifted with forceps; in handling the stinger, avoid putting pressure on the venom sac.

Honey is the best-known and most valuable product of the Honey Bee. It is manufactured by the bees from flower nectar, and it consists mainly of glucose and fructose sugars. Honey is fed to the larvae and is an important supplemental food for adult bees. Nectar, however, is the adults' primary carbohydrate food source; they also store and eat pollen, which supplies protein. Honey Bees often pass excess ingested pollen, and the fecal drops may fall on objects, defacing cars with small brownish "bee spots."

For man honey is a centuries-old food and elixir. Much has been said about the cosmetic use of another bee product, royal jelly. As yet, however, no definite evidence exists to support claims that when applied topically it benefits the human skin. The bees use it as special food for young worker larvae and for the queen in both her larva and adult stages.

Beeswax, the construction material of the comb, is secreted by special glands in the abdomen of the bee. It has many legitimate industrial and home applications.

At this writing, the so-called Africanized or Killer Bees *(Apis mellifera scutellata,* formerly *A. m. adonsoni)* have just reached the United States on their slow expansion out of South America. Because they are very agressive and easily agitated, these bees require special handling by the beekeeper: they are more apt to sting en masse than are bees of the more docile European strains. The effects on local apiculture and agriculture of Africanized bees—should they become established here—cannot be predicted.

REFERENCES. Butler, C. 1955. *The world of the Honeybee.* New York: Macmillan.

Ribbands, C. 1953. *The behaviour and social life of honeybees.* London: Bee Research Association Ltd. [Available in Dover reprint edition.]

Root, A. I., and E. Root. 1983. *The ABC and XYZ of bee culture: An encyclopedia pertaining to scientific and practical culture of bees,* 39th edition. Medina, Ohio: A.I. Root.

Winston, M. L. 1987. *The biology of the honey bee.* Cambridge, Mass.: Harvard University.

_______. 1992. *Killer bees: The Africanized honey bee in the Americas.* Cambridge, Mass.: Harvard University.

# Part III
# THE ARACHNIDS

THE SPIDERS, mites, ticks, scorpions, and related arthropods belong to the class Arachnida, a large group whose members are biologically quite diverse. The majority are free-living solitary predators who capture and kill other arthropods; many arachnids feed on plants, however, and some are parasitic on humans.

As a group, the arachnids are quite ancient: fossil scorpions are known from the Silurian Period, over 400 million years ago. In contrast, the early insects appeared later, around 350 million years ago.

Although arachnids are often mistaken for insects, they differ in fundamental and rather obvious ways. Arachnids possess six rather than three pairs of jointed appendages, and only the four pairs to the rear of the body usually function as legs (thus the arachnid has eight "legs" to the insect's six). Unlike insects, arachnids have neither mandibles nor antennae; instead, the first of the six pair of jointed structures is modified into "scissors" or needle-like mouthparts used for chewing or sucking food—in spiders, these are the fangs. The second pair of arachnid appendages has a variety of shapes and functions—in scorpions, for example, they are claws used for grasping prey; in male spiders, they are copulatory organs. Eyes, when present in an arachnid species, are simple (not compound) and sessile (not on a stalk).

The body of an arachnid is unlike that of an insect in that it has only two divisions instead of three: its abdomen is comparable to the insect abdomen, but the arachnid's *cephalothorax* is a combined head and thorax in structure and function. The main parts of the arachnid body may or may not be segmented; in some (for example, the mites), even the cephalothorax and abdomen are fused, so that the body is undivided.

Arachnids develop gradually, increasing in size and sexual maturity after each molt. Molting often continues throughout life, and many arachnids are capable of regenerating lost limbs, which are replaced incrementally over several molts. Although arachnids lack metamorphosis, mites and ticks have rather different appearing immature stages, which include a "larva" with only six legs and one or more stages that are called "nymphs" even though they are like the adults except for smaller size and undeveloped external genitalia.

A few arachnids are rightfully considered harmful because they are agricultural pests or because they bite, sting, or carry human disease. However, the group as a whole is a very beneficial one from our standpoint. The vast majority of species are predatory and thus help to control populations of injurious insects in fields, gardens, and even in the home. Because they do good work and are inoffensive and for the most part rarely seen, their presence around buildings and yards should be tolerated and indeed encouraged.

# 16 SCORPIONS, SUN SPIDERS, AND DADDY LONG-LEGS

## SCORPIONS (Order Scorpionida)

THE SCORPION is an invertebrate that has prompted odd beliefs from earliest times, and it is the only arachnid having a place, as "Scorpio," among the signs of the Zodiac. The popular belief that, when exposed to fire, scorpions sting themselves to death and thus commit suicide is without validity.

Scorpion anatomy is common knowledge: large "lobster" claws, eight legs, and an elongate body with a slender segmented tail tipped by a bulbous sting. A pair of conspicuous comb-like structures on the belly (one is visible in Figure 426), which are held out parallel to the ground, further characterize these animals; the function of these organs is not definitely known, but they probably serve to perceive ground vibrations.

Scorpions are generally nocturnal, moving about at night in search of soft-bodied insects for food. They usually feed on beetles, cockroaches, crickets, centipedes, spiders, sun spiders, and other ground dwellers.

In the basin we are most likely to encounter scorpions in hilly areas being developed for residences. When their natural hiding places are disturbed or destroyed, they often seek shelter around or within homes.

In nature, scorpions live in burrows, under rocks and logs lying on the ground, in sand, and under loose tree bark. Around human habitation, they thrive in lumber piles, in accumulations of boxes, rags, brush, and other rubble, and in crevices and seldom-used areas such as cellars, garages, and the like. Scorpion control around the home consists mainly of keeping the premises clean and free of litter.

Courtship in scorpions takes the form of an exotic dance. Upon finding a female, the male grasps her claws with his and promenades sideways or backward with her. Copulation occurs after an extended period of such activity. The fertilized eggs develop within the female, and the young are born alive. Until their first molt the young are carried upon the mother's back and subsist on embryonic yolk; they are not fed by the adult.

None of our indigenous scorpions is considered dangerous, although any may inflict a wound that is temporarily painful. However, there is a potentially lethal species *(Centruroides exilicauda)* in southern Arizona and portions of the neighboring states. In Arizona, during a twenty-year period, *C. exilicauda* has been responsible for the majority of the deaths due to venomous creatures, many more than rattlesnakes and all other types put together. The species has turned up in recent years in parts of Orange County; it has been particularly common in Irvine, where it poses a health hazard.

The stings of our scorpions usually cause only a local reaction similar to that of a bee sting, consisting of pain and a burning sensation, with swelling that lasts from a few minutes to over an hour. First-aid treatment involves immersing the affected area in ice water or applying an ice pack. If symptoms persist, a physician should be consulted.

The species described here, all members of the family Vejovidae, are the most common scorpions in the Los Angeles Basin.

REFERENCES. Ennik, F. 1972. A short review of scorpion biology, management of stings, and control. *California Vector Views,* vol. 19, pp. 69-80.

Geck, R. 1980. Introduction of scorpions to Orange County. Proceedings and Papers of the 48th Annual Conference of the California Mosquito and Vector Control Association, p. 136.

_______. 1989. Orange County Vector Control District, Vector Ecology Program Report, Report of District Activities for January 13, 1989, p. 1.

Parrish, C. 1966. The biology of scorpions. *Pacific Discovery,* vol. 19, pp. 2-11.

Russell, F. E., and M. B. Madon. 1984. Introduction of the scorpion *Centruroides exilicauda* into California and its public health significance. *Toxicon,* vol. 22, pp. 658-664.

Stahnke, N. L. 1966. Some aspects of scorpion behavior. *Bulletin of the Southern California Academy of Sciences,* vol. 65, pp. 65-80.

Williams, S. C. 1976. The scorpion fauna of California. *Bulletin of the Society of Vector Ecology,* vol. 3, pp. 1-14.

## ■ VEJOVID SCORPIONS (Family Vejovidae)

■ BURROWING SCORPION *(Anuroctonus phaiodactylus)* Figure 425

This is a common species, distinguished from the other species described here by its bulky claws and large size (the length of the adults from the front of the stout body to the tip of the extended tail is 2 1/2 in., or 65 mm). The sting is naked and shiny; in mature males, there is frequently a second small inflated bulb between the base and tip of the sting. The Burrowing Scorpion is light yellow brown to dark brown.

These scorpions dig burrows, often in colonial aggregations; the male's is usually 6 to 8 inches (15 to 20 cm) deep, but the female may dig to a depth of up to 2 feet (60 cm). The scorpion ambushes prey from the mouth of its burrow during the night, retreating into its shelter during the day or when startled or threatened. It seldom uses its sting.

REFERENCE. Williams, S. C. 1966. Burrowing activities of the scorpion *Anuroctonus phaiodactylus*. *Proceedings of the California Academy of Sciences* (series 4), vol. 34, pp. 419-428.

---

■ STRIPE-TAILED SCORPION
*(Paruroctonus silvestrii)*
Figure 426

The body and claws of this scorpion are slender, and it is medium sized (adults are up to 1¾ in., or 45 mm, long). It is generally mottled dark gray-brown in color, except for the last segment, which is pale; the species has four conspicuous longitudinal dark brown lines on the underside of its tail. The claws or pincers are scalloped or toothed along the inner margins. The species stings readily but without doing lasting harm.

425. Burrowing Scorpion. Photograph by B. Hebert

426. Stripe-tailed Scorpion. Photograph by C. Hogue.

The Stripe-tailed Scorpion lives in the Palos Verdes Hills and other foothills and mountains surrounding the basin. It is a burrower and prefers dry areas in chaparral, scrub, and grassland.

■ COMMON SCORPIONS
*(Vejovis* species)
Figure 427

These somewhat flattened scorpions are the smallest of the local types (the body length of adults is 1¼ in. or 32 mm). There are four basin species, which vary in shape and color. Their sting causes an immediate burning sensation followed by temporary swelling and slight numbness of the general area. The effects subside after two or three days and cause no lasting damage.

REFERENCES. Russell, F. E., C. B. Alender, and F. W. Buess. 1968. Venom of the scorpion *Vejovis spinigerus. Science,* vol. 159, pp. 90-91.

Williams, S. C. 1970. The effects on man of a natural sting by the scorpion *Vejovis confusus* Stahnke. *Pan-Pacific Entomologist,* vol. 46, pp. 77-78.

## SUN SPIDERS (Order Solpugida)

THESE HAIRY ARACHNIDS (Figure 428) are greatly feared because of their ferocious appearance. Although they possess a formidable pair of jaws *(chelicerae)* and can pinch with some force, they are actually harmless because there are no poison glands or stingers anywhere on their bodies. They are small- to medium-sized arachnids, with body lengths ranging from ½ to 2 inches (13 to 50 mm).

The group is known by several additional names—"wind scorpions," "scorpion spiders," "false spiders," and "hunting spiders." Although somewhat similar in appearance to spiders or scorpions, they belong to a distinct order. They also bear a resemblance to Jerusalem crickets, for which local residents sometimes mistake them.

Suckers at the tip of each of the two slender appendages (the *pedipalps)* at the front of the body enable the sun spider to climb smooth surfaces—even glass—and also are used in feeding, drinking, and battling. The remaining appendages are true legs. Each of the legs in the hind two pairs has a row of wedge-shaped structures called "racket organs"; although their function is unknown, they are believed to be related to the perception of ground vibrations.

Sun spiders usually inhabit hot dry desert areas. They are not common in the basin, although they occasionally occur under stones or in sandy soil and may be attracted indoors by lights on warm nights.

They are ground dwellers and excellent excavators. When digging, they push the soil ahead of them like a bulldozer, the pedipalps holding the pile from the sides and the chelicerae pushing from the rear. All of our species are nocturnal, wandering by night in search of the small invertebrate animals that are their prey. They are extremely voracious carnivores and crush and tear captive organisms to shreds with their huge jaws.

All of the several local species that have been identified are in the family Eremobatidae (none has a common name): *Hemerotrecha marginata, Therobates morrisi,* and *Eremobates scopulatus, E. purpusi, E. zinni, E. mormonus,* and *E. gracilidens.*

REFERENCE. Muma, M. H. 1970. A synoptic review of North American, Central American, and West Indian Solpugida (Arthropoda: Arachnida). *Arthropods of Florida and neighboring land areas,* vol. 5.

## DADDY LONG-LEGS (Order Opiliones)

ALTHOUGH THEY RESEMBLE SPIDERS, daddy long-legs (or "harvestmen" as they are sometimes called) belong to a separate order of arachnids. Unlike spiders, they have no fangs and do not bite. The extremely long

427. Common scorpion *(Vejovis* species). Photograph by C. Hogue.

428. Sun spider. Photograph by C. Hogue.

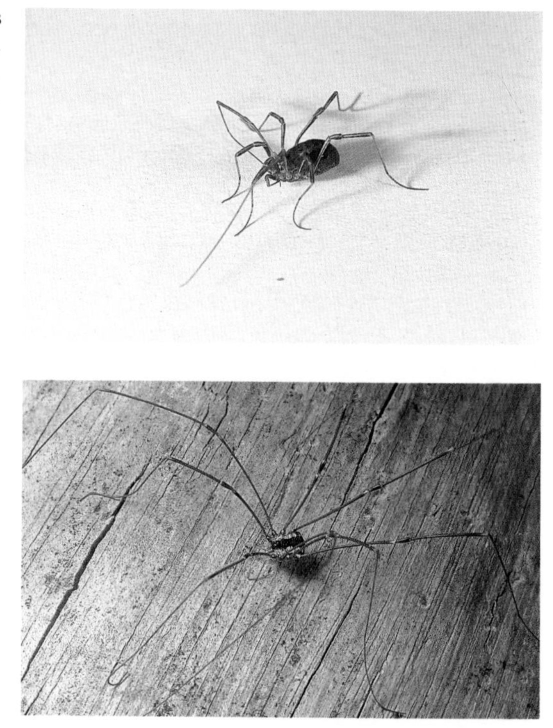

429. Daddy long-legs *(Protolophus singularis)*. Photograph by C. Hogue.

430. Daddy long-legs *(Leuronychus pacificus)*. Photograph by C. Hogue.

slender legs and an abdomen fused to the cephalothorax are characteristic. Also, there are usually only two eyes, one looking out to each side; these are located on a single tubercle atop the head region.

Daddy long-legs live in low-growing vegetation and frequently inhabit secluded niches such as ground cavities under stones, boards, or brush; sometimes they congregate in masses. These creatures use their crushing mouthparts to feed primarily on the carcasses of invertebrates that have recently died—insects, pillbugs, and spiders, for example. They may scavenge almost any kind of organic matter.

There are several local species. Perhaps the commonest are *Protolophus singularis* (Figure 429), which is grayish and has relatively short legs and toothed tubercles on its back, and *Leuronychus pacificus* (Figure 430), a very long-legged light-brown species with a broad brown band running down the middle of its smooth back.

REFERENCES. Banks, N. 1911. The Phalangida of California. *Pomona College Journal of Entomology*, vol. 3, pp. 412-421.

Goodnight, C. T., and M. L. Goodnight. 1942. The genus *Protolophus* (Phalangida). American Museum Novitates, no. 1157, 7 pp. New York: American Museum of Natural History.

# 17 SPIDERS (Order Araneae)

SPIDERS ARE SUCH FAMILIAR ARTHROPODS that they really require no description, although several characteristics should be pointed out. Like all arachnids, spiders have a pair of leg-like appendages near the front of the body called *pedipalps,* which serve as sensory structures. But in male spiders, these appendages also carry the copulatory organs. The genital opening is at the rear of the abdomen in its normal position, but the sperm are transferred to the pedipalps, which in turn are used to transfer them to the female during the mating process.

Spiders feed in a special way. All are predaceous and subdue their victims with poison injected by a pair of fangs *(chelicerae)* situated just in front of the mouth. Food is not swallowed whole by the spider but is first liquified externally using digestive enzymes regurgitated from the mouth; it is then slowly imbibed.

The only evidence of a head on the spider's cephalothorax is the mouth, the associated fangs, and the eyes. In spiders, there are several eyes—usually eight—lying atop the cephalothorax; their number and position are important in spider identification.

Spiders are as well known by their webs as by their body form. Each kind of spider spins its own distinctive web, and a species may create several types for different purposes. Usually the web functions as a snare for prey, but it can also serve as a home for the spider, a place to rest and find protection from the elements, or a place to raise young. Often the female spins a tight silken bag or ball around the egg mass and then carries it about until the eggs hatch. The silk is produced by special glands within the abdomen that open to the outside of the body through a set (usually three pairs) of finger-like appendages called *spinnerets* at the rear end.

Spiders figure in some of what Los Angeles newspaper columnist Jack Smith calls "urban myths." The story about the Black Widow spiders that hid in a woman's coiffure has been around for a long time. More recently I heard about a couple who bought a large potted saguaro cactus. They had had it in their home for a few days when they noticed that it was moving. The police bomb squad was called in after someone advised them that tarantulas breed inside this kind of cactus and that their plant would soon explode and send the spiders everywhere!

Many species of spiders live in the basin. Only a very few of the most notable can be described here.

REFERENCES. Comstock, J. H. 1940. *The spider book,* 2nd edition (revised by W. Gertsch). New York: Doubleday.

Gertsch, W. W. 1979. *American spiders,* 2nd edition. New York: Van Nostrand.

Kaston, B. J. 1978. *How to know the spiders,* 3rd edition. Pictured Key Nature Series. Dubuque, Iowa: W. C. Brown.

Smith, J. 1990. Spiders in the spines. *Los Angeles Times Magazine* (6 May 1990), vol. 6, no. 18, p. 6.

Thompson, M. E. 1979. Common house and garden spiders of the Los Angeles area. *Terra,* vol. 17, no. 4, pp. 23-29.

---

## ■ SPIDER BITES

THE BITES of only two kinds of local spiders are usually considered dangerous—those of the black widow *(Latrodectus hesperus)* and the violin spider *(Loxosceles* species). The venom of the former is a neurotoxin that often (but not always) causes debilitating systemic disorders, sometimes leading to death. Violin spider venom is haemotoxic, with local affects only; it can cause disfigurement and sometimes extensive tissue loss, but it is rarely lethal. (Many bites from black widows and violin spiders are completely innocuous because for some reason the spider has not injected venom with its bite.)

Because the medical importance of these two spiders is best known, almost all spider bites are blamed on them. There is evidence, however, that a number of other species can and often do inflict bites on people, sometimes with serious consequences. In the basin such overlooked offenders are mainly orb weavers *(Araneus* and *Neoscona* species), running spiders (genus *Cheiracanthium),* a dysderid *(Dysdera crocota),* false black widows *(Steatoda* species), and wolf spiders (genus *Lycosa).* The bites of large individuals of all these types can cause pain, necrosis, and septicemia. Physicians should be aware of this fact and not be too ready to diagnose a bite as originating with the two famous culprits. In fact, it is rarely possible to definitely determine the type of spider that has bitten a person from a skin reaction or the appearance of the bite alone, although sometimes the twin punctures from the fangs of the biter may persist.

Spiders bite humans only in self-defense, when

they are molested. This commonly occurs when the creature—hiding among bedding, in clothes, or among stored items—is compressed by an unwitting person rolling on it or handling the infested materials.

REFERENCES. Russell, F. E. 1986. A confusion of spiders. *Emergency Medicine,* vol. 18, no. 11, pp. 8-9, 13.

Russell, F. E., and W. J. Gertsch. 1982. Last word on araneism. *American Arachnology,* vol. 25, pp. 7-10.

_______. 1983. Letter to the editor. *Toxicon,* vol. 21, pp. 337-339. [Comments on spider bites.]

## TARANTULAS (Family Theraphosidae)

■ CALIFORNIA TRAPDOOR SPIDER
*(Bothriocyrtum californicum)*
Figures 431-433

Trapdoor spiders are novelties in the Los Angeles Basin today, although they were commonplace a few years ago. They were even collected and sold as curios in the Los Angeles area at the beginning of the twentieth century. Their rarity now is another example of human expansion destroying the habitat of a local animal. The spider prefers to build its nest on sunny south-facing dry hillsides, which in the spring bear a thick covering of short grasses and low herbs. Such areas are becoming increasingly rare in the basin (they are also the habitat of our local tarantulas, and both types of spiders can be found living on the same hillside).

The nest of the trapdoor spider (Figure 433) consists of a tube or burrow running perpendicular to the surface and into the ground to a depth of 6 to 10 inches (15 to 25 cm). The walls of the burrow are lined with a thick smooth papery silken sheet, and the ground-level entrance is capped with a very tight-fitting lid, which is hinged on one (the uphill) side. The lid opens like a trap door (hence the name of the spider) and has beveled sides that fit into the tube like a cork into a bottle. The spider holds the door shut with its fangs (fang marks may be seen on the inside of the lid).

Females of this large shiny slow-walking spider (its body length is 1 to 1 1/4 in., or 25 to 32 mm) possess heavy digging spines on their chelicerae and on the first two pairs of legs to assist them in the task of excavating the burrow. The smaller darker males (body length 3/4 to 1 in., or 20 to 25 mm) are sometimes seen crawling over the ground on rainy evenings in the fall or winter, searching for the burrows of females. The legs of the male are much longer and more slender in proportion to its body size than those of the female. It also has a tan to orangish abdomen.

Food consists of other terrestrial arthropods, especially ground-dwelling insects happening to pass the burrow. Rarely does this nocturnal spider stray from the immediate vicinity of the burrow opening—it usually prefers to "keep one foot in the door." When the trap is shut, the nest is well-camouflaged by its dirt construction and by the moss and surface debris spun into the outer wall of the door to enhance its resemblance to the surrounding terrain. It is said the female has a difficult time in opening the trap from the outside, especially when it is tightly closed.

Like its relative the tarantula, the trapdoor spider is hunted by spider wasps, most commonly by *Psorthaspis planata.* The wasps bore a hole in the trapdoor to reach their quarry. Trapdoor spiders also have other enemies, including a small-headed fly parasitoid (family Acroceridae) and skunks, which dig up the nests and eat the spiders.

REFERENCES. Davidson, A. 1905. An enemy of the trap door spider. *Entomological News,* vol. 16, pp. 233-234.

Holder, C. F. 1901. A singular industry in the poisonous insects of California. *Scientific American,* vol. 85, pp. 202-203.

Jenks, G. E. 1938. Marvels of metamorphosis. *National Geographic,* vol. 74, pp. 807-828.

Passmore, L. 1933. California trapdoor spider performs engineering marvels. *National Geographic,* vol. 64, pp. 195-211.

Van Riper, W. 1946. How strong is the trapdoor spider? *Natural History,* vol. 55, pp. 68-71.



■ CALIFORNIA TARANTULAS *(Aphonopelma* species) Figures 434, 435

Local hill residents are sometimes shocked to find a giant hairy spider crawling about their patios on a late summer's eve. Few Angelenos realize that tarantulas are permanent inhabitants of the dry grass- and brush-covered hillsides of the basin.

Tarantulas usually live in groups or "colonies"; their presence in an area is revealed by the numerous ground burrows in which they live and lay their eggs. For their homes, the spiders utilize existing holes, such as abandoned rodent burrows or natural cavities under stones, which they enlarge and modify, using their fangs and legs as digging tools. The burrows are 1 to 2 inches (3 to 5 cm) in diameter at the mouth and wider below and are 1 to 2 feet (30 to 60 cm) deep. They are always lined at least partially by silk, which both the males and females produce with the two

431. California Trapdoor Spider female. Photograph by C. Hogue.

432. California Trapdoor Spider male. Photograph by C. Hogue.

433. California Trapdoor Spider nest opening. Photograph by C. Hogue.

pairs of finger-like spinnerets located at the end of the abdomen. The burrow occupant usually weaves a collar of webbing around the edge of the opening and often creates a thin veil of silk across the entrance; at various times, they close the burrow with a plug of silk, dirt, and debris.

Adults and immatures are virtually identical except for size and sexual structures; the adults may attain a body length of up to 2½ inches (65 mm) and a leg spread of 6 to 7 inches (16 to 18 cm). Mature males may be recognized by the hard sharp-pointed bulb—the copulatory organ used in mating—at the end of each of the pedipalps. Males also have proportionately longer legs and a hook-like appendage underneath the third from the last segment of the front legs that is used to hold the female during mating. Males have relatively smaller bodies than females.

Tarantulas usually have a bald spot on the back of the otherwise very hairy abdomen, a result of the hind legs being frequently rubbed over this area, loosening the hairs and causing them to fly out. The loosened hairs can embed in and inflame the skin or eyes of an enemy, and the action of loosening the hairs is presumed to be a protective one.

The tarantula's food consists of other small arthropods such as beetles, ground crickets, sow bugs, and other spiders. The hunter sits and waits at the burrow entrance for its meals to wander by. It kills its prey by crushing and injecting poison with the large fangs that project downward from the front end of the body. Although deadly to small creatures, the tarantula's bite is of little consequence to human beings: its effect has been described as a mildly painful burning sensation accompanied by slight swelling. Furthermore, the spider must be strenuously provoked to cause it to bite.

Because they are really not very poisonous and because their life span is long (females may live up to twenty years or more), tarantulas make interesting pets. They should be kept in a large container with an inch or two of soil; they need a live insect once a week or so and a bottlecap full of water for sustenance. Even without food (but with water), these spiders may survive a considerably long time, up to several months.

Some pet shops sell imported tarantulas, especially the Red-kneed Tarantula *(Brachypelma smithi)* from Mexico. This species is protected by international law and can only be imported under permit. Other species are not legally controlled but should only be taken sparingly so as not to put their populations in danger of extinction.

It is appropriate here to repeat a curious story from the Middle Ages about the origin of the word "tarantula." Near the town of Taranta in southern Italy, there was a strange malady called "tarantism" that was erroneously believed to result from the bite of a large spider. The most conspicuous symptom of the disease was an uncontrollable urge to dance wildly and violently to any light tune until totally exhausted and—at the same time—cured. Rooted in tarantism is the origin of the Tarantella, a lively and passionate Neapolitan folk dance.

The wolf spider *(Lycosa tarantula),* which supposedly caused tarantism, can be considered the original and "true" tarantula, even though it belongs to a quite different family, the Lycosidae. But it does resemble the spiders we know as tarantulas (family Theraphosidae) and thus has lost its name to our spiders. A further complication in names results from the use of the generic term "tarantula" for a group of Old World whipless whip scorpions belonging to the order Pedipalpida.

We have two tarantula species, one with males maturing in the fall *(Aphonopelmus eutylenum;* Figure 434) and the other maturing in midsummer *(A. reversum;* Figure 435). In both species the carapace and outer ends of the legs are lighter than the abdomen and legs; this pattern is most obvious in females of *A. eutylenum,* less conspicuous in the males of this species, and scarcely noticeable in either sex of *A. reversum. Aphonopelma eutylenum* is also recognizable by its overall brownish color and relative hairiness *(A. reversum* tends to be gray-black to black and less hairy). A useful and apparently consistent structural distinction is in the extent of the hair pads (scopulae) on the undersides of the penultimate segment of the hind legs: in *A. eutylenum* they cover about three-fourths of

434. Tarantula *(Aphonopelmus eutylenum).* Photograph by C. Hogue.

435. Tarantula *(Aphonopelmus reversum).* Photograph by C. Hogue.

the segment, in *A. reversum* only about a fourth.

REFERENCES. Baerg, W. J. 1958. *The tarantula.* Lawrence, Kans.: University of Kansas.

Browning, J. G. 1981. Tarantulas. Neptune, N.J.: TFH Books. [Care and feeding of pets.]

Gabel, J. R. 1972. Further observations of theraphosid tarantula burrows. *Pan-Pacific Entomologist,* vol. 48, pp. 72-73.

Marer, P. J. 1972. An eye deformity in a tarantula spider (Araneae: Theraphosidae). *Pan-Pacific Entomologist,* vol. 48, pp. 221-225.

______. 1972. Preening behavior in the tarantula spider *Aphonopelma reversum* Chamberlain (Araneae: Thera–phosidae). *Wasmann Journal of Biology,* vol. 30, pp. 167-168.

Randall, T. B., and W. H. Whitcomb. 1976. Notes on the molting of a tarantula, *Brachypelma smithi* (Araneae: Theraphosidae). *Florida Entomologist,* vol. 59, pp. 438.

Schultz, S. A. 1984. *The tarantula keeper's guide.* New York: Sterling.

Woodson, W. 1951. Tarantula-tarantella. *Pacific Discovery,* vol. 4, pp. 25-28.

## DYSDERID SPIDERS (Family Dysderidae)

■ SOW BUG KILLER
*(Dysdera crocota)*
Figure 436

This fairly common local spider, which is cosmopolitan in distribution, is probably an immigrant from Europe. It lives under bark or under stones or debris on the ground and is recognized by its shiny brick-red carapace and ovoid purplish-cream abdomen and its medium brown legs. The front two pair of legs are considerably longer than the rest and are habitually held strongly forward. Females are medium sized, measuring about 1/2 inch (13 mm) in length.

This spider does not spin a web for entrapping prey but instead uses its extra large powerful fangs to feed on hard-shelled sow bugs and pill bugs, which it catches from ambush. Thus, in spite of the fact that it sometimes bites people, this species should be considered beneficial.

OTHER SCIENTIFIC NAMES. The species name is sometimes spelled "*crocata.*"

REFERENCES. Cook, J. A. L. 1965. Spider genus *Dysdera* (Araneae, Dysderidae). *Nature,* vol. 205, no. 4975, pp. 1027-1029.

______. 1965. A contribution to the biology of the British spiders belonging to the genus *Dysdera. Oikos,* vol. 16, pp. 20-25.

______. 1968. Factors affecting the distribution of some spiders of the genus *Dysdera* (Araneae, Dysderidae). *Entomologists Monthly,* vol. 103, pp. 221-225.

Petrankevitch, A. 1910. Courtship in *Dysdera crocata. Biological Bulletin,* vol. 19, pp. 127-129.

## COMB-FOOTED SPIDERS (Family Theridiidae)

■ WESTERN BLACK WIDOW *(Latrodectus hesperus)* Figure 437

The female's occasional practice of devouring the male after mating has earned this species its name.

The sexes differ greatly in appearance: males are small ($^{1}/_{2}$ in., or 13 mm, in body length) and grayish black in color, with variable marks of red and yellow spots and stripes; females are large (1 to $1^{1}/_{4}$ in., or 25 to 32 mm, in body length) with very bulbous shiny black abdomens. Both sexes usually display a red spot shaped like an hourglass on the underside of the abdomen, although this mark and the rest of the color pattern varies from individual to individual.

The Black Widow is all too common in the basin and likely to be found in any cool dry protected hideout. Favorite haunts are hollow stumps, woodpiles, crawlspaces under houses, and refuse stored in garages and attics. The web is made of rather coarse silk and is without definite form.

436. Sow Bug Killer female. Photograph by C. Hogue.

437. Underside of female Western Black Widow, showing characteristic hourglass mark. LACM photograph.

Although they are not at all aggressive, males and females occasionally bite people, often causing serious and even fatal consequences. The venom is reputed to be proportionately fifteen times as poisonous as that of the rattlesnake. For this reason, one should be very careful to avoid the spider and should immediately call a physician if bitten. First-aid treatment consists of placing an ice bag on the wound and keeping the patient calm until the doctor arrives.

Two local parasites help keep Black Widow populations in check. These are the chloropid fly *Pseudogaurax signatus* and the scelinoid wasp *Baeus latrodecti* (formerly known as *B. californicus*).

REFERENCES. Barton, C. 1938. How it feels to be bitten by a Black Widow—A case history. *Natural History,* vol. 42 (June), pp. 43-51.

Jenks, G. E. 1938. The birth of a baby black widow and its enemy, a small parasitic fly. *Natural History,* vol. 42 (June), pp. 52-57.

Kaston, B. J. 1970. Comparative biology of American black widow spiders. *Transactions of the San Diego Society of Natural History,* vol. 16, pp. 33-82.

Pierce, W. D. 1938. The black widow spider and its parasites. *Bulletin of the Southern California Academy of Sciences,* vol. 37, pp. 101-104.

_______. 1942. Fauna and flora of the El Segundo Sand Dunes, part 12: Utilization of the black widow parasite, and further data on spiders and parasites. *Bulletin of the Southern California Academy of Sciences,* vol. 41, pp. 14-26.

Thorp, R. W., and W. D. Woodson. 1945. *Black Widow, America's most poisonous spider.* Chapel Hill, N.C.: University of North Carolina.

■ FALSE WIDOW
*(Steatoda grossa)*
Figure 438

The False Widow is very abundant locally and probably suffers considerable undeserved abuse because of its general similarity to the Black Widow, upon which it is reported to prey; it also eats sow bugs. It lacks the red hourglass mark of the Black Widow and has a marbled purplish-brown rather than black abdomen. Females are just over 1/2 inch (13 mm) long.

False Widows are found in and around houses, under the loose bark of trees, and in rock and wood piles; these spiders are more tolerant of outdoor conditions than are Black Widows.

REFERENCE. Barmeyer, R. A. 1975. Predation on the isopod crustacean *Porcellio scaber* by the theridiid spider *Steatoda grossa. Bulletin of the Southern California Academy of Sciences,* vol. 74, pp. 30-36.

438. False Widow female. Photograph by C. Hogue.

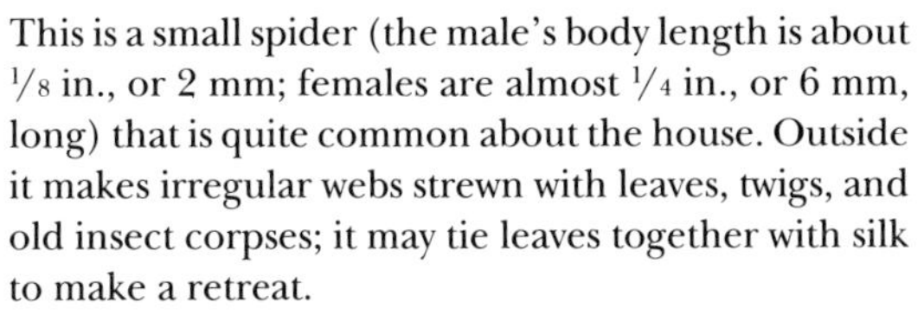

439. House Spider female. Photograph by C. Hogue.

■ House Spider *(Achaearanea tepidariorum)* Figure 439

This is a small spider (the male's body length is about 1/8 in., or 2 mm; females are almost 1/4 in., or 6 mm, long) that is quite common about the house. Outside it makes irregular webs strewn with leaves, twigs, and old insect corpses; it may tie leaves together with silk to make a retreat.

Its abdomen is very large and bulbous relative to the rest of the body and is generally cream colored with irregular dark blotches.

## PHOLCID SPIDERS (Family Pholcidae)

■ Cobweb Spider *(Pholcus phalangioides)* Figure 440

This strictly domestic spider is another species imported from Europe. It is the major contributor to ceiling cobwebs and is common under eaves of homes in tree-shaded neighborhoods.

The spider is drab gray-brown, with an elongate abdomen; it is a little less than 1/4 inch (6 mm) long. Its very long slender legs and small rounded body give it a superficial resemblance to a daddy long-legs. When disturbed, it gyrates wildly in its web.

REFERENCE. Platel, T. G. H. 1989. The egg laying and larval development of *Pholcus phalangioides* (Fuesslin) (Araneae: Pholcidae). *Tijdschrift voor Entomologie,* vol. 132, pp. 135-147.

## FOLIAGE SPIDERS (Family Clubionidae)

■ Cream House Spiders (*Chiracanthium* species) Figure 441

There are two locally common species in this genus, *Chiracanthium mildei* and *C. inclusum.* The former is introduced from Europe and is the more common; it often enters homes, where it builds a sack-like web in corners and crevices (even in household appliances). *C. inclusum* is a native and more of an outdoor species.

Both are small (body length about 3/8 in., or 9 mm) and pale translucent brown in color with an oval cream-colored abdomen. When disturbed they draw the fore pair of legs back and in, forming a cage around the body. As they walk, these spiders often wave the fore legs about or thrust them forward as if testing the path.

These spiders have relatively strong long fangs and have been known to bite humans, causing a wound that is painful and slow to heal.

## GROUND SPIDERS (Family Gnaphosidae)

■ Mouse Spider (*Scotophaeus blackwalli*) Figure 442

This is a small spider (body length to 1/4 in., or 6 mm) that is elongate and dull brown in color; it has short legs and two pairs of spinnerets that project straight backwards. The abdomen is clothed with silky hair. It is found indoors, walking on walls, ceilings, and curtains. The species is another immigrant from Europe.

## GRASS SPIDERS (Family Agelenidae)

■ Corner Spider (*Hololena curta*) Figure 443

This spider prefers dark shade and occasional moisture and often takes up residence in ornamental plantings near watered lawns. It builds its sheet webs on shrubs, tree branches, and structures (patio furniture and neglected walls and barbecues); each web is equipped with a funnel-like retreat running into a nearby crevice.

Adults vary in size; the body length may be 1/4 to 3/4 inch (6 to 19 mm). They are medium brown with a pair of parallel dark stripes that run the length of the body but are most distinct on the carapace.

## BROWN SPIDERS (Family Loxoscelidae)

■ Violin Spiders (*Loxosceles* species) Figure 444

In 1969 the discovery of a population of *Loxosceles laeta* in Sierra Madre stirred recognition of the "spider menace" to public health in the Los Angeles area. And in April 1991 the same species was found again in buildings in downtown Los Angeles. Although there have not been any local cases of a bite from this species

440. Cobweb Spider female. Photograph by C. Hogue.

441. Cream house spider *(Chiracanthium* species). Photograph by C. Hogue.

442. Mouse Spider. Photograph by C. Hogue.

443. Corner Spider with egg sac (left) and wrapped prey. LACM photograph.

or from the closely related Brown Recluse Spider, *L. reclusa* (which, like *L. laeta,* is an introduced species), there is evidence that a third member of the genus, *L. unicolor,* which is native to the Mojave Desert, has been responsible for a few instances of spider bite resulting in serious consequences. The venom of these spiders acts on the tissues locally, rather than on the nervous system in general, and causes a troublesome sore, which may grow in size and be so resistant to healing that plastic surgery is indicated.

These are small pale brown long-legged spiders (body length 3/16 to 1/4 in., or 4 to 6 mm). The back of the carapace has a faint violin-shaped marking (the neck of the violin points toward the rear), and this is the source of the common name for the group.

Violin spiders build their small loose webs in dark recesses. Common habitats outdoors are wood piles, spaces in and under stones and wood debris loosely set on the ground, and piles of broken concrete. Indoors, they occupy packing crates, piles of old books and newspapers, and other accumulations of trash. They are extremely rare in the basin.

REFERENCES. Foil, L. D., J. L. Frazier, and B. R. Norment. 1979. Partial characterization of lethal and neuroactive components of the Brown Recluse Spider *(Loxosceles reclusa)* venom. *Toxicon,* vol. 17, pp. 347-354.

Gertsch, W. T., and F. Ennik. 1983. The spider genus *Loxosceles* in North America, Central America and the West Indies (Araneae, Loxoscelidae). *Bulletin of the American Museum of Natural History,* vol. 175, pp. 264-360.

Keh, B. 1970. *Loxosceles* spiders in California. *California Vector Views,* vol. 17, pp. 29-34.

Levi, H. W., and A. Spielman. 1964. The biology and control of the South American brown spider, *Loxosceles laeta* (Nicolet), in a North American focus. *American Journal of Tropical Medicine and Hygiene,* vol. 13, pp. 132-136.

Waldron, W. G., M. B. Madon, and T. Suddarth. 1975. Observations on the occurrence and ecology of *Loxosceles laeta* (Araneae: Scytodidae) in Los Angeles County, California. *California Vector Views,* vol. 22, no. 4, pp. 29-36.

## ORB WEAVERS (Family Araneidae)

■ Golden Orb Weaver
*(Argiope aurantia)*
Figure 445

Large size (a body often over 1 in., or 25 mm, long) and a black abdomen with broad yellow bars along the sides characterize the female of this spider. It is frequently seen hanging head down in the center of its large strong web. The web is constructed among shrubs and is of the orb type (like a wheel, with strands radiating out like spokes from the center and ringed

444. Violin spider *(Loxosceles laeta)*. Photograph by C. Hogue.

445. Golden Orb Weaver female in web; note zigzag stabilmentum. Photograph by C. Hogue.

with concentric circles of silk); below the center there is a zigzag band of silk (the *stabilmentum;* Figure 445) that apparently helps camouflage the spider, whose abdomen is similarly patterned. This spider and other orb weavers are sometimes called "writing spiders" because of the appearance of the stabilmentum, which seems to some imaginative people to contain letters of the alphabet.

The male of the species, which is one-fourth the size of the female, lives inconspicuously in a feeble imperfect web beside that of his mate. The adults mature and mate in late summer. The female places her eggs in a large ovoid cocoon (3/4 in., or 20 mm, in diameter) that has a brown paper-like surface. The egg sac is suspended in a niche among leaves by a stiff web framework. The young hatch during the winter and remain in the cocoon until the warm days of the following spring, when they escape and disperse.

The Golden Orb Weaver is also known as the Common Garden Spider and the Black and Yellow Argiope. Two other orb weavers in the genus *Argiope* are occasionally encountered locally. The Banded

Orb Weaver *(Argiope trifasciata)* is similar to the Golden in shape, but its abdomen is cream with narrow circular black rings. The abdomen of another species, the Silver Orb Weaver *(A. argentata)*, is flattened and strongly lobed posteriorly, and the carapace and front half of its abdomen are shiny silver.

All three *Argiope* species have similar habits; however, the Silver Orb Weaver constructs four stabilmenta, one at each point where the tips of the outstretched pairs of front and rear legs contact the web. All of these species are mature in the fall and can be seen in their conspicuous webs in August and September.

REFERENCES. Enders, F. 1973. Selection of habitat by the spider *Argiope aurantia* Lucas (Araneidae). *American Midland Naturalist,* vol. 90, pp. 47-55.

Tolbert, W. W. 1975. Predator avoidance behaviors and web defensive structures in the orb weavers *Argiope aurantia* and *Argiope trifasciata* (Araneae, Araneidae). *Psyche,* vol. 82, pp. 29-52.

---

■ COMMON ORB WEAVER *(Neoscona oxacensis)* Figure 446

This is our most common orb weaver; in late summer and fall, its moderate-sized webs adorn gardens everywhere in the basin. The webs lack stabilmenta.

The spider itself is moderate sized; the body length of females is 1/2 to 3/4 inch (13 to 19 mm). It is identified by two pairs of yellow spots on the underside of the abdomen; the top of the purplish-brown bulbous abdomen has two broken zigzaging yellow stripes that converge toward the rear.

REFERENCE. Berman, J. D., and H. W. Levi. 1971. The orbweaver genus *Neoscona* in North America (Araneae: Araneidae). *Bulletin of the Museum of Comparative Zoology, Harvard University,* vol. 141, pp. 465-500.

---

■ JEWELED ARANEUS *(Araneus gemmus)* Figure 447

One of several *Araneus* species in the basin, this spider is distinguished by broad shoulder humps on its fat abdomen, which is medium brown in color and has a white arrowhead design in the center that is flanked on each side by a small round spot. Its body length is up to 3/4 inch (20 mm), and its abdomen and legs are hairy.

The females spin loose orb webs that are very expansive—up to 6 to 8 feet (1.8 to 2.4 meters) across—and of very strong silk strands. Webs are often placed in trees or bushes or on bridges, buildings, and other manmade structures. The spiders are active in spinning webs and tending prey mainly during the night.

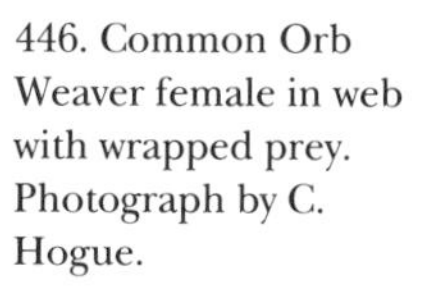

446. Common Orb Weaver female in web with wrapped prey. Photograph by C. Hogue.

447. Jeweled Araneus female. Photograph by C. Hogue.

448. Trash-web Spider's trash; the abdomen and hind legs of the spider are visible at the bottom of the trash pile. Photograph by M. Badgley.

■ TRASH-WEB SPIDER
*(Cyclosa turbinata)*
Figure 448

The resting place of the female of this spider in its web is distinctive: it is a loose line of thick webbing upon which the female collects an odd assortment of trash, mainly the carcasses of old prey wrapped in defunct cocoons. All of the web, except for the debris string, is dismantled each day, and the old web material is added to the trash pile.

The female spider itself is small (body length about 1/4 in., or 6 mm) and may vary in color from a mixture of gray and white to almost solid black. It has a disproportionately large bulbous abdomen with a prominent rearward protruding bump; the spider's eyes are on tubercles.

REFERENCE. Levi, H. W. 1977. The American orb-weaver genera *Cyclosa, Metazygia* and *Eustala* north of Mexico (Araneae, Araneidae). *Bulletin of the Museum of Comparative Zoology, Harvard University,* vol. 148, no. 3, pp. 61-127.

## JUMPING SPIDERS (Family Salticidae)

■ Red Jumping Spider
*(Phidippus formosus)*
Figure 449

The brilliant red abdomen of this species frequently attracts attention in the spring, when it is most active, and the undiscerning sometimes mistake it for the Black Widow. It is quite different, however, because of its much smaller size (length 3/8 in., or 9 mm), short legs, and hairy body. Furthermore, the red color on this spider covers the entire abdomen and is not restricted to an hourglass-shaped mark on the underside, as it is in the Black Widow. The Red Jumping Spider is not considered dangerous, although its bite may be painful to sensitive persons. Like all jumping spiders, it has a pair of very large eyes.

This is a hunting spider and thus does not use a permanent web for trapping prey. Adult females do construct a funnel-like web that is usually in contact with the soil; this structure is used as a retreat for the adults and a safe repository for the eggs. Both sexes spend the daylight hours wandering over the ground and vegetation in search of small invertebrates, upon which they may leap from some distance. Because they are so active, they often wander into houses, where they attract immediate attention (and cause undue concern) with their bright color and rapid movements.

There are many other species of jumping spiders in the basin; although they vary in color pattern and size, their habits are similar to those of the Red Jumping Spider.

## LYNX SPIDERS (Family Oxyopidae)

■ Green Lynx Spider
*(Peucetia viridans)*
Figure 450

This is a conspicuous large green spider (its body may be up to 3/4 in., or 20 mm, long) that lives on low bushes and herbaceous vegetation throughout the basin. Adults are hunters that spring on their prey from ambush and do not form a web for entrapment. They often feed on moths, but they also catch and eat a wide variety of other kinds of insects.

Males wander in search of mates from May through July. In the fall the female constructs a rounded light green to straw-colored egg sac that is 1/2 to 1 inch (13 to 25 mm) in diameter and flattened on one side. It is usually placed in the upper branches of woody plants and contains 25 to 600 bright orange eggs. The spiderlings, which are also colored orange, emerge in about two weeks. Some maternal care is displayed by the female: she aids the emerging young by opening the egg sac along a seam with her chelicerae.

The back of the adult's abdomen is marked with a series of cream-colored chevrons bounded on either side by an obscure cream stripe; both chevrons and stripes are partially margined with russet.

REFERENCES. Fink, L. S. 1987. Green Lynx Spider egg sacs: Sources of mortality and the function of female guarding (Araneae, Oxyopidae). *Journal of Arachnology,* vol. 15, pp. 231-239.

Hall, R. E., and M. B. Madon. 1973. Envenomation by the Green Lynx Spider *Peucetia viridans* (Hentz 1832), in Orange

449. Red Jumping Spider. LACM photograph.

450. Green Lynx Spider female. Photograph by C. Hogue.

County, California. *Toxicon,* vol. 11, pp. 197-199.

Randall, J. B. 1977. New observations of maternal care exhibited by the Green Lynx Spider, *Peucetia viridans* Hentz (Araneida: Oxyopidae). *Psyche,* vol. 84, pp. 286-291.

Turner, M. 1979. Diet and feeding phenology of the Green Lynx Spider, *Peucetia viridans* (Araneae: Oxyopidae). *Journal of Arachnology,* vol. 7, pp. 149-154.

Whitcomb, W. H., M. Hite, and R. Eason. 1966. Life history of the Green Lynx Spider, *Peucetia viridans* (Araneida: Oxyopidae). *Journal of the Kansas Entomological Society,* vol. 39, pp. 259-267.

## CRAB SPIDERS (Family Thomisidae)

THE COMMON NAME of this family of small spiders refers to the crab-like appearance and behavior of the majority of species: the body is flattened and short, and the legs—the first two pair of which are enlarged—are strongly oriented to the sides. The spiders have the ability to move forward, sideways, and backward with facility.

Crab spiders capture their prey from ambush, without the use of a web. The colors of many species match those of the flowers that they sit on as they wait for visiting flies and bees. Individual spiders can change their color to some degree to mimic their background, although the process requires several days and they cannot deviate greatly from their predominating hue of white, yellow, bluish, or red.

One of the commonest species in our area is *Misumenoides formosipes* (Figure 451). It has a compressed triangular white to bright yellow abdomen and is most abundant in the spring on garden flowers. The adult's body is about 3/8 inch (10 mm) long.

REFERENCE. Schick, R. 1965. The crab spiders of California (Araneida, Thomisidae). *Bulletin of the American Museum of Natural History,* vol. 129, pp. 1-180.

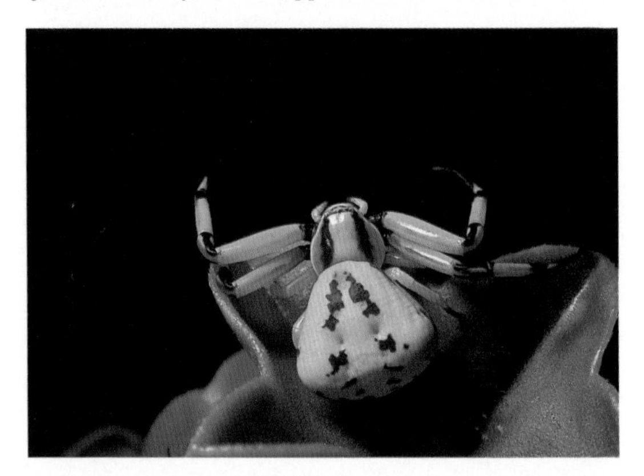

451. Crab spider female *(Misumenoides formosipes).* Photograph by C. Hogue.

# 18 MITES AND TICKS (Order Acari)*

## MITES

"MITE," MEANING A TINY OBJECT or creature, is an apt description of these eight-legged relatives of spiders; indeed, most species are barely visible to the naked eye.

We know of a large number of species of mites, and the ranks of the order increase each year with further discoveries. Acarids live in every imaginable habitat—for discussion purposes, it is useful to divide them into the following seven categories:

- Plant-feeding mites (spider mites; Clover Mite; eriophyid mites).
- Soil-inhabiting mites (oribatid or beetle mites; velvet mites).
- Food-infesting mites—these occasionally bite people (*Acarus* and *Tyrophagus* species; culture mites).
- Nest-inhabiting species that are parasitic on urban birds and rodents and occasionally bother people (Tropical Rat Mite; Mouse Mite; Chicken Mite).
- Mites that are parasitic on wild animals and occasionally bother human beings or their pets (chiggers; scale mites).
- Species directly parasitic on man (Hair Follicle Mite; Scabies Mite).
- Domestic mites (house dust mites).

REFERENCES. Baker, E. W. 1956. *A manual of parasitic mites of medical or economic importance.* New York: National Pest Control Association.

Baker, E. W., and G. W. Wharton. 1952. *An introduction to acarology.* New York: Macmillan.

Hughes, T. E. 1959. *Mites or the Acari.* London: Athlone.

Jeppson, L. R., H. H. Keifer, and E. W. Baker. 1975. *Mites injurious to economic plants.* Berkeley, Calif.: University of California.

MacNay, C. 1963. Control of mites in the home. Canada Department of Agriculture Publication, no. 934.

Senff, W. A., and J. R. Gorham. 1979. The Food and Drug Administration and regulatory acarology. In *Recent advances in acarology,* part 1, edited by J. Rodriguez. New York: Academic.

*This order has also been known as Acarida and Acarina.

## ■ SPIDER MITES (Family Tetranychidae)

SPIDER MITES acquired the name from their habit of spinning fine silken carpets on the undersides of leaves (Figure 452), where they spend their lives and lay their eggs. The mites themselves are minute and whitish, greenish, or reddish in color, often with faint spots on their backs. They suck the sap of economically important plants and, in our area, are especially injurious to citrus. Infested leaves become blotched and sickly, then gradually die and drop.

Perhaps the most important species is the Citrus Red Mite *(Panonychus citri,* also called the Red Spider; Figure 453), a velvety red species that is mainly a pest of lemon trees. Two other species that are quite injurious are the light yellow Six-spotted Mite *(Eotetranychus sexmaculatus),* which infests citrus and avocado, and the greenish Two-spotted Mite *(Tetranychus urticae*—also known as *T. telarius* or *bimaculatus),* which feeds on weeds and cover crops in citrus orchards, sometimes attacking the trees as well.

REFERENCE. Helle, W., and M. W. Sabelis, editors. 1985. *Spider mites: Their biology, natural enemies and control.* Amsterdam: Elsevier.

■ CLOVER MITE
*(Bryobia praetiosa)*
Figure 454

This species is a member of the spider mite family (Tetranychidae) but does not spin silk. Although its external coloring is reddish brown, it leaves a conspicuous blood-red stain if crushed. A plant feeder that is found on a wide variety of herbaceous hosts in the garden, it tends to migrate and readily enters homes, where it can create great concern for the residents. It is a nuisance only, since it does not bite people. But a determined attack by an irate householder will result in a permanent sign of its passing, in the form of a small but ugly stain on a wall or window sill.

The species is less than 1/32 inch (0.75 mm) long and has feather-like projections on its body and front legs that are much longer than its rear legs.

REFERENCES. Anonymous. 1958. Clover mites—how to control them around the home. United States Department of Agriculture Leaflet, no. 443.

English, L. L., and R. Snetsinger. 1957. The habits and control of the clover mite in dwellings. *Journal of Economic Entomology,* vol. 50, pp. 135-141.

## ■ ERIOPHYID MITES (Family Eriophyidae)

THESE ARE SO TINY—they average about 1/100 inch (0.2 mm) long—that they are almost never seen directly,

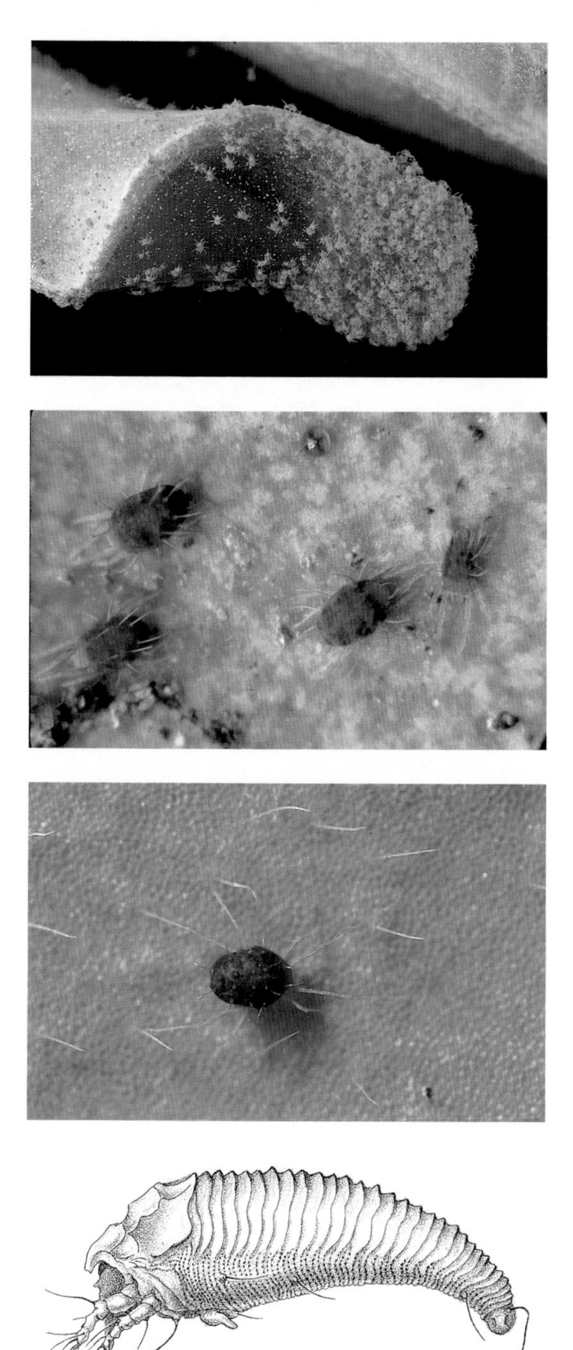

452. Spider mites massed in webbing on eggplant leaf. Photograph by C. Hogue.

453. Citrus Red Mite. Photograph courtesy of Los Angeles County Agricultural Commissioner's Office.

454. Clover Mite. Photograph by M. Badgley.

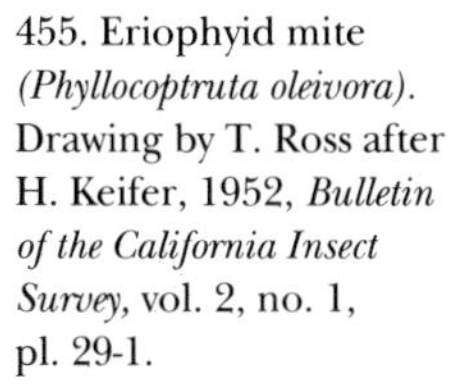

455. Eriophyid mite *(Phyllocoptruta oleivora)*. Drawing by T. Ross after H. Keifer, 1952, *Bulletin of the California Insect Survey*, vol. 2, no. 1, pl. 29-1.

and it is only their effects on plants that announce their presence. Many species are serious agricultural pests, causing leaf curling and abnormal tissue growths, galls, and blisters. The mite's body is elongate and worm-like, with numerous fine ring-like striations. Only the front two pairs of legs are developed.

A local species that is a pest is the Citrus Rust Mite *(Phyllocoptruta oleivora;* Figure 455), which infests the leaves, stems, and fruit of lemon and orange trees.

REFERENCES. Keifer, H. H. 1952. The eriophyid mites of California. *Bulletin of the California Insect Survey,* vol. 2, no. 1, pp. 1-123.

Keifer, H. H., E. W. Baker, G. Kono, M. Delfinado, and W. E. Styer. 1982. An illustrated guide to plant abnormalities caused by eriophyid mites in North America. U.S.D.A., Agricultural Research Service, Agricultural Handbook, no. 573, pp. 46-47.

## ■ TARSONEMID MITES (Family Tarsonemidae)

■ CYCLAMEN MITE
*(Stenotarsonemus pallidus)*
Figure 456

This is a common local plant-feeding mite that causes considerable damage to a variety of ornamentals, particularly cyclamen but also snapdragon, geranium, begonia, fuchsia, and strawberries. Infestations cause leaves to distort, buds to fall, and flowers to develop unsightly discolorations.

The mite is very small—around 1/100 inch (0.2 mm) long—and shiny orangish-pink. The hind pairs of legs in the female are slender and simple at the tips; those of the male are stout with hooked tips. The hind legs of both sexes are tipped with a long filamentous hair.

REFERENCES. Munger, F. 1933. Investigations in the control of the Cyclamen Mite *(Tarsonemus pallidus* Banks). University of Minnesota, Agricultural Experiment Station Technical Bulletin, no. 93, 20 pp.

Smith, L. M., and E. V. Goldsmith. 1936. The cyclamen mite, *Tarsonemus pallidus,* and its control on field strawberries. *Hilgardia,* vol. 10, pp. 53-94.

## ■ ACARID MITES (Family Acaridae)

■ FOOD-INFESTING MITES
*(Acarus* and *Tyrophagus* species)
Figure 457

These mites are very small (less than 1/64 in., or 0.4 mm), colorless, and covered with long spine-like hairs. They commonly infest stored and dried food materials, sometimes straying to stuffed furniture, mattresses, and pillows. They may be present in such quantities that food and mites form a single moving mass.

Occasionally their bite causes an itching irritation called "grocer's" or "baker's itch." When abundant, some species known as "culture mites" can also become a nuisance in food cultures maintained for animals in zoos, laboratories, and vivaria.

REFERENCE. Solomon, M. E. 1943. Tyroglyphid mites in stored products, part 1: A survey of published information. London: Department of Scientific and Industrial Research.

## ■ PARASITIC MITES (Family Macronyssidae)

■ TROPICAL RAT MITE
*(Ornithonyssus bacoti)*
Figure 458

The hosts of this parasitic blood-sucking mite are the Domestic Rat and House Mouse, in whose nests it may be found in large numbers. On occasion, particularly after the hosts die or are eradicated from the pre-

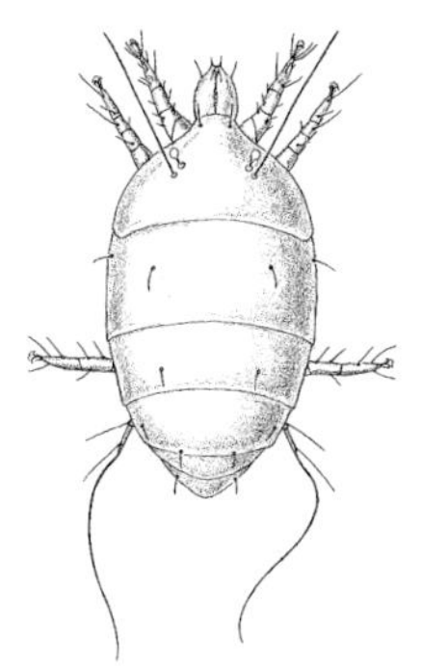

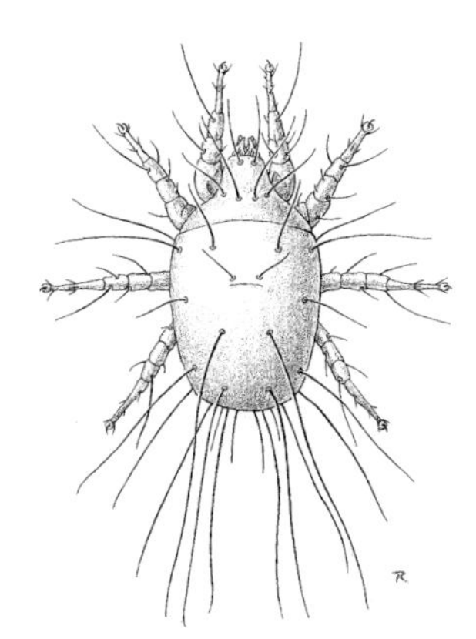

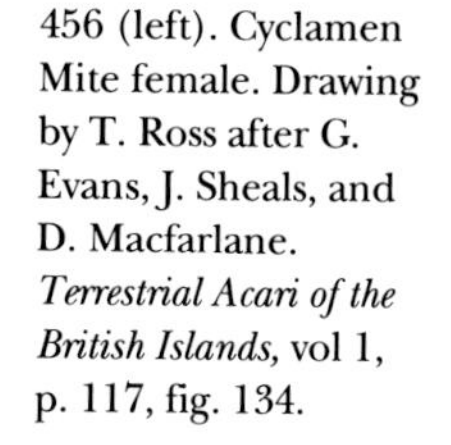

456 (left). Cyclamen Mite female. Drawing by T. Ross after G. Evans, J. Sheals, and D. Macfarlane. *Terrestrial Acari of the British Islands,* vol 1, p. 117, fig. 134.

457 (right). Food-infesting mite *(Tyrophagus* species). Drawing by T. Ross after E. Essig and W. Hoskins, 1944, University of California Agriculture Experiment Station Circulars, no. 87, fig. 135.

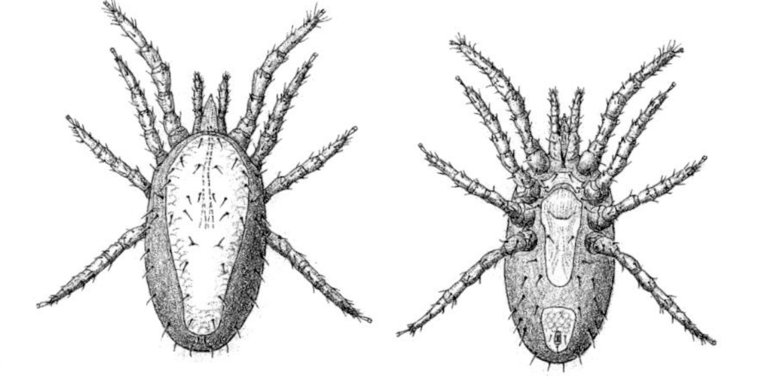

458. Tropical Rat Mite. Drawing by T. Ross after E. Baker and others, 1956, *Manual of parasitic mites of medical and economic importance,* fig. 5 (New York: National Pest Control Association).

mises, these mites may attack the human occupants, causing itching inflammations with their bites. They bite primarily at night and after feeding may migrate to warm places indoors—near water heaters or hot water pipes, for example.

The species is pale grayish yellow in color and tiny (slightly over 1/32 in., or 1 mm, long). Its occurrence in the basin is sporadic.

Although it is the most common local species, this is only one of several widespread parasitic mites that may attack man; the others are difficult to distinguish except by the specialist. The related species *Ornithonyssus sylviarum* and *O. bursa* infest the nests of urban birds, such as House Sparrows and pigeons, and may invade homes and buildings in droves.

OTHER SCIENTIFIC NAMES. *Liponyssus bacoti.*

REFERENCES. Larson, B. 1973. Some observations on rat mite dermatitis. *California Vector Views,* vol. 20, pp. 34-35.

Shelmire, B., and W. Dove. 1931. The Tropical Rat Mite, *Liponyssus bacoti* Hirst. *Journal of the American Medical Association,* vol. 96, pp. 579-584.

Skaliy, P., and W. J. Hayes, Jr. 1949. The biology of *Liponyssus bacoti* (Hirst, 1913) (Acarina, Liponyssidae). *American Journal of Tropical Medicine,* vol. 29, pp. 759-772.

---

■ MOUSE MITE
*(Allodermanyssus sanguineus)*
Figure 459

The most obvious difference between this species and the very similar Tropical Rat Mite is that this mite has two back plates rather than one. It is a blood-sucking parasite of house mice and domestic rats and therefore may infest homes and occasionally bite people. Outbreaks of rickettsialpox, a mild human disease similar to chicken pox, have been traced to this mite in other parts of the country. The House Mouse has acted as a reservoir of the disease, with the mite acting as the vector to transmit it to human beings.

---

■ CHICKEN MITE
*(Dermanyssus gallinae)*
Figure 460

This is a cosmopolitan parasitic species that attacks not only chickens but many kinds of wild, feral, and domestic birds. It is an average-sized mite (its body length is less than 1/32 in., or 0.9 mm) that is bright red after engorgement with blood, which it takes from its host with long piercing-sucking mouthparts. The legs are long, and the body has a large back plate and two belly plates.

The Chicken Mite occasionally bites humans, causing dermatitis but fortunately transmitting no diseases.

CHIGGERS, OR "RED BUGS" as they are often called, are not as great a problem locally as in more humid climates. Nevertheless, there is one pestiferous species in the basin that occasionally attaches to man, causing the typical itching welts known so well to residents of the South and Midwest. Belkin's Chigger (or California Pest Chigger, *Eutrombicula belkini;* Figure 461) occurs in the foothills of the San Gabriel Mountains, in the Palos Verdes Hills, and in our other natural areas. This

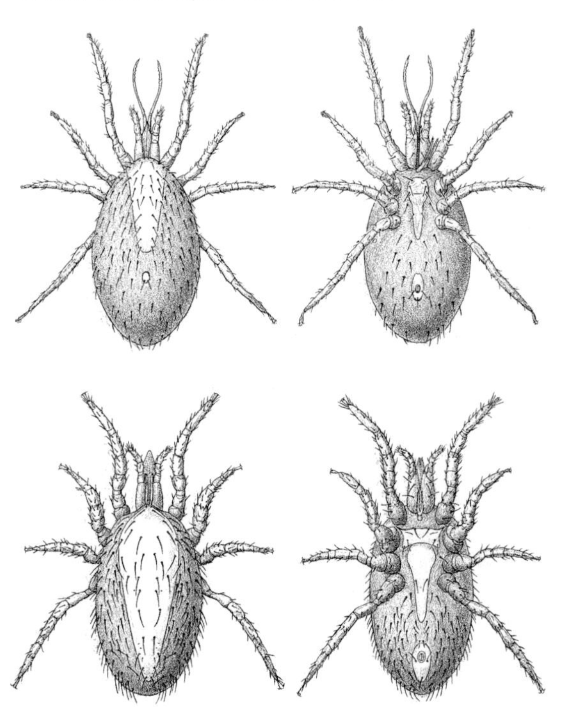

459. Mouse Mite. Drawing by T. Ross after E. Baker and others, 1956, *Manual of parasitic mites of medical and economic importance,* fig. 4 (New York: National Pest Control Association).

460. Chicken Mite. Drawing by T. Ross after E. Baker and others, 1956, *Manual of parasitic mites of medical and economic importance,* fig. 3 (New York: National Pest Control Association).

461. Belkin's Chigger. Drawing by T. Ross after D. Gould, 1950, *Wasmann Journal of Biology,* vol. 8, p. 368.

species normally infests lizards and occasionally birds and small mammals. Man is only an accidental host.

The chigger does not burrow into the skin, as is commonly believed, but attaches only with its mouthparts. Furthermore, it does not suck blood. Instead, when firmly attached in the skin, the mite injects a digestive fluid from its salivary glands, causing a disintegration of the skin cells. The resultant fluid is then ingested. The irritating action of this digestive fluid may persist for some time after the mite has been removed or has dropped off. Chiggers do not transmit any human diseases in our area, but they do act as vectors of serious illnesses, especially scrub typhus, in parts of Asia.

The irritating chigger is the parasitic six-legged larval stage of this family of otherwise free-living terrestrial mites (the adults and nymphs have eight legs). Adults are relatively large (1/16 in., or 1.5 mm, long) compared to the larvae (which are less than 1/64 in., or 0.5 mm) and are conspicuously covered with a dense furry coat of bright red or orange feathery hairs. Like the nymphs, which are somewhat similar in structure, they live on or in loose soil and feed on the eggs of other small terrestrial arthropods, probably most often those of springtails.

In addition to the pest species mentioned above, there are several other local species of chiggers, especially in the genera *Euschoengastia, Neotrombicula,* and *Odontacarus.* Their larval hosts are mainly lizards and small mammals, and they never bite people.

In a celebrated case in nearby Ventura County, a suspect was shown to have been in the area of a murder by the presence on his body of bites of Belkin's Chigger, which was found to be abundant at the scene of the crime.

REFERENCES. Gould, D. J. 1956. The trombiculid mites (Acarina) of California. *University of California Publications in Entomology,* vol. 11, pp. 1-116.

Webb, J. P., R. B. Loomis, M. B. Madon, S. G. Bennett, and G. E. Greene. 1983. The chigger species *Eutrombicula belkini* Gould (Acari: Trombiculidae) as a forensic tool in a homicide investigation in Ventura County, California. *Bulletin of the Society of Vector Ecologists,* vol. 8, pp. 141-146.

## ■ VELVET MITES (Family Trombidiidae)

■ Angelitos
*(Angelothrombium* species)
Figure 462

There is probably more than one species of giant red velvet mite in the deserts of southern California. But at least one occasionally emerges in the dry eastern margins of the basin in large numbers, usually follow-

462. Angelito *(Angelothrombium* species). Photograph by C. Hogue.

ing a rain. These creatures never fail to attract attention because of their large size (the body length of adults is about 1/4 to 3/8 in., or 5 to 8 mm) and brilliant crimson furry bodies.

The larvae are parasites of grasshoppers, and the adults are predators on subterranean termites. The adults remain in the soil most of the year and spend only a few hours above ground, probably to feast on their prey, which also respond to rains by emerging in numbers. Little else is known of their biology.

Our commonest species is one named only recently, in 1960, as *Angelothrombium pandorae.* It had previously been labeled *Dinothrombium tinctorum,* which is actually the name of an African mite.

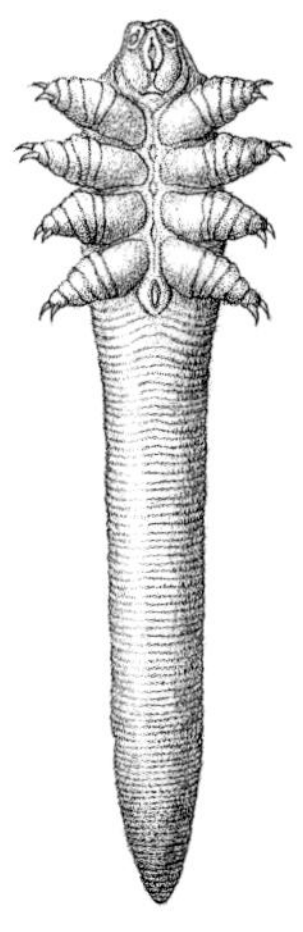

463. Underside of Hair Follicle Mite. Drawing by T. Ross after various sources.

REFERENCE. Newell, I. M., and L. Tevis, Jr. 1960. *Angelothrombium pandorae* n. g., n. sp. (Acari, Trombidiidae), and notes on the biology of the giant red velvet mites. *Annals of the Entomological Society of America,* vol. 53, pp. 293-304.

## ■ FOLLICLE MITES (Family Demodicidae)

■ **Hair Follicle Mite**
*(Demodex follicularum)*
Figure 463

This extremely minute arthropod lives, of all places, in the follicles of human hair. Here it no doubt goes unnoticed in a fairly high percentage of the population, because its presence is not felt and generally causes no ill effects (it does not cause blackheads and pimples, but infestations of this mite are sometimes found to be associated with diabetes in older people).

Peculiar to this mite, in addition to its extremely small size (it is less than 1/250 to 1/64 in., or 0.1 to 0.4 mm, long), are its elongate worm-like body and stout legs.

Closely related species are found in the hair follicles of dogs, cats, horses, cattle, and other mammals.

REFERENCE. Clifford, G. W., and G. W. Fulk. 1990. Association of diabetes, lash loss, and *Staphylococcus aureus* with infestion of eyelids by *Demodex follicularum* (Acari: Demodicidae). *Journal of Medical Entomology,* vol. 27, pp. 467-470.

## ■ PYROGLYPHID MITES (Family Pyroglyphidae)

■ HOUSE DUST MITES *(Dermatophagoides* species) Figure 464

In recent years we have begun to appreciate the role of certain mites in allergies that have been erroneously attributed to house dust. A normal habitat of various species in the genus *Dermatophagoides* seems to be accumulations of dust in human living quarters. The mites can be inhaled with the dust and may generate severe allergic reactions in sensitive persons.

Because the mites are extremely small (up to 1/50 in., or 0.5 mm, long) and pale colored, they are virtually invisible. Their food consists of bits of organic detritus, such as dandruff or flakes of dead skin that become mixed with the dust.

REFERENCES. Furumizo, R. T. 1975. Geographical distribution of house-dust mites (Acarina: Pyroglyphidae) in California. *California Vector Views,* vol. 22, no. 11, pp. 89-95.

_______. 1978. Seasonal abundance of *Dermatophagoides farinae* Hughes 1961 (Acarina: Pyroglyphidae) in house dust in southern California. *California Vector Views,* vol. 25, pp. 13-19.

Lang, J. D., and M. S. Mulla. 1978. Spatial distribution and abundance of house dust mites, *Dermatophagoides* spp., in homes in southern California. *Environmental Entomology,* vol. 7, pp. 121-127.

Spieksma, F., and M. Spieksma-Boezeman. 1967. The mite fauna of house dust with particular reference to the house-dust mite *Dermatophagoides pteronyssinus* (Trouessart, 1897) (Psoroptidae: Sarcoptiformes). *Acarologia,* vol. 9, pp. 226-241.

## ■ SKIN MITES (Family Sarcoptidae)

■ SCABIES MITE *(Sarcoptes scabiei)* Figure 465

These mites, which cause human scabies, burrow into the horny layer of the skin and produce an intense itching sensation. Skin between the fingers, on the inside of the elbow or the back of the knee, or around the genitals, breasts, and shoulder blades is the usual site of infestation, although any part of the body is vulnerable. At first, symptoms may be mild, but serious eruptions, scabbing, and hardening (mange) of the skin may develop when the wounds are aggravated by scratching and secondary infection.

The mites themselves are tiny (about 1/64 in., or 0.4 mm, long), nearly round in outline, and usually difficult to find even when involved with the symptoms just mentioned. The species occurs very sporadically in the basin, usually manifesting itself only under crowded or unsanitary conditions.

REFERENCE. Mellanby, K. 1943. *Scabies.* London: Oxford University.

■ SCALE MITES
*(Knemidokoptes pilae* and *mutans)*
Figure 466

These mites live beneath the scales of a bird's feet and legs or around the bill, feeding on keratin and often causing a disease called "scaly leg," to which most kinds of poultry and pet birds are vulnerable (the comb, neck, and head of the bird may be similarly affected). The condition is very contagious and can cause considerable deformity and crippling. It is a common affliction of caged canaries and parakeets and when severe may result in the death of the pet.

The mites are minute (body length 1/80 in., or 0.3 mm, long) and virtually invisible to the naked eye. The body is rotund, and the legs are short and stubby and

464. House dust mite *(Dermatophagoides farinae)*. Scanning electron photomicrograph courtesy of G. Wharton, Ohio State University.

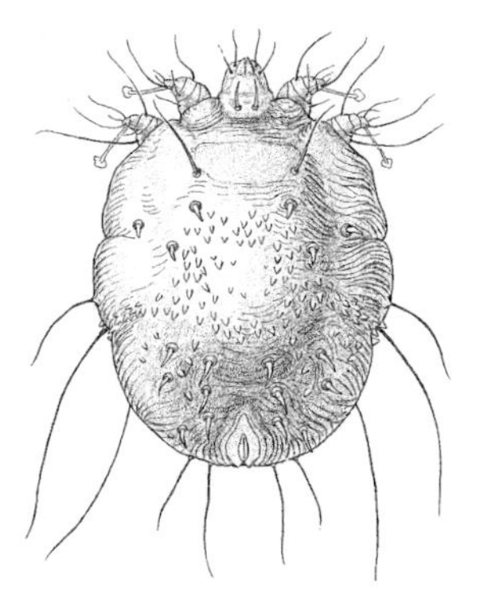

465. Scabies Mite. Drawing by T. Ross after K. Mellanby, 1943, *Scabies,* fig. 1 (London: Oxford University).

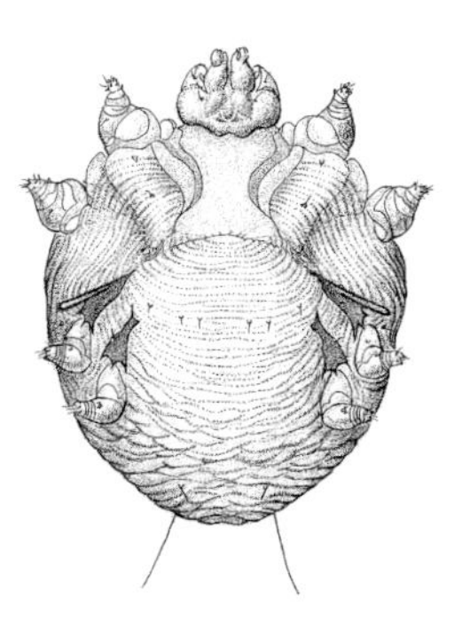

466. Underside of scale mite *(Knemidokoptes* species). Drawing by T. Ross after E. Baker and others, 1956, *Manual of parasitic mites of medical and economic importance,* fig. 46 (New York: National Pest Control Association).

without claws. A pair of long hairs is present at the rear of the body.

REFERENCES. Kaschula, V. R. 1950. "Scaly-leg" of the canary [*Serinus canaria* (Linn.)]. *Journal of the South African Veterinary Medicine Association*, vol. 21, pp. 117-119.

Keymer, I. R. 1982. Knemidokoptic mites. Pages 583-587 in *Diseases of cage and aviary birds*, edited by M. L. Petrak. Philadelphia: Lea & Febiger.

## ■ SOIL MITES (Group Oribatei)

ORIBATID OR "beetle" mites are abundant in mountain canyons and wherever leaf litter and deep soils occur in the basin; they make up a large percentage of the fauna of humic soils. They are important in promoting soil fertility through their digestion of organic matter, but little else is known of their biology.

Members of the group are characterized by a hard dark integument and often heavily plated body, sometimes with wing-like extensions on its sides (Figure 467). Most of the species look like bird seeds with legs. An additional recognition feature is that the front portion of the body is separated from the rear and forms a plate that covers the head region, which is frequently jointed to the remainder of the thorax-abdomen.

A representative local species is *Galumna humida* (Family Galumnidae). Several other species in this genus are vectors of tapeworms.

REFERENCE. Marshall, V. G., R. M. Reeves, and R. A. Norton. 1987. Catalog of the Oribatida (Acari) of continental United States and Canada. *Memoirs of the Entomological Society of Canada*, vol. 139, pp. 1-418.

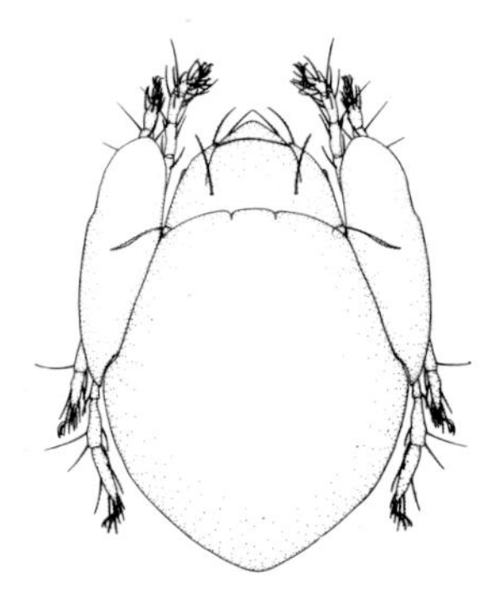

467. Soil mite (*Notaspis pectinata*). Drawing from F. Cox and others, 1921, *Journal of Entomology and Zoology*, Pomona College, vol. 13, p. 31.

## TICKS

There are two general types of ticks—hard ticks and soft ticks. The hard ticks (Family Ixodidae; also called wood ticks or, as larvae, seed ticks) are characterized by a thick leathery scutum or plate on the back. In the male, this plate completely covers the back and is often ornamented with a harlequin pattern of dark and light (frequently silvery) areas; the plate is reduced in the female and covers only a small region behind the "head." In the soft ticks (Family Argasidae), both sexes lack this plate, and as a result the whole body is a pliable bag.

The ticks found in the basin range in size from 3/32 to 1/2 inch (2.6 to 12 mm). However, females of both soft and hard types are capable of unbelievable distention during the course of feeding on blood. An unfed individual the size of an apple seed may, after engorging, swell to the size of a grape.

Ticks differ from insects in that the body is undivided; it is a simple sack-like structure in which the thorax, abdomen, and part of the head are fused together. There is no true head, but there are mouthparts arising from a small square base. Although nymphs and mature ticks bear four pairs of legs, the larvae have six legs, which could result in them being confused with certain insects.

Ticks are very hardy creatures and may survive for months or years without food. In all stages they feed on the blood of vertebrates, but like many other ectoparasitic arthropods they spend only a portion of their life on the host. Fully engorged females usually drop to the ground, where they deposit their eggs. The six-legged larvae hatch from the eggs and attach themselves to the first suitable host coming within their reach. After feeding, they drop to the ground and transform into eight-legged nymphs. After undergoing one to five molts, and engorging on a host at each intermediate stage, the nymphs mature. As in the case of immature stages, adult ticks again seek a passing host. New adult hard ticks climb to the tips of blades of grass or twigs to await passing animals. Vibration or movement of their perch, or rising carbon dioxide levels, cause them to wave their legs in the air, groping for a lucky grab at a passerby. Adult soft ticks usually visit the nests or roosts of potential hosts to obtain their first blood meal.

Ticks frequently attach themselves to humans or pets and can cause harm. The feeding action of an attached hard tick can sometimes induce paralysis ("tick paralysis"). Normal function of the afflicted body part, which is most often a limb, usually returns after the tick is removed, but death can be a consequence, particularly if the attachment site is near the head or vital nerve centers. The area around the bite may become very sensitive or painful. The bite of the Pajahuello may cause tissue decay and the development of a severe wound.

Lyme Disease, which is caused by a spirochaete that is transmitted by the Western Black-legged Tick, is a possible new public health concern in the basin. Another disease caused by a spirochaete, Relapsing Fever, is carried by a species of soft tick and occasionally breaks out in the nearby mountains.

The California Department of Public Health (Anonymous, 1963) recommends the following procedure for removal of an attached tick:

> Use a pair of flat-tipped tweezers or a piece of paper. These are recommended because the engorged tick may carry disease organisms . . . which can penetrate the unbroken skin.
>
> Grasp the body of the attached tick with tweezers or paper and exert a slight, steady pull. It may take some time before the tick releases its hold.
>
> Applications of a small amount of kerosene sometimes make it easier to remove the tick.
>
> Apply an antiseptic to the area of the bite. Washing with soap and water will serve the same purpose.
>
> See your physician if you are unable to completely remove the tick.
>
> Report to your physician should you become ill following a tick bite.
>
> Vaccination as protection against Rocky Mountain Spotted Fever is recommended only for those people who are frequently exposed to ticks in heavily infested areas where this disease occurs. There are no vaccines available to the public for the prevention of the other tick-borne diseases.

The following are our major ticks (they may be identified only with the aid of technical literature). The measurements given for these species are for unengorged individuals.

HARD TICKS (Family Ixodidae)

■ Western Black-legged Tick *(Ixodes pacificus)*. This is a small species (3/32 in., or 2.6 mm) with a dull reddish body and dark legs. It is usually found on deer, although it commonly crawls onto humans, and its bite may be painful. It is the vector for Lyme Disease spirochaetes.

■ Pacific Coast Tick *(Dermacentor occidentalis;* Figure 468). The back plate and legs of this tick are silvery-gray; its length is 3/16 inch (4.5 mm). It is found on a wide variety of hosts, including man.

■ American Dog Tick *(Dermacentor variabilis;* Figure 469). The adult is similar to that of the Pacific Coast Tick, but there are large dark areas on its back plate; it is 3/16 inch (4.5 mm) long. The dog is its preferred host, but this tick will also attack people and many wild and domesticated large mammals.

■ Brown Dog Tick *(Rhipicephalus sanguineus)*. This is another pest of dogs, and it is more common locally than the American Dog Tick. Its length is 3/32 inch (2 mm).

■ Rabbit Tick *(Haemaphysalis leporispalustris)*. This species is small (3/32 in., or 2 mm, long) and is often found on rabbits. It rarely bites people.

## SOFT TICKS (Family Argasidae)

■ Pajahuello *(Ornithodoros coriaceus;* Figure 470). The bite of this tick can result in tissue decay and the development of a very serious wound. Cattle and deer are the usual hosts. The tick is found in chaparral and scrub oak areas of the basin's foothills; it is 3/8 to 1/2 inch (10 to 12 mm) long.

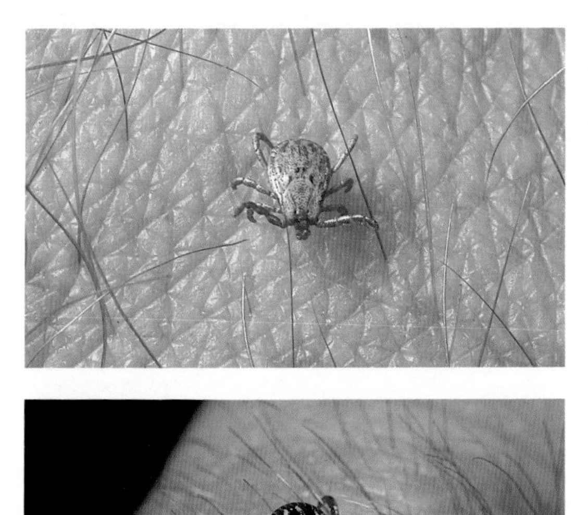

468. Pacific Coast Tick male biting human arm. Photograph by C. Hogue.

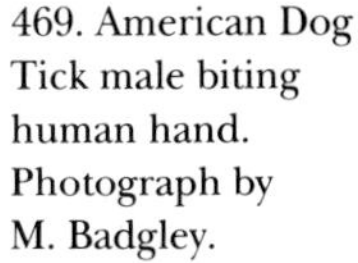

469. American Dog Tick male biting human hand. Photograph by M. Badgley.

■ Fowl Tick *(Argas persicus)*. This tick, which is found on chickens and other birds, is very flat and disk-shaped; its length is 1/8 to 3/16 inch (3 to 4 mm). (It is not officially recorded from the Los Angeles Basin but probably occurs here occasionally on chicken ranches.)

■ Spinose Ear Tick *(Otobius megnini)*. The adults are large, up to 1/3 inch (8 mm) long; their round bodies are constricted slightly at the middle and bear numerous pock marks. The nymph is spiny beneath. The species is a pest of livestock and horses, feeding primarily in the host's ears.

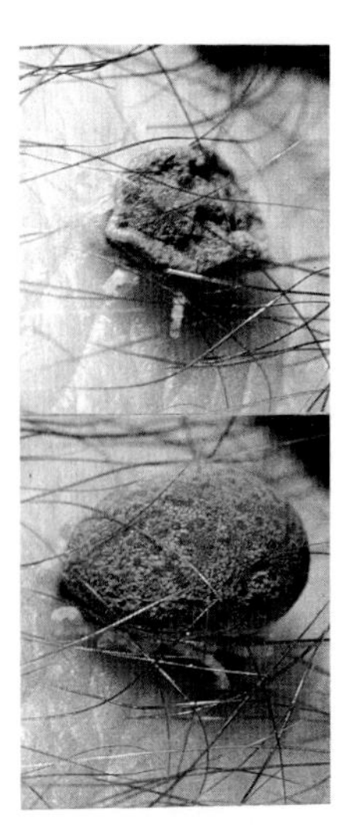

470. Pajahuello unfed (top) and after engorgement with blood. Photograph by R. Pence.

REFERENCES. Anonymous. 1963. Wood ticks in California. California State Department of Public Health, Berkeley.

Arthur, D. R., and K. R. Snow. 1968. *Ixodes pacificus* Cooley and Kohls, 1943: Its life history and occurrence. *Parasitology*, vol. 58, pp. 893-906.

Camin, J. H., and R. W. Drenner. 1978. Climbing behavior and host finding of larval rabbit ticks *(Haemaphysalis leporispalustris)*. *Journal of Parasitology*, vol. 64, pp. 905-909.

Furman, D. P., and E. C. Loomis. 1984. The ticks of California (Acari: Ixodida). *Bulletin of the California Insect Survey*, vol. 25, pp. 1-239.

Lane, R. S. 1984. Tick paralysis, an underreported disease of dogs in California. *California Veterinarian*, vol. 38, pp. 14-16.

Loomis, E. C., E. T. Schmidtmann, and M. N. Oliver. 1974. A summary review on the distribution of *Ornithodoros coriaceus* Koch in California (Acarina: Argasidae). *California Vector Views*, vol. 21, no. 10, pp. 57-62.

Loye, J. E., and R. S. Lane. 1988. Questing behavior of *Ixodes pacificus* (Acari: Ixodidae) in relation to meterological and seasonal factors. *Journal of Medical Entomology*, vol. 25, pp. 391-398.

Waldron, W. G. 1962. Notes on the occurrence, observations and public health significance of the Pajaroello tick *Ornithodorus coriaceus* Koch, in Los Angeles County. *Bulletin of the Southern California Academy of Sciences*, vol. 61, pp. 241-245.

# Part IV
# OTHER INSECT-LIKE ANIMALS

SMALL TERRESTRIAL creatures that crawl and are pesky are sometimes thought of as insects even though they belong to various other invertebrate groups. I have included a few of them here because they are common or conspicuous.

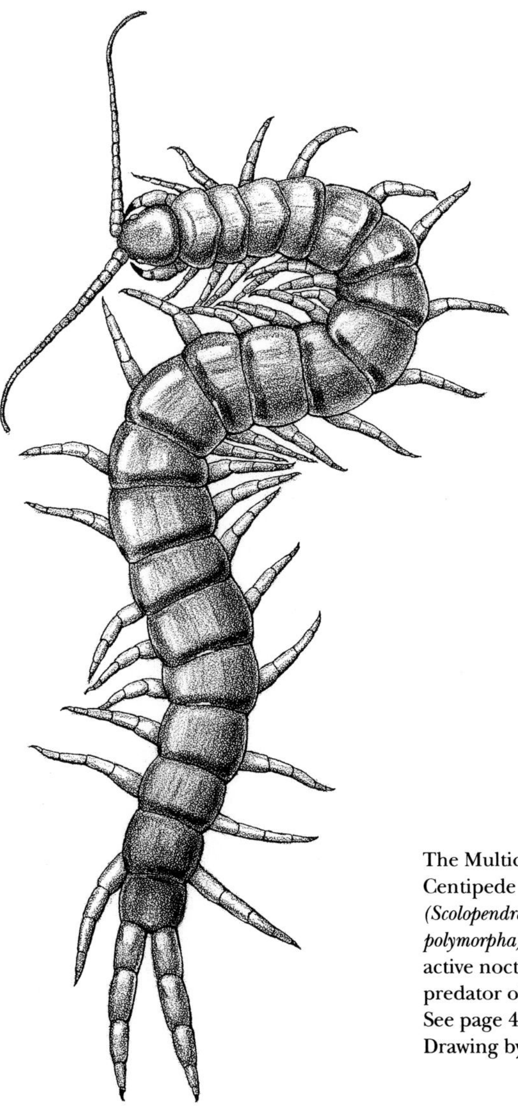

The Multicolored Centipede *(Scolopendra polymorpha)*, an active nocturnal predator of insects. See page 401. Drawing by T. Ross.

# 19 CENTIPEDES AND MILLIPEDES

## CENTIPEDES (Class Chilopoda)

ALTHOUGH ITS NAME MEANS "hundred-legged," no centipede can have exactly that number because there is always an odd number of pairs of legs, ranging from 15 to well over 200 pairs in adults. The legs are distributed one pair per segment, an arrangement that distinguishes these arthropods from millipedes, in which the legs appear in double pairs.

Centipedes are nocturnal and carnivorous, hunting at night for other small creatures, which they subdue with a dose of poison administered by a pair of fangs that are actually modified front legs. During the day, they hide in a damp dark obscure places—under stones and bark, in leaf litter and rotten logs, and in cracks and crevices in the soil.

Though their poison is lethal to their small prey, there is no evidence that the bites of all but the largest species of centipedes cause more than mild discomfort to people. A number of species are known to occur in the basin; the most common and conspicuous are the House and Multicolored Centipedes.

REFERENCE. Chamberlain, R. V. 1910-1912. Chilopoda of California. *Pomona College Journal of Entomology and Zoology*, vol. 2, pp. 363-374; vol. 3, pp. 470-479; vol. 4, pp. 651-672.

■ HOUSE CENTIPEDE *(Scutigera coleoptrata)* Figure 471

This centipede is about an inch (25 mm) long and distinctive because of its very long slender legs; there are only fifteen pairs of legs in adults, and the last pair is much longer than the rest (in the female they are twice as long as the body). The entire body is exceedingly frail and pale translucent-bluish in general color.

The species is very active and moves rapidly in its attempts to escape capture and to snatch the small insects it requires as food. Individuals are commonly found indoors, darting across the floor or clinging to a wall. They are particularly attracted to bathtubs, washbasins, and damp basements. Outdoors they are active in the summer and fall; most summer evenings, a certain House Centipede visits my porch light to catch the insects that it attracts.

It is doubtful that this species of centipede can

even inflict a wound to the human skin, so it should not be considered dangerous. It is actually beneficial in that it preys on many insect housepests, such as silverfish.

OTHER SCIENTIFIC NAMES. The species name is sometimes misspelled as *cleopatra.*

■ MULTICOLORED CENTIPEDE *(Scolopendra polymorpha)* Figure 472

This is a fairly large centipede, attaining a maximum length of 3 to 4 inches (7 to 10 cm). It varies in color from clear or dark olive yellow to greenish brown; the rear borders of the back plates are mostly dark green.

Practically nothing is known about its biology, other than that its general habitat is the same as for most centipedes—secluded places in contact with logs, rocks, or the ground.

The bite of this species may be painful. Although there are no data on the effects of its poison on humans, it is probably harmless. Contrary to popular belief, the sharp claws on the legs are not poisonous, although the last pair of legs is capable of pinching.

## MILLIPEDES (Class Diplopoda)

THE WORD "DIPLOPODA" means "double-footed," and refers to the two close-set pairs of legs on each "apparent" segment of these worm-like arthropods (each segment actually consists of two coalesced true segments).

Millipedes prefer moist conditions, and they abound in damp leaf litter and under rocks, logs, and loose bark; however, in their nocturnal wandering, they may be seen climbing on tree trunks or rocky cliffs, and they burrow into the ground when the weather is cold or dry. Their food consists of humus—rotting leaves, wood, and bark.

If disturbed, a millipede will coil up like a watch spring; many species also exude fluids that stain the skin (Figure 473) and have a repugnant odor that has been compared to iodine, quinine, or chlorine. The fluids commonly are benzoquinones and other chemicals that evaporate rapidly and act as repellents against predators. Certain millipedes even produce cyanide fumes.

Very little is known about the biology of these interesting creatures. They are easily collected in the habitats listed above, especially after rains; a few are luminescent and others fluoresce in the presence of ultraviolet light, making them easy to find at night with an ultraviolet or black light. Tightly closed jars containing several handfuls of moist humus are good quarters for captive specimens. All diplopods molt a number of times as they develop, and during each molt the number of legs and body segments of the individual is increased.

Several species live in the basin; most are small and inconspicuous. But two closely related species, *Hiltonius pulchrus* (Figure 473) and *Tylobolus claremontus,* are very large (exceeding 3 in., or 8 cm, in length), and a third species, *Atopetholus californicus (=angelus),* is only slightly smaller (up to 2 in., or 50 mm). All are otherwise similar, with cylindrical shiny black, dark gray, or brown bodies.

REFERENCE. Keeton, W. T. 1960. A taxonomic study of the milliped family Spirobolidae (Diplopoda: Spirobolida). *Memoirs of the American Entomological Society,* vol. 17, pp. 1-146.

---

■ **GREENHOUSE MILLIPEDE**
*(Oxidus gracilis)*
Figure 474

This millipede (order Polydesmida, family Paradoxosomatidae) is found throughout the tropical and temperate regions of the world but is thought to be native to Asia. Although it may be very abundant and a nuisance by its numbers in hothouses or outdoor gardens during warm weather, it does not eat living plant tissues and should not be considered a plant pest. It actually eats soil and decomposing organic matter. During the warm months, enormous swarms of the species may develop in beds filled with potting mix.

This is a small species (the body length of adults is about 1/2 to 3/4 in., or 13 to 20 mm) with a somewhat flattened body. Each segment has plate-like expansions along the side that are about one-fourth the width of the body. There are thirty to thirty-one pairs of legs in adults. Recently molted specimens are creamy white; as individuals mature they become medium to dark brown.

Females deposit their eggs in soil crevices. The young molt seven times.

OTHER SCIENTIFIC NAMES. *Orthomorpha gracilis.*

REFERENCE. Causey, N. B. 1943. Studies on the life history and the ecology of the hothouse millipede, *Orthomorpha gracilis* (C. L. Koch 1847). *American Midland Naturalist,* vol. 29, pp. 670-682.

471. House Centipede. Photograph by C. Hogue.

472. Multicolored Centipede. Photograph by C. Hogue.

473. Large common millipede (probably *Hiltonius pulchrus);* note the stains from noxious fluids secreted by the animal as a defense mechanism. Photograph by C. Hogue.

474. Greenhouse Millipede. Photograph by C. Hogue.

# 20 CRUSTACEANS

## ISOPODS (Order Isopoda)

■ Sow Bugs and Pill Bugs
Figures 475, 476

Sow bugs and pill bugs, which are in the Subphylum Crustacea, are very common in damp places around the yard and are considered a nuisance by gardeners and householders. The two most common local species were introduced. The Common Pill Bug *(Armadillidium vulgare;* Figure 475) is 3/8 inch (10 mm) long; it and other pill bugs are so called because of their habit of rolling up into a tight ball when disturbed. The Dooryard Sow Bug *(Porcellio laevis;* Figure 476) grows to 9/16 inch (15 mm) in length (and thus is considerably larger than the Common Pill Bug) and has two prominent tail-like appendages. It is incapable of rolling itself into a ball.

Both kinds are omnivorous and feed on young and decaying plant material. Unless very numerous, they do little damage except perhaps to tender seed-

475. Common Pill Bug. Photograph by C. Hogue.

476. Dooryard Sow Bug. Photograph by C. Hogue.

lings, but population explosions do sometimes occur. These crustaceans are probably eaten by few predators because they produce distasteful secretions from glands in the integument. One enemy is a dysderid spider, the Sow Bug Killer *(Dysdera crocota)*.

Both pill bugs and sow bugs are most abundant during the wet season, in winter and spring. As the dry summer approaches, these isopods take refuge in the soil, migrating down to lower levels to maintain body moisture and survive until the rains return.

REFERENCES. Gunter, G. 1979. Death on a plane surface, a result of enrollment in the common pill-bug *Armadillidum vulgare* Latreille (Isopoda). *Louisiana Academy of Sciences,* vol. 42, pp. 16-18.

Paris, O. H. 1963. The ecology of *Armadillidum vulgare* (Isopoda: Oniscoidea) in California grassland: Food, enemies, and weather. *Ecological Monographs,* vol. 33, pp. 1-22.

Paris, O. H., and F. A. Pitelka. 1962. Population characteristics of the terrestrial isopod *Armadillidium vulgare* in California grassland. *Ecology,* vol. 43, pp. 229-248.

## AMPHIPODS (Order Amphipoda)

■ HOUSE HOPPER (Family Talitridae; *Talitroides sylvaticus)* Figure 477

These black-eyed, shrimp-like terrestrial crustaceans are small (length about 5/16 in., or 8 mm); they are gray in life but turn orange after dying. Ivy ground cover and leaf mold beneath shrubbery apparently offer suitable conditions for their development.

During or just after a rain, residents in various parts of Los Angeles County are sometimes startled to find a number of these amphipods in their houses. The creatures are usually dead when found and are a nuisance merely by their presence. It is likely that the House Hoppers seek the dryness of buildings when their natural habitats become flooded.

REFERENCE. Mallis, A. 1942. An amphipod household pest in California. *Journal of Economic Entomology,* vol. 35, p. 595.

477. House Hopper. Photograph by R. Pence.

# 21 SNAILS, SLUGS, AND PLANARIANS

## HELIX SNAILS (Family Helicidae)

■ Brown Garden Snail *(Helix aspersa)* Figure 478

This snail was purposely introduced into California in the 1850s by European immigrants for use as food. It turned out to be a poor substitute for the snail served in restaurants (Edible Snail, *Helix pomatis),* and its cultivation was abandoned. It has since become well established and has assumed the status of a major plant pest. Its diet includes living and decaying vegetables, flowers, ground cover, citrus leaves and fruit, and many other plants—and even carrion and cardboard.

More money is probably spent by California homeowners in controlling this creature than in eradicating all other invertebrate pests combined.

REFERENCE. Basinger, A. J. 1931. The European brown snail in California. *Bulletin of the University of California, Agriculture Experiment Station,* vol. 515, pp. 1-22.

## COMMON SLUGS (Family Limacidae)

■ Gray Garden Slug *(Agriolimax reticulatum)* Figure 479

There are several species of slugs in our area, but this seems to be the commonest. Adults are buff to light brown in general color and about 2 inches (50 mm) long when fully extended. The back is keeled near the rear end, and the mantle covering a portion of the front of the body has irregular dark spots.

The slug was introduced from Europe and is a very destructive garden pest here as there, feeding on a wide variety of plants.

These slugs require moist surroundings to be active. Consequently, they feed at night and on over-

478. Brown Garden Snail. Photograph by M. Badgley.

cast or misty days. During dry periods, they hide beneath objects lying on the ground or in protected soil or rock niches.

OTHER SCIENTIFIC NAMES. *Deroceras reticulatum.*

## LAND PLANARIANS (Class Turbellaria; Order Tricladida)

■ ARROW-HEADED FLATWORM (Family Bipaliidae; *Bipalium kewensis*) Figure 480

This land planarian is slender and brown, with five dark longitudinal stripes; it can be large, up to 10 inches (25 cm) in length. The species is "hammer-headed": the head is shovel-shaped (wider than body), and there are numerous minute eyes along its border.

The species was discovered in 1878 in the greenhouses of Kew Gardens near London, hence its scientific name. It has a wide distribution in warm climates. It needs a moist habitat and is usually encountered near outdoor water faucets, where the soil often remains wet. Its original home is unknown but is possibly the Indo-Malayan region.

Flatworms are hermaphroditic, and copulation involves mutual insemination; they may also reproduce asexually by fragmentation. The eggs are encapsulated and affixed to objects in damp places.

These are benign creatures—they do not damage plants or cause any medical problems.

REFERENCE. Hyman, L. H. 1943. *Endemic and exotic land planarians in the United States.* American Museum Novitates, no. 1241. New York: American Museum of Natural History.

479. Gray Garden Slug. Photograph by C. Hogue.

480. Arrow-headed Flatworm. Photograph by J. Hogue.

A Crab Louse *(Pthirus pubis)* and its egg (top) attached to a human hair, the louse near the hair root. See pages 103,104. Photograph courtesy of Los Angeles County Agricultural Commissioner's Office, R. Garrison.

Appendixes

# Appendix A
# INSECT AND ARACHNID PESTS IN THE LOS ANGELES BASIN

Unwelcome guests within our homes, stealing our food and wrecking our possessions, and sometimes repaying us with bites and stings—these are the insect pests. It is convenient to recognize three primary categories here: household pests, plant pests, and health pests.

## HOUSEHOLD PESTS

A need for shelter, food, or a nesting place, attraction to lights, or perhaps the enticement of shade and food odors—and not a conscious desire to bother us—bring these guests to our door. In spite of screening, entry is easy for most insects: their small size permits them to squeeze through small cracks in the flooring, around baseboards, and under doors and through other imperfections in construction.

REFERENCE. Wilson, G. W., A. V. Bennett, and A. V. Provonsha. 1977. *Insects of man's household and health.* Prospect Heights, Ill.: Waveland.

### ■ FABRIC PESTS

The principal offenders here are the carpet beetles and clothes moths. Larvae of both these types normally use animal hair and skin as food, and they make no distinction between dead carrion in the out-of-doors and a fur coat or wool sweater hanging in a closet.

Usually only natural products—such as wool, cotton, and silk—are eaten, but occasionally synthetics such as nylon and rayon are attacked. Mixed weaves of natural and synthetic fibers are more susceptible. Fabrics stained with food are especially attractive to these pests.

REFERENCES. Mallis, A. 1959. The attraction of stains to three species of fabric pests. *Journal of Economic Entomology,* vol. 52, pp. 382-384.

Mallis, A., A. C. Miller, and R. C. Hill. 1958. Feeding of four species of fabric pests on natural and synthetic textiles. *Journal of Economic Entomology,* vol. 51, pp. 248-249.

### ■ WOOD AND PAPER PESTS

Several kinds of insects—most notably termites, cockroaches, silverfish, and larvae of several wood-boring beetles—eat and digest cellulose in its original wood form or in paper products.

When wood in the structure of the house is eaten away to the extent that it no longer can perform its support function, expensive repairs will be needed. Hardwood furniture, paneling, or picture frames may be riddled. Wallpaper, paper keepsakes, and books may be shredded or mined out. And flooring may deteriorate and cave in.

---

## ■ PANTRY PESTS

Two categories of insects—pantry moths and pantry beetles—are unwanted additives of stored food. The damage caused by these pests is usually confined to dried food products in storage. But fresh fruit or vegetables may occasionally be ruined by other species, such as fly maggots or beetle grubs, that really belong to the out-of-doors rather than in the household.

REFERENCES. Anonymous. 1953. Stored grain pests. U.S. Department of Agriculture, Farmers' Bulletin, no. 1260, pp. 1-46.

Gorham, J. R., editor. 1981. Principles of food analysis for filth, decomposition, and foreign matter, 2nd edition. Food and Drug Administration, Technical Bulletin, no. 1.

Kurtz, O. L., and K. L. Harris. 1962. *Microanalytical entomology for food sanitation control.* Washington, D.C.: Association of Official Agricultural Chemists.

Strong, R. G., and G. T. Okumura. 1958. Insects and mites associated with stored foods and seeds in California. *California Department of Agriculture Bulletin,* vol. 47, no. 3, pp. 233-249.

---

## ■ ANNOYANCES

Housekeepers will not hesitate to place in this group any creeping crawling creature that graces their doorstep. No one enjoys having a cockroach run across his face at night, and the persistent buzz of a fly about one's ears is not really appreciated. Insects such as these, which cause no real damage or harm but are simply undesirable house guests, are annoyances. In addition to those that are adapted to living with human beings, there are other species—such as gnats, owlet moths, Jerusalem crickets, and spiders—that join the "annoying" category as we invade new realms in our real estate ventures.

REFERENCE. Pence, R. J. 1955. Is new home construction upping the invasion of "strangers"? *Pest Control Magazine,* vol. 23, no. 10, pp. 30f.

## PLANT PESTS

A wide variety of insects prey on ornamental, fruit, and vegetable plants. They may be classified into two groups according to their method of feeding:

- Chewing insects. With mandibulate mouthparts (Figure 2 left) that bite, crush, or grind, these pests consume the solid matter of garden plants. The category includes cutworms, grasshoppers, grubs, weevils, and stem borers.
- Sucking insects. Piercing mouthparts (Figure 2 right) rob plants of their vital sap. Aphids, scale insects, mealybugs, and leafhoppers are of this type.

In the face of urban sprawl, agricultural acreage is steadily diminishing in the Los Angeles Basin. Yet dairy, citrus, truck crop, nursery stock, and cut-flower farming still occupy a surprisingly broad territory. Basin agriculture contributes substantially to Los Angeles County's position within the top ten in the United States in total crop production.

And what devotee of gardening, the popular hobby of many Angelenos, has not felt ardor dampened upon finding prize roses curled under masses of aphids, or tomato vines denuded by giant green caterpillars?

It is not possible to characterize all the pests of plants, even in our limited area, in this book. Many species are found in both fields and gardens, and those of major importance are included in the main text of this book. We are fortunate at least in that the basin is not home to some of the infamous offenders of other parts of the United States, such as the Mexican Bean Beetle *(Epilachna varivestis)*, migratory locusts *(Melanoplus* species), European Corn Borer *(Ostrinia nubilalis)*, Mormon Cricket *(Anabrus simplex)*, Colorado Potato Beetle *(Leptinotarsa decemlineata)*, Japanese Beetle *(Popilia japonica)*, and Gypsy Moth *(Porthetria dispar)*.

REFERENCES. Brown, L. R., and C. O. Eads. 1965. A technical study of insects affecting the oak tree in southern California. California Agricultural Experiment Station Bulletin, no. 810.

_______. 1965. A technical study of insects affecting the sycamore tree in southern California. California Agricultural Experiment Station Bulletin, no. 818.

_______. 1966. A technical study of insects affecting the elm tree in southern California. California Agricultural Experiment Station Bulletin, no. 821.

_______. 1967. Insects affecting ornamental conifers in southern California. California Agricultural Experiment Station Bulletin, no. 834.

______. 1969. Unnamed and little known insects attacking cottonwood in southern California. *Journal of Economic Entomology,* vol. 62, pp. 667-674.

Ebeling, W. 1959. *Subtropical fruit pests.* Berkeley, Calif.: University of California.

Johnson, W. T., and H. H. Lyon. 1976. *Insects that feed on trees and shrubs: An illustrated practical guide.* Ithaca, N.Y.: Cornell University Press.

Michelbacher, A. E., and J. C. Ortega. 1958. A technical study of insects and related pests attacking walnuts. California Agricultural Experiment Station Bulletin, no. 764, 86 pp.

Michelbacher, A. E., J. Swift, C. David, D. Hall, and R. Raabe. 1959. Ridding the garden of common pests. California Agricultural Experiment Station Extension, Service Circular, no. 479.

Oatman, E. R., and G. R. Platner. 1972. An ecological study of lepidopterous pests affecting lettuce in coastal southern California. *Environmental Entomology,* vol. 1, pp. 202-204.

Okumura, G. T. 1959. Illustrated key to the lepidopterous larvae attacking lawns in California. *Bulletin of the California Department of Agriculture,* vol. 48, no. 1, pp. 15-21.

Tashiro, H. 1987. *Turfgrass insects of the United States and Canada.* Ithaca, N.Y.: Cornell University.

## HEALTH PESTS

THIS CATEGORY OF PESTS may be organized to separate those that inflict us and our animals with direct bodily injury (direct agents) and those that act as vectors of pathogenic microorganisms (disease vectors).

### ■ MEDICAL PESTS

The basin is inhabited by many venomous insects and arachnids that sting us (bees, ants, wasps, scorpions), bite us (spiders, centipedes, horse flies, cone-nose bugs, ticks), or burrow into our skin and bodies (scabies mites, flesh flies). Stings and bites are seldom serious unless the venom injected is very toxic (primarily in the case of the Black Widow spider) or capable of eliciting a shock reaction in sensitive individuals (as is that of the Western Cone-nose Bug).

Contrary to popular belief, the perpetrator of a bite or sting cannot be readily identified by the appearance of the wound because bite symptoms are so variable (sensitive persons may exhibit severe reactions to bites that produce no noticeable effects in other people). A wound may offer some clues, by its position on the body (swollen lips from kissing bugs) or by a characteristic feature such as necrosis (usually associated with spider bites) or blood spots in series (often produced by fleas). Circumstancial evidence regarding the prevalance of a pest or a specific sighting of it is also of interest. And because the outward features of a bite itself are seldom diagnostic, it is always wise for sensitive persons to catch and save possible offending species for later diagnosis, should complications arise.

Fortunately, few active vectors live in our region, although a surprising number of arthropod-borne diseases are indigenous. Occasionally there are cases of encephalitis (St. Louis or Western Equine varieties) resulting from the bite of mosquitoes infected with the virus, or of Relapsing Fever, which is carried by a soft tick, *Ornithodorus hermsi,* that lives in neighboring mountains. A new threat is Lyme Disease, which is transmitted by the Western Black-legged Tick; although this disease has not yet been recorded in the basin, it has recently been found for the first time in southern California.

Fleas may transmit plague and Murine Typhus, two serious diseases that are latent in our wild rodent populations. Even malaria remains a potential threat to our local health and welfare; although they are rare, mosquito vectors of the genus *Anopheles* live in the basin and are capable of transmitting this ailment should they acquire the protozoans that cause it as they feed on the blood of an infected person from abroad.

The local status of Chagas' disease is puzzling. Microorganisms that are apparently identical to those causing symptoms in the tropics, but without the virulence to do so, exist here in wild rodents and in the Western Cone-nose Bug; thus, Chagas' disease must also be classified as potential threat to the local public health.

The incidence of dwarf tapeworms *(Hymenolepis diminuta* and *H. nana)* in the Los Angeles area is apparently growing as a result of the commonness of their intermediate hosts, granary beetles of the genera *Tribolium* and *Tenebrio.*

The comfort of many is destroyed by pesky insects and mites whose bites, although not venomous, can cause allergic reactions, itching welts, and skin rash; the list of such pests includes mosquitoes, lice, bedbugs, black flies, and punkies.

REFERENCES. Andrews, M. 1977. *The life that lives on man.* New York: Taplinger.

James, M. T., and R. F. Harwood. 1969. *Herm's medical entomology,* 6th edition. London: MacMillan.

Rau, M. E. 1979. The frequency distribution of *Hymenolepis diminuta* cysticercoids in natural, sympatric populations of *Tenebrio molitor* and *T. obscurus*. *Journal of Parasitology*, vol. 9, pp. 85-87.

## ■ PSYCHOLOGICAL PESTS

Skin rashes are frequently psychosomatic, and a rash may sometimes be attributed to insects in the absence of clearcut evidence of their involvement. Some mentally or emotionally ill people have intense emotional upset over imagined "bug infestations" of the body or home *(delusory parasitosis* or *delusory cleptoparasitosis);* these delusions are not illnesses in themselves but instead are symptoms of other mental disorders, such as depression or organic brain syndrome.

An extreme aversion to insects *(entomophobia)* seems to be relatively common even among otherwise healthy people.

REFERENCES. Grace, J. K., and D. L. Wood. 1987. Delusory cleptoparasitosis: Delusions of arthropod infestation in the home. *Pan-Pacific Entomologist*, vol. 63, pp. 1-4.

Keh, B. 1983. Cryptic arthropod infestations and illusions with delusions of parasitosis. Pages 165-185 in *Urban entomology: Interdisciplinary perspectives*, edited by G. W. Frankie and C. S. Koehler. New York: Praeger.

Pierce, W. D. 1944. Entomophobia. *Bulletin of the Southern California Academy of Sciences*, vol. 43, pp. 78-80.

St. Aubin, F. [No date.] Ectoparasites; real or delusory? How to recognize and cope with either. Reformatted reprint from *Pest Control Technology*, 26 pp.

Waldron, W. G. 1963. Psychiatric and entomological aspects of delusory parasitosis. *Journal of the American Medical Association*, vol. 186, pp. 213-214.

## ■ PESTS OF PETS AND LIVESTOCK

Dairy cattle, horses, swine, and poultry suffer the attacks of a variety of parasitic and blood-sucking insects. Veterinary practice must still deal with these pests, although modern insecticides and drugs render their effects much less severe than in the past.

Many veterinary pests are specifically adapted to pet animals. Dogs and cats are usually infested by fleas and often suffer from ticks and chiggers and other parasitic mites. The latter, especially the knemodectic or "itch" mites, can seriously affect the health of caged birds.

An outbreak of the flea-borne disease Murine Typhus affected cats in northern sections of Los Angeles in 1984 and 1985.

REFERENCES. Keymer, I. F. 1982. Parasitic diseases. Pages

535-598 in *Diseases of cage and aviary birds,* edited by M. L. Petrak. Philadelphia: Lea & Febiger.

Williams, R. E. 1985. *Livestock entomology.* New York: Wiley.

## QUARANTINE

LOCAL AGRICULTURE needs protection from hitchhiking foreign plant pests and animal disease organisms. Exotic pests that become established cost much more to control than those that have been here a long time, because newcomers are not kept in check by their natural enemies, which usually get left behind.

Three agencies are concerned with the undesired importation of potential pest insects and arthropods from other countries: the Los Angeles County Agricultural Commissioner's Office, the California Department of Food and Agriculture, and the U.S. Department of Agriculture, Animal and Plant Health Inspection Service, Plant Protection and Quarantine (known as "USDA-APHIS-PPQ," for short). Officials and inspectors from each bureau monitor travelers' baggage and vehicles at border crossings, including Los Angeles International Airport and Harbor. Quarantine zones controlled by county and state personnel may also be set up within the basin, as has been done in the Medfly control program. These agencies cooperate in cases of major outbreaks of pest species.

Literally thousands of shipments of prohibited meats, fruits, vegetables, animals, and soil are intercepted each year by agricultural inspectors. It is believed that imported fruit that passed undetected into Los Angeles started an infestation of Medflies in orchards near the airport in 1975. Other such immigrant pests that have turned up in the basin in recent years are the Japanese Beetle, Mexican Fruit Fly, Oriental Fruit Fly, Eucalyptus Borer, Gypsy Moth, Ash Whitefly, and Eugenia Psyllid.

Even products that have been legally imported are frequently infested with known pests or possibly injurious species. Food inspectors find great numbers of mites, beetles, and other foreign insects thriving in the holds of arriving ships or in import warehouses onshore.

Thus by land, sea, and air, there is a constant threat of introduction of new pests. It is important for residents to realize the dangers posed by these would-be immigrants and to cooperate with authorities by not transporting potentially infested material from place to place.

Permits are required from both the state and federal offices for the legal importation of live insects, spiders, and scorpions that are to be used for scientific or other legitimate purposes (for example, to be sold as pets, as food for pets, or as biological pest control agents). Permits must be obtained in advance of the time of importation.

REFERENCES. Anonymous. 1980. Traveler's tips, revised edition. U.S. Department of Agriculture, Animal and Plant Inspection Service, Program Aid, no. 1083.

Olsen, A. R. 1981. List of stored-product insects found in imported foods entering United States at southern California ports. *Bulletin of the Entomological Society of America,* vol. 27, pp. 18-20.

_______. 1983. Food-contaminating mites from imported foods entering the United States through southern California. *International Journal of Acarology,* vol. 9, pp. 189-193.

Olsen, A. R., J. R. Bryce, J. R. Lara, J. J. Madenjian, R. W. Potter, G. M. Reynolds, and M. L. Zimmerman. 1987. Survey of stored-product and other economic pests in import warehouses in Los Angeles. *Journal of Economic Entomology,* vol. 80, pp. 455-459.

Zimmerman, M. L. 1990. Coleoptera found in imported stored-food products entering southern California and Arizona between December 1984 through December 1987. *Coleopterists' Bulletin,* vol. 44, pp. 235-240.

# Appendix B
# KEEPING INSECTS IN CAPTIVITY

To satisfy their curiosity about the habits and activities of insects or just to have an unusual pet, many people like to keep live specimens. Most species are easy to maintain, and the pet owner can learn much about feeding, life cycle, reproduction, and other aspects of the biology of our local insects.

Some kind of enclosure is needed for terrestrial types. Although a screened box, cardboard carton, or glass-covered aquarium makes the best cage, a glass jar with holes punched in the lid will suffice. It is important in any case to follow a few simple rules regarding the proper care and feeding of the captives:

- Provide good ventilation. Insects do not need much oxygen—even a closed jar contains more than enough air for an insect's lifetime use. But fresh air is required to dissipate heat and to control humidity in the cage and prevent harmful growths of mold.
- Offer water at regular intervals by sprinkling it lightly about the cage or placing a water-filled vial stoppered with a wad of cloth in the cage.
- Provide proper food in adequate amounts, and keep it fresh. Remember that many species have very specific food requirements; these can be determined by reading about a species or observing the insect under natural conditions. If the exact food is not known, generally successful substitutes are available. For herbivorous chewing types (but not caterpillars), try fresh apple or potato slices or dog food (caterpillars always need fresh leaves of their own specific foodplant species). For carnivorous chewing captives, give mealworms, crickets, or caterpillars. For nectar-feeding types, offer saturated sugar or honey water on a cotton wad. Some herbivorous sucking types can be maintained on fresh prunes or raisins.
- Try to make the insect's surroundings conform as closely as possible to those of the environment in which it was found. For example, provide rocks under which light-shy types can hide, damp soil for burrowers, twigs or other supports for climbers.

Some aquatic insects require rapidly flowing water, but most can be cared for much like tropical fish. Temperature control is not usually a factor, but it is important to keep the water level in the container low to provide air space for those that take in oxygen at the surface or transform into flying adults. Aquatic insects are generally not as particular in their food requirements as are terrestrial types: a piece of meat tied to a string and suspended in the water satisfies gnawing carnivores; those requiring living prey will take mosquito larvae or minnows; and green algal masses and common aquarium plants will do for most aquatic herbivores.

Excellent books that provide specific instructions on how to maintain insects in captivity are listed below and in the bibliography under "Insect Observation." Instructions for keeping and rearing praying mantids, ants, caterpillars, tarantulas, and millipedes are given in the text of this book (see the index for page numbers).

REFERENCES. Evans, A.V. 1992. Bug husbandry at the Insect Zoo. *Terra,* vol 30, no. 3, pp. 16-25

Siverly, R. E. 1962. *Rearing insects in schools.* Dubuque, Iowa: W. C. Brown.

# Appendix C
# SIMPLE WAYS TO MAKE AN INSECT COLLECTION

Insect collecting is not only an excellent learning experience for the serious entomology student, it is an exciting, healthful, enjoyable, and satisfying hobby for anyone. Endless new additions to the collection are possible because of the enormous variety of insects (the world's species are many times the number of insect postage stamp designs that have been produced to date!).

To be more than a helter-skelter assemblage of "dead bugs," a collection must be properly prepared and cared for. The following instructions will serve the beginner; but after gaining a little experience, the collector should refer to the references cited at the end of this section.

There are four phases of collecting activities—finding the specimens, catching and killing them, keeping them temporarily, and providing permanent storage.

Finding insects requires only diligence and a sharp eye, for they occur virtually everywhere. The twenty selected microhabitats of insects given in Chapter 2 are good places to begin the search. An excellent way to obtain night-flying insects is to take advantage of their tendency to be drawn to lights (especially on moonless nights): an outdoor porch light, street lamp, battery-powered camp light, or gas lantern set on a white bed sheet on the ground may yield hundreds of specimens. Most collectors nowadays use bulbs that emit ultraviolet light, such as the "Black Light," the GE "BL"-15-watt unfiltered fluorescent tube, or the mercury vapor bulb, because ultraviolet light is particularly attractive to nocturnal species.

The human forefinger and thumb are the best instruments for snaring specimens. However, a net is essential for catching fast-flying and stinging species. Many kinds of nets are available; a simple one can be made of a bag of curtain material sewn to a wire coat hanger that has been bent out into a circle and wired to a handle. The bag should be deep enough to fold completely across the face of the ring, leaving a small pocket in which the specimen is trapped. Aquatic insects can be taken with a small dip net (the kind used in tending aquaria) or a wire kitchen strainer.

Captured specimens can be removed from the net with fingers, if they are not stinging or biting types, or coaxed directly into the killing bottle. Dry potassium cyanide crystals may be used as the killing agent; however, this is a dangerous substance, and safer chemicals may be preferred, especially for children collectors. A very good killing bottle can be made by pouring 1/4 inch (6 mm) of wet plaster into the bottom of a small jar and, when the plaster is dry, soaking it with ethyl acetate, fingernail polish remover, or lacquer thinner; these volatile fluids are capable of killing insects with their fumes but are relatively safe to handle. You can also kill insects by placing them in plastic bags or jars and putting them in the freezer. Soft-bodied types such as caterpillars can be killed by dropping them in alcohol (ordinary "hospital" or isopropyl alcohol is good, but ethyl alcohol is preferred).

The final step is to preserve the insect for display. A soft-bodied type may be kept in the alcohol in which it was killed. But other insect specimens are usually "mounted"—each insect is placed on a pin passed through its body. The pin acts as a handle so that the specimen can be moved or examined from all angles without touching it and possibly breaking appendages or otherwise damaging the delicate body.

After a specimen dies it should be removed from the killing bottle or jar and stored in a safe place until it can be mounted. Many collectors use glassine envelopes kept in a cardboard box for this purpose. If possible, a day's catch should be processed that evening for permanent preservation or display.

Dead insects quickly dry and become brittle; if your catch cannot be mounted before this happens, you must relax the specimens in a humidifier (moisture box) so that they can be handled. A plastic or glass box (a sandwich box or refrigerator dish) works as a humidifier: simply cover the bottom with a pad of paper towelling that is thoroughly damp but not dripping wet, and place the dry specimens on the towelling for one or two days. Mold may be inhibited by keeping the box in the refrigerator or by adding two or three drops of carbolic acid to the paper pad.

Mounting procedures vary according to the kind of insect. Most large specimens are pinned directly through the body with special steel pins ("insect pins"; the popular sizes are #1, 2, or 3), leaving about 3/8 inch (1 cm) below the head as finger room for holding the specimen while piercing it with the pin. Tiny specimens are glued to the tip of a small triangle of paper, which is then pierced at the broad end with a #3 insect pin. The wings of butterflies and moths should be spread open (this requires special mounting boards and techniques, which are described in the references cited below).

Pinned and mounted specimens require no preservation treatment other than thorough drying through exposure to air. Never spray them with plastic fixatives or lacquers; this is unnecessary and obscures anatomical details. Kept dry and free of pests in an air-tight box with a few moth crystals (paradichlorobenzene) or moth flakes (napthalene), a collection of insects will last indefinitely.

The recording of data is a very important part of maintaining an insect collection. A specimen has little scientific value without notes on the exact place, date, and circumstances of the insect's capture. The accepted procedure is to print this information finely and carefully with India ink onto a small piece of heavy paper (dimensions no greater than 1/3 by 1/2 inch, or 8 by 13 mm); this label is then placed on the pin beneath the mounted specimen, or in the jar of alcohol with "pickled" specimens.

In California, collection equipment and supplies may be obtained from several sources, including BioQuip Products, 17803 La Salle Avenue, Gardena 90248; Ward's Natural Science Establishment, Inc., 11850 East Florence Avenue, Santa Fe Springs 90670-4490; and Tri-Ess Sciences, 1020 West Chestnut Street, Burbank 91506. Other sources of materials and supplies are given in the bibliography and resources section at the back of this book.

REFERENCES. Anonymous. [No date.] How to make an insect collection. Rochester, N.Y.: Ward's Natural Science Establishment.

BioQuip. [No Date.] How to make an insect collection. Gardena, Calif.: BioQuip. [A revised edition is in preparation.]

Dirig, R. 1977. Labeling and storing an insect collection. 4H Member's Guide, no. M-6-7. Ithaca, N.Y.: New York State College of Agriculture.

Martin, J. E. H. 1977. Collecting, preparing and preserving insects, mites and spiders. In The insects and arachnids of Canada, part 1. Biosystematics Research Institute Publication, Agriculture Canada, no. 1643, 182 pp.

Oman, P. 1948. Collection and preservation of insects. U.S. Department of Agriculture, Miscellaneous Publications, no. 601.

# GLOSSARY

AMETABOLOUS: Without metamorphosis; proceeding in the life cycle from egg to adult with increase in size but without change in form. Compare with hemimetabolous and holometabolous.

APICULTURE: Beekeeping.

BALANCER: See halter.

CAECUM: See gastric caecum.

CALYPTER: Lobes at the base of the wings in the Diptera (gnats, midges, and flies).

CAPRIFICATION: The process of producing caprifigs.

CAPRIFIGS: Figs fertilized by pollen carried from the wild fig *Ficus carica sylvestris* by fig wasps (*Blastophaga* species).

CARAPACE: The dorsal (back) portion of the cephalothorax of spiders and other arachnids and of crustaceans.

CARNIVORE: A flesh-eating animal. Compare with herbivore and omnivore.

CEPHALOTHORAX: The fused head and thorax of arachnids and some other arthropods.

CHELICERAE; singular chelicera: The first set of appendages of an arachnid, which are modified into fangs (spiders) or mouthparts.

CHITIN: A strong resistant chemical substance found in the cuticle or body wall of insects and related arthropods.

CHRYSALID, or CHRYSALIS: The pupa of butterflies.

CLEPTOPARASITOSIS: Infestation by cleptoparasites.

CLEPTOPARASITE, or KLEPTOPARASITE: A species that steals food from other insects or spiders.

COCOON: A protective cover of silk for the eggs (spiders) or pupa (many insects).

COXA; plural coxae: The first or base segment of an insect's leg, by which the leg articulates with the body.

CREPITATION: Insect sound production by wing snapping (grasshoppers) or explosive discharges (bombardier beetles). Compare with stridulation.

CREPUSCULAR: Active in the twilight. Compare with diurnal and nocturnal.

CURATOR: A person responsible for caring for collections in museums.

CYSTICERCOID: An immature stage of certain tapeworms.

DELUSORY CLEPTOPARASITOSIS: A mental condition in human beings in which afflicted people imagine their home (or "nest") to be infested with insects. See cleptoparasitosis, delusory parasitosis.

DELUSORY PARASITOSIS: A mental condition in human beings in which afflicted people imagine themselves to be infested with insects. See delusory cleptoparasitosis.

DIURNAL: Active in the daytime. Compare with crepuscular and nocturnal.

Ectoparasite: A parasite that lives on the exterior of its host. Compare with endoparasite.

Elytra; singular elytron: The hardened fore wings of insects (especially beetles).

Emergence: Exit of an adult insect from an immature stage. Compare with hatching.

Endoparasite: A parasite that lives in the internal organs or tissues of its host. Compare with ectoparasite.

Endemic: Confined to a geographic area and found nowhere else.

Entomologist: A scientist who studies insects.

Entomophobia: An extreme fear of or aversion to insects.

Extinction: The complete disappearance or death of a species from its total range. Compare with extirpation.

Extirpation: The disappearance of a species from a particular geographic area but not from its total range. Compare with extinction.

Filature: The unwinding of or reeling out of silk from cocoons.

Free-living: Not fixed to a substrate or host and thus capable of locomotion. Not parasitic.

Frenulum: A bristle-like structure on the leading edge of the rear wing of many moths that interlocks with a hook-like process on the front wing and thus unites the two wings so that they function as one.

Furcula: A forked appendage on the underside of the fourth abdominal segment of a springtail that may be snapped against the substratum to propel the insect into the air.

Ganglion; plural ganglia: A mass of nerve cells on the ventral (bottom) nerve cord of an arthropod.

Gastric caecum (plural caeca): One of several sac-like digestive organs associated with the insect stomach.

Halter; also called "balancer": In dipterans (gnats, midges, or flies), hind wing modified to form an elongate knob-shaped organ for sensing flight orientation.

Hatching: Exit of an immature insect from the egg stage. Compare with emergence.

Hemimetabolous: Undergoing incomplete metamorphosis; proceeding through not four but three life stages (egg, nymph, and adult) with relatively slight change in body form and habits during development. Compare with ametabolous and holometabolous.

Herbivore: A plant-eating animal. Compare with carnivore and omnivore.

Hermaphrodite: An organism that is normally equipped with both male and female reproductive organs.

Holometabolous: Undergoing a complete or four-stage metamorphosis (egg, larva, pupa, and adult) and exhibiting dramatic change in body form and habits at each stage. Compare with ametabolous and hemimetabolous.

Hyperparasite: An organism that is a parasite of another parasite.

Hypopharynx: Sensory or tongue-like structure in the mouthparts of an insect.

IMAGO: An insect in its sexually mature adult—and usually winged—stage.

INSTAR: The immature insect between molts during development.

LABIUM: The lower lip of an insect's mouthparts.

LABRUM: The upper lip of an insect's mouthparts.

LACM: Acronym for Natural History Museum of Los Angeles County.

LARVA; plural larvae: The immature and wingless form that hatches from the egg of an holometabolous insect and that will eventually transform into a pupa, prior to reaching adulthood. Also, sometimes, the immature form (or nymph) of a hemimetabolous insect.

LARVIFORM: Term used to describe an adult insect that retains the appearance of the larva.

MALPIGHIAN TUBULES: The kidneys of an insect.

MANDIBLES: The first of the paired mouth appendages in insects and other arthropods; usually jaw-like (in chewing forms) or needle-like (in sucking forms).

MAXILLAE; singular maxilla: The second of the paired mouth appendages of insects.

METAMORPHOSIS: A series of marked and more or less abrupt changes in the form of a developing insect. See ametabolous, hemimetabolous, and holometabolous.

MESOTHORAX: The second or middle of the three segments in the thorax of an insect. Compare with metathorax and prothorax.

METATHORAX: The third or posterior of the three segments in the thorax of an insect. Compare with mesothorax and prothorax.

MYIASIS: The invasion of human or animal tissue or body cavities by fly maggots.

NECROSIS: Death and decomposition of tissue, often following a wound or venomous bite or sting.

NOCTURNAL: Active at night. Compare with crepuscular and diurnal.

NYMPH: The immature stage that hatches from the egg of a hemimetabolous insect.

OMNIVORE: An animal that eats a broad variety of foods, often both animals and plants. Compare with carnivore and herbivore.

OVIPOSITION: Egg laying.

OVIPOSITOR: An organ used by insects for depositing eggs in a place suitable for their development.

PALPUS; plural, palpi: A segmented sensory appendage of the labium and maxilla of an insect's mouthparts.

PARASITE: An organism that lives on or in another organism, obtaining food from but not killing its host. Compare with parasitoid.

PARASITOID: A parasite that eventually kills its host.

PARASITOSIS: Infestation with or disease caused by parasites.

PARTHENOGENESIS: Reproduction without fertilization.

PEDIPALPS (or PEDIPALPI); singular pedipalp (or pedipalpus): The second set of paired appendages of an arachnid. Compare with chelicerae.

Polymorphism: Existence of varied body forms among individuals of a given species.

Proboscis: Elongate, often extensile, mouthparts of insects that take liquid food.

Prolegs: Unsegmented abdominal legs of caterpillars.

Pronotum: Back of the prothorax of an insect.

Prothoracic shield: Pronotum enlarged or expanded into a broad plate.

Prothorax: The first or anterior of the three segments in the thorax of an insect. Compare with mesothorax and metathorax.

Pubescence: Hairiness; fuzziness.

Pupa; plural pupae: An intermediate, usually quiescent, stage in the life cycle of an holometabolous insect in which the insect is usually enclosed in a hardened cuticle (chrysalid or puparium) or in a cocoon and from which the adult will eventually emerge.

Puparium: In some flies, the thickened, hardened capsule surrounding the pupa; formed from the last larval skin, which has been shed but not discarded.

Sclerites: Plate-like regions of an arthropod's integument bounded by sutures or membranes.

Sclerotin: A rigid dark chemical found in the cuticle or body wall of insects and related arthropods.

Scopula, plural scopulae: A brush of hairs on the lower surfaces of the feet of some spiders.

Scutellum: A triangular sclerite at the rear of the back of the thorax in many insects.

Sericulture: The cultivation of silkworms.

Spinnerets: Appendages at the rear end of the spider's abdomen, from which silk is emitted.

Spiracles: Openings of the tracheal system of arthropods through which they breathe.

Stabilmentum; plural stabilmenta: A thick band of silk in the center of the web of some orb weaver spiders that apparently helps camouflage the spider.

Stridulation: Insect sound production by rubbing a roughened part of the body against another part. Compare with crepitation.

Subimago: Winged next-to-last stage in the development of a mayfly.

Symbiosis: The intimate association of two dissimilar and unrelated organisms in a mutually beneficial relationship.

Tarsus; plural tarsi: The segmented foot of an insect.

Thorax: The middle or second of the three main regions (head, thorax, and abdomen) of the insect's body; bears the legs and wings.

Trochanter: The segment of the insect leg between the coxa and the femur.

Vector: An agent (sometimes an arthropod) capable of carrying and transmitting a disease organism from one plant or animal to another.

Vestiture: Body covering (scales, hairs, etc.).

Viviparity: The bearing of live young (as opposed to egg laying).

Wild: Living under natural conditions; not domestic or domesticated.

# GENERAL BIBLIOGRAPHY AND RESOURCE LIST

## LITERATURE

READERS HAVING QUESTIONS about the biology of specific insects, pest control, or other aspects of entomology may consult the sources listed here. Some of these publications are considered technical literature and are not available in outlying branches of the local public libraries. (The main Los Angeles City Library may have some of these references, but the best source is a university, college, or museum library.)

The books mentioned here that are still in print may be obtained through the Book Shop of the Natural History Museum of Los Angeles County or through any major book store (ask to have a book ordered from the publisher if you cannot find it on the shelves). New, used, and out of print insect literature may be ordered through the mail from second-hand and specialty book dealers.

### ■ GENERAL ENTOMOLOGY

Daly, H. V., J. T. Doyen, and P. R. Ehrlich. 1978. *An introduction to insect biology and diversity*. New York: McGraw-Hill.

Farb, P., and Editors of Life. 1962. *The insects*. Life Nature Library. New York: Time Life.

Imms, A. D. 1957. *A general textbook of entomology*, 9th edition. London: Methuen.

Klots, A. B., and E. B. Klots. 1959. *Living insects of the world*. Garden City, N.Y.: Doubleday.

Milne, L., and M. Milne. 1980. *The Audubon Society field guide to North American insects and spiders*. New York: A. Knopf.

Ross, H. H., C. A. Ross, and J. R. P. Ross. 1982, *A textbook of entomology*, 4th edition. New York: Wiley and Sons.

Wigglesworth, V. B. 1964. *The life of insects*. Cleveland, Ohio: World.

### ■ INSECT FAUNA OF THE LOS ANGELES BASIN

Assis de Moraes, A. P. 1977. Flies (Diptera) attracted to blacklight at the Anaheim Bay salt marsh, California. Master's thesis, California State University, Long Beach.

Augustson, G. F. 1939. The fauna and flora of the El Segundo Sand Dunes: Some of the mites (Acarina) of the dunes. *Bulletin of the Southern California Academy of Sciences*, vol. 28, pp. 191-197.

Daly, H. V. 1975. Orders of intertidal insects; Collembola, Hemiptera. Pages 432-435 in *Light's Manual: Intertidal invertebrates of the central California coast*, 3rd edition, edited by R. I. Smith and J. T. Carlton. Berkeley, Calif.: University of California.

Doyen, J. T. 1975. Intertidal insects: Order Coleoptera. Pages 446-452 in *Light's Manual: Intertidal invertebrates of the central California coast,* 3rd edition, edited by R. I. Smith and J. T. Carlton. Berkeley, Calif.: University of California.

Evans, W. G. 1980. Insecta, Chilopoda, and Arachnida: Insects and allies. Pages 641-658 in *Intertidal invertebrates of California,* edited by R. H. Morris, D. P. Abbott, and E. C. Haderlie. Stanford, Calif.: Stanford University.

Goeden, R. D., and D. W. Ricker. 1975. The phytophagous insect fauna of the ragweed, *Ambrosia confertiflora,* in southern California. *Environmental Entomology,* vol. 4, pp. 301-306.

_______. 1987. Phytophagous insect faunas of the native thistles, *Cirsium brevistylum, Cirsium congdonii, Cirsium occidentale,* and *Cirsium tioganum,* in southern California. *Annals of the Entomological Society of America,* vol. 80, pp. 152-160.

Gustafson, J. F., and R. S. Lane. 1978. An annotated bibliography of literature on salt marsh insects and related arthropods in California. *Pan-Pacific Entomologist,* vol. 44, pp. 327-331.

Ingles, L. G. 1929. The seasonal and associational distribution of the fauna of the upper Santa Ana River Wash. *Pomona College Journal of Entomology and Zoology,* vol. 21, pp. 1-48, 57-96.

Illingworth, J. F. 1927. Insects attracted to carrion in southern California. *Proceedings of the Hawaiian Entomological Society,* vol. 6, pp. 397-401.

Legner, E. F., G. S. Olton, R. E. Eastwood, and E. J. Dietrick. 1975. Seasonal density, distribution and interactions of predatory and scavenger arthropods in accumulating poultry washes in coastal and interior southern California. *Entomophaga,* vol. 20, pp. 269-283.

McFarland, N. 1965. The moths (Macroheterocera) of the chaparral plant association in the Santa Monica Mountains of southern California. *Journal of Research on the Lepidoptera,* vol. 4, pp. 43-74.

Nagano, C. D., C. L. Hogue, R. R. Snelling, and J. P. Donahue. 1981. The insects and related terrestrial arthropods of Ballona. Pages E1-E89 in The biota of the Ballona Region, Los Angeles County, edited by R. W. Schreiber. Report to the L.A. County Department of Regional Planning.

Perkins, P. D. 1976. Psammophilous aquatic beetles in southern California: A study of microhabitat preferences with notes on responses to stream alteration (Coleoptera: Hydraenidae and Hydrophilidae). *Coleopterists Bulletin,* vol. 30, pp. 309-324.

Pierce, W. D. 1939. The dodder and its insects. *Bulletin of the Southern California Academy of Sciences,* vol. 38, pp. 43-53.

Schlinger, E. I. 1975. Intertidal insects: Order Diptera. Pages 436-446 in *Light's Manual: Intertidal invertebrates of the central California coast,* 3rd edition, edited by R. I. Smith and J. T. Carlton. Berkeley, Calif.: University of California.

Sweet, H. E. 1930. An ecological study of the animal life associated with *Artemisia californica* Less, at Claremont, Calif. *Pomona College Journal of Entomology and Zoology,* vol. 22, pp. 57-70, 75-103.

Wallwork, J. A. 1972. Distribution patterns and population dynamics of the micro-arthropods of a desert soil in southern California. *Journal of Animal Ecology*, vol. 41, pp. 291-310.

## ■ IDENTIFICATION

Arnett, R. H., Jr. 1985. *American insects: A handbook of the insects of America north of Mexico.* Florence, Ky.: Van Nostrand Reinhold.
Borror, D. J., and R. E. White. 1970. *A field guide to the insects of America north of Mexico.* Boston: Houghton Mifflin.
Borror, D. J., C. A. Teriplehorn, and N. F. Johnson. 1989. *An introduction to the study of insects,* 6th edition. New York: Holt, Rinehart and Winston.
Brues, C. T., A. L. Melander, and F. M. Carpenter. 1954. Classification of insects. *Bulletin of the Museum of Comparative Zoology, Harvard University,* vol. 108.
Essig, E. O. 1958. *Insects and mites of western North America,* 2nd edition. New York: Macmillan.
Jaques, H. E. 1978. *How to know the insects,* 3rd edition. Dubuque, Iowa: W. C. Brown. [Others books in the series treat true bugs, beetles, spiders, grasshoppers and their allies, and mites and ticks.]
Powell, J. A., and C. L. Hogue. 1981. *California insects.* Berkeley, Calif.: University of California.
Zim, H. S., and C. Cottam. 1951. *Insects.* Golden Nature Guide. New York: Simon and Schuster. [Other titles in the series include *Butterflies and Moths; Insect Pests; Spiders and Their Kin.]*

## ■ URBAN ENTOMOLOGY

Ebeling, W. 1975. *Urban entomology.* Berkeley, Calif.: University of California.
Hickin, N. 1986. *Pest animals in buildings.* Somerset, N.J.: Wiley and Sons.

## ■ SPECIAL TOPICS

Andrews, M. 1977. *The life that lives on man.* New York: Taplinger.
Blum, M. S. 1985. *Fundamentals of insect physiology.* New York: Wiley and Sons.
Chapman, R. F. 1982. *The Insects: Structure and function,* 3rd edition. New York: American Elsevier.
Chu, H. F. 1949. *How to Know the Immature Insects.* Dubuque, Iowa: W. C. Brown.
Cooper, E. K. 1963. *Insects and plants, the amazing partnership.* New York: Harcourt, Brace and World.
Matthews, R. W. 1978. *Insect behavior.* New York: Wiley and Sons.
Merritt, R. W., and K. W. Cummings, editors. 1984. *An introduction to the aquatic insects of North America,* 2nd edition. Dubuque, Iowa: Kendall/Hunt.
Michener, C. E., and M. H. Michener. 1951. *American social insects.* Toronto: Van Nostrand.
Price, P. W. 1984. *Insect ecology,* 2nd edition. New York: Wiley and Sons.
Snodgrass, R. E. 1935. *Principles of insect morphology.* New York: McGraw Hill.
Stehr, R., editor. 1987. *Immature insects,* 2 volumes. Dubuque, Iowa: Kendall/Hunt.
Sterling, D. 1954. *Insects and the homes they build.* Garden City, N.Y.: Doubleday.
Usinger, R. L., editor. 1956. *Aquatic insects of California.* Berkeley, Calif.: University of California.
Wigglesworth, V. B. 1984. *Insect physiology,* 8th edition. London: Methuen.

■ PESTS AND PEST CONTROL

Anonymous. 1983. The pesticide manual. London: British Crop Protection Council.

Carson, R. 1962. *Silent spring.* Boston: Houghton Mifflin.

Davidson, R. M., and W. F. Lyon. 1986. *Insect pests of farm, garden and orchard,* 8th edition. New York: Wiley and Sons.

Graham, F., Jr. 1970. *Since silent spring.* Greenwich, Conn.: Fawcett World.

James, M. T. 1969. *Herm's medical entomology,* 6th edition. New York: Macmillan.

Kettle, D. S. 1984. *Medical veterinary entomology.* New York: Wiley and Sons.

Kono, T., and C. S. Papp. 1977. *Handbook of agricultural pests.* California Department of Food and Agriculture, Laboratory Services, Entomology, Sacramento.

Mallis, A. 1982. *Handbook of pest control,* 6th edition. New York: MacNair-Dorland.

Metcalf, C. L., W. Flint, and R. L. Metcalf. 1962. *Destructive and useful insects: Their habits and control,* 4th edition. New York: McGraw Hill.

Papp, C. S., and L. A. Swan. 1983. *Ouch! A guide to biting and stinging insects and other arthropods.* Sacramento, Calif.: Entomography.

Whitten, J. L. 1966. *That we may live.* Princeton, N.J.: Van Nostrand.

Further valuable sources are the innumerable circulars, pamphlets, bulletins, newsletters, and other publications of the U.S. Department of Agriculture (available from Government Printing Office, Washington, D.C.) and the California Agricultural Experiment Station, Extension Service (University of California, Berkeley); agricultural pest control bulletins and other pertinent information is available from the County of Los Angeles and from local city, inter-agency, and public health offices. Private pest control operators, laboratories, insecticide companies, and other businesses concerned with insects also issue useful publications on pest species.

■ INSECT OBSERVATION

Brown, V. 1968. *How to follow the adventures of insects.* Healdsburg, Calif.: Naturegraph.

Cummins, K. W., L. D. Miller, N. A. Smith, and R. M. Fox. 1964. *Experimental entomology.* New York: Reinhold.

Headstrom, R. 1973. *Your insect pet.* New York: David McKay.

Kalmus, H. 1960. *101 simple experiments with insects.* Garden City, N.Y.: Doubleday.

Tweedie, M. 1968. *Pleasure from insects.* New York: Taplinger.

Villiard, P. 1973. *Insects as pets.* New York: Doubleday.

■ ARACHNIDS AND OTHER ARTHROPODS

Baker, E. W., J. H. Camin, F. Cunliffe, T. A. Woolley, and C. E. Yunker. 1958. Guide to the families of mites. Contributions of the Institute of Acarology, no. 3. University of Maryland.

Cloudsley-Thompson, J. L. 1968. *Spiders, scorpions, centipedes and mites,* 2nd edition. New York: Pergamon.

Comstock, J. H., and W. J. Gertsch. 1940. *The spider book.* New York: Doubleday.

Gertsch, W. J. 1979. *American spiders,* 2nd edition. New York: Van Nostrand Reinhold.

Kaston, B. J., and E. Kaston. 1972. *How to know the spiders,* 2nd edition. Dubuque, Iowa: W. C. Brown.
Levi, H. W., and L. R. Levi. 1968. *A guide to spiders and their kin.* A Golden Nature Guide. New York: Golden Press.
Savory, T. 1964. *Arachnida.* London: Academic Press.

---

■ FACTS AND FOLKTALES

Clausen, L. W. 1954. *Insect fact and folklore.* New York: Macmillan.
Klots, A. B., and E. B. Klots. 1961. *1001 questions answered about insects.* New York: Dodd, Mead.
Neider, C., editor. 1954. *The fabulous insects.* New York: Harper.
Teale, E. W., editor. 1949. *The insect world of J. Henri Fabre.* New York: Dodd, Mead.
Verrill, A. 1937. *Strange insects and their stories.* New York: Grosset Dunlap.

---

■ CHILDREN'S BOOKS

Baranowski, R. 1964. *Golden bookshelf of natural history: Insects.* New York: Golden.
Conklin, G. 1966. *The bug club book.* New York: Holiday House.
Daly, K. 1977. *A child's book of insects.* Garden City, N.Y.: Doubleday.
Griffen, E. 1967. *A dog's book of bugs.* New York: Atheneum.
Rood, R. 1963. *The how and why Wonder Book of butterflies and moths.* New York: Wonder Books. [Other titles in the series concern insects in general and ants and bees.]
Selsam, M. E. 1981. *Back-yard insects.* New York: Four Winds.
Shuttlesworth, D., and S. Z. N. Swain. 1959. *The story of spiders.* Garden City, N.Y.: Doubleday.
Sterling, D. 1961. *Caterpillars.* Garden City, N.Y.: Doubleday.
Teale, E. W. 1953. *The junior book of insects.* New York: Dutton.

## CONSULTANTS

AN EASY WAY TO GET AN ANSWER to a question about insects is to ask a person who knows about them. A surprising number of public agencies in our area employ entomologists who are available for impartial consultation.

The local Yellow Pages telephone directory is the key to additional consultant sources among private firms involved with insects in some way, among them insectaries (which sell insects for biological control), pest control operators, testing laboratories, independent entomologists, insect specimen and book dealers, beekeepers, nurseries, and chemical (pesticide) companies.

---

■ GENERAL INFORMATION

The following are sources of information and identifications but do not make control recommendations.

Entomology Section, Natural History Museum of Los Angeles County, 900 Exposition Boulevard, Los Angeles 90007.
Department of Biology, University of California, 405 Hilgard Avenue, Los Angeles 90024.
Biology Departments, California State Universities: Los Angeles, Long Beach, Fullerton, and Northridge.
Los Angeles County Department of Parks and Recreation, Natural Areas and Nature Centers: McCurdy (Eaton Canyon), 1750 N. Altadena Drive, Pasadena 91107; Placerita Canyon, 19152 Placerita Canyon Road, Newhall 91321; Whittier Narrows, 1000 North Durfee Avenue, South El Monte, California 91733; Charmlee, 2577 S. Encinal Canyon Road, Malibu, California 90265

■ QUARANTINE

U.S. Department of Agriculture, Plant Protection and Quarantine, Animal and Plant Health Inspection Service ("APHIS") offices:
Harbor Office: U.S. Customs House, 300 South Ferry Street, Terminal Island, San Pedro 90731.
Airport Office: Bradley Terminal, L.A. International Airport.
Plant Inspection Station: 9650 South La Cienega Boulevard, Inglewood 90301.

■ HEALTH AND HOUSEHOLD PESTS

Vector Borne Disease Surveillance and Entomology Program, Bureau of Consumer Protection, Los Angeles County Department of Health Services, 2525 Corporate Place, Monterey Park, California 91754.
Structural Pest Control Board, California Department of Consumer Affairs, 1422 Howe Avenue, Sacramento, California 95825. [This office regulates and takes complaints against pest control operators.]
Vector Control Districts: West Valley, 13355 Elliot Avenue, Chino, California 91710. Orange County, 13301 Garden Grove Boulevard, Garden Grove, California 92643.
Local Mosquito Abatement District Offices: Los Angeles County West, 12107 W. Jefferson Boulevard, Culver City, California 90230; Southeast District, 9510 South Garfield Avenue, South Gate, California 90280; Compton Creek, 1224 South Santa Fe Avenue, Compton, California 90221.

■ VENOMOUS INSECTS AND ARACHNIDS

Dr. Robert Gallia, Department of Emergency Medicine, Olive View Medical Center, 14445 Olive View Drive, Sylmar, California 91342.
Poison Information Center, Los Angeles County Medical Association, telephone 213/222-3212.

■ AGRICULTURAL AND GARDEN PESTS

Bee Inspector, Los Angeles County Agricultural Commissioner's Office, 3400 La Madera Avenue, El Monte, California 91732.
Agricultural Entomologist, Los Angeles County Agricultural Commissioner's Office, 3400 La Madera Avenue, El Monte, California 91732.
County and State Arboreta and Botanic Gardens, 301 North Baldwin Avenue, Arcadia, California 91006.
Farm Advisor, University of California Extension, 2615 S. Grand Avenue, Los Angeles, California 90007.

Los Angeles District Office, U.S. Food and Drug Administration, 1521 West Pico Boulevard, Los Angeles, California 90015.

Food and Drug Branch, California Department of Health Services, 1449 West Temple Street, Suite 224, Los Angeles, California 90020.

## SUPPLIES

MATERIALS AND EQUIPMENT for use in collecting, raising, and teaching about insects may be obtained from the following sources.

The Biology Store, 275 Pauma Place, Escondido, California 92029.

BioQuip Products, 17803 La Salle Avenue, Gardena, California 90248.

Insect Lore Products, 132 S. Beech Street, Shafter, California 93263.

Carolina Biological, 2700 York Road, Burlington, North Carolina 27215.

Tri-Ess Sciences, 1020 West Chestnut Street, Burbank, California 91506.

Ward's Natural Science Establishment, Inc., 5100 W. Henrietta Road, Henrietta, New York 14586 (P.O. Box 92912, Rochester, New York 14692-9012); 815 Fiero Lane, San Luis Obispo, California 93501 (P.O. Box 5010, San Luis Obispo, California 93403).

## ENTOMOLOGY ORGANIZATIONS

PERSONAL CONTACT WITH OTHERS interested in entomology is informative and enjoyable. Two societies dedicated to the study of insects are active in our local area:

Lorquin Entomological Society. Open to persons of all ages, including amateur collectors, interested in any aspect of entomology. Meets on the fourth Friday evening of each month, except July, August, and December; sponsored by and normally meets at the Natural History Museum of Los Angeles County.

Entomological Association of Southern California. Concerned primarily with economic entomology. Meets on first Friday in March, June, September, and December at the Los Angeles County Arboretum, Arcadia (contact the Los Angeles County Agricultural Commissioner's Office for further information).

## INSECT ZOO

ONE OF THE FEW PUBLIC EXHIBITIONS of living insects in the United States opened in Los Angeles in spring 1992:

Ralph M. Parsons Insect Zoo at the Natural History Museum of Los Angeles County. A 3,000-square-foot hall featuring live arthropods—about half of which are from southern California and southeastern Arizona—and preserved and fossil specimens as well as interactive exhibits. 900 Exposition Boulevard, Los Angeles. Open 10 a.m. to 4 p.m., Tuesday through Sunday (closed Thanksgiving Day, Christmas Day, New Year's Day, and Mondays).

# INDEX

Boldface numerals indicate illustrations

*Edited by*
Robin A. Simpson

*Design and typography by*
Dana Levy, Perpetua Press

*Design assistance and mechanicals by*
D.J. Choi

*Assistance in computer typography by*
Brenda Johnson-Grau

*Production administration by*
Letitia Burns O'Connor

*Index prepared by*
Kathy Talley-Jones

*Printed in Hong Kong*